# Test Quality for Construction, Materials and Structures

# Other RILEM Proceedings available from Chapman and Hall

1 Adhesion between Polymers and Concrete. ISAP 86
Aix-en-Provence, France, 1986
Edited by H.R. Sasse

2 From Materials Science to Construction Materials Engineering
Proceedings of the First International RILEM Congress
Versailles, France, 1987
Edited by J.C. Maso

3 Durability of Geotextiles
St Rémy-lès-Chevreuse, France, 1986

4 Demolition and Reuse of Concrete and Masonry
Tokyo, Japan, 1988
Edited by Y. Kasai

5 Admixtures for Concrete
Improvement of Properties
Barcelona, Spain, 1990
Edited by E. Vázquez

6 Analysis of Concrete Structures by Fracture Mechanics
Abisko, Sweden, 1989
Edited by L. Elfgren and S.P. Shah

7 Vegetable Plants and their Fibres as Building Materials
Salvador, Bahia, Brazil, 1990
Edited by H.S. Sobral

8 Mechanical Tests for Bituminous Mixes
Budapest, Hungary, 1990
Edited by H.W. Fritz and E. Eustacchio

9 Test Quality for Construction, Materials and Structures
St Rémy-lès-Chevreuse, France, 1990
Edited by M. Fickelson

**Publisher's Note**
This book has been produced from camera ready copy provided by the individual contributors. This method of production has allowed us to supply finished copies to the delegates at the Symposium.

# Test Quality for Construction, Materials and Structures

Proceedings of the International Symposium held by RILEM (The International Union of Testing and Research Laboratories for Materials and Structures) and ILAC (The International Laboratory Accreditation Congress): Test Quality and Quality Assurance in Testing Laboratories for Construction Materials and Structures/Qualité des Essais et Assurance de Qualité des Laboratoires d'Essais pour les Matériaux de Construction et les Ouvrages, organized by AFREM (Association Française de Recherches et d'Essais sur les Matériaux et les Constructions), co-sponsored by ASTM (American Society for Testing and Materials), CEN (European Committee for Standardization), ISO (International Standards Organization), OIML (Organisation Internationale de Métrologie Légale).

Saint-Rémy-lès-Chevreuse, France
October 15–17, 1990

EDITED BY

**M. Fickelson**

CHAPMAN AND HALL

LONDON · NEW YORK · TOKYO · MELBOURNE · MADRAS

| | |
|---|---|
| UK | Chapman and Hall, 2–6 Boundary Row, London SE1 8HN |
| USA | Van Nostrand Reinhold, 115 5th Avenue, New York NY10003 |
| JAPAN | Chapman and Hall Japan, Thomson Publishing Japan, Hirakawacho Nemoto Building, 7F, 1-7-11 Hirakawa-cho, Chiyoda-ku, Tokyo 102 |
| AUSTRALIA | Chapman and Hall Australia, Thomas Nelson Asutralia, 480 La Trobe Street, PO Box 4725, Melbourne 3000 |
| INDIA | Chapman and Hall India, R. Seshadri, 32 Second Main Road, CIT East, Madras 600 035 |

First edition 1990



Printed in Great Britain
at the University Press, Cambridge

ISBN 0 412 39450 2 0 442 31285 7 (USA)



**British Library Cataloguing in Publication Data**
Available

**Library of Congress Cataloging-in-Publication Data**
Available

# International Scientific Committee
# Conseil Scientifique

# Contents

# Preface

Technical events, publications, congresses or conferences may well fit into the mainstream of development in a major topic, but they can also be staged to respond to a subject of current interest at a point in time and a concurrence of events.

Undoubtedly, the reliability of measurement and the quality of testing form one of these basic themes which we must keep constantly in mind, with implications reaching well beyond laboratories alone. But this subject has now become particularly topical with a combination of events where a new order is being set up for international trade.

The first International conference in 1978 on Accreditation of Test Laboratories, since followed by a regular series known under the abbreviation of ILAC, prefigures the emergence of questions which also concerned RILEM as an international union of testing and research laboratories. So it was logical that RILEM and ILAC should come together to hold a symposium on Test Quality and Quality Assurance in Testing Laboratories.

An opportune symposium, well-timed, which thereby benefited from the best of co-operation from the Organisation Internationale de Métrologie Légale (OIML), the International Standards Organization (ISO), the European Committee for Standardization (CEN) and the American Society for Testing and Materials (ASTM).

I am most grateful to these authoritative institutions of great renown for their association with the RILEM/ILAC initiative.

The specific vocation of RILEM made us limit the field of experience covered to that of construction materials, building and engineering structures. But even if there are certain specific characteristics at the application stage, the various aspects of measurement and testing, also of the organization of laboratories to ensure the necessary level of quality, are of common interest to all branches of industry.

Consequently, it will be very satisfying to know that this volume, which brings together the essential part of the contributions presented and discussed at the RILEM/ILAC symposium, will become a source of reference for both laboratories and users of test methods and results.

Maurice Fickelson
*Secretary General of RILEM*

# PART ONE

# INTRODUCTION

# 1 QUALITY IN TESTING FOR QUALITY

C. BANKVALL
Swedish National Testing and Research Institute (SP),.
Borås, Sweden

**Abstract**
This paper discusses the right quality level for testing, and indicates the factors that are of importance for industry and public authorities when evaluating testing services. It also discusses the most important elements in achieving the quality that we need. The role of quality assurance in the work of testing is described, and the need for a functional approach is underlined. The most important factors influencing the 'right' quality and the 'right' results in testing are set out in the final summary.
Keywords: Testing, Quality, Verification, Traceability, Comparison, References, Standards.

## 1 Introduction

What do we mean by quality of testing; how do we ensure it, and how do we know that we are attaining it? By attempting to answer these questions, I hope to indicate some of the most important factors in ensuring Quality in Testing for Quality.

## 2 What is the 'right' quality level for testing?

Evaluation of technology, testing and measurement are important parts of research and development, manufacture and marketing. Technical investigations and test results are also often called for by authorities having jurisdiction when products or systems have to comply with specifications.

To a very considerable extent, testing is a specialised service, of which the content and objectives are determined by the needs of industry, authorities, consumers and other parties. It is these needs, of the recipient of the testing, that determine what is the 'right' quality. At one end of the scale, we have the consumer who buys a punnet of strawberries in the market, whose needs stretch no further than a fairly rough idea of the weight or volume (provided, of course, that the strawberries are fresh). At the other end of the scale, we have the control room operator in a nuclear power station, who needs considerably more accurate information concerning, say, the cooling water flow through the reactor core. Developing and manufacturing building materials, designing the structures of buildings, building a house, a bridge or a nuclear power station - all these applications require differing extents of testing and evaluation in the laboratory or at site, with specific and appropriate levels of quality requirements, depending on the particular needs. For the laboratory, quality in testing is a matter of responsibility for the results and their acceptance by the market.

Essentially, quality in testing is a question of reliable and useable results, i.e. of actually answering given questions or solving specific problems. The overall concept of Quality in Testing also includes factors such as delivery time and the cost of the specialised, expert services. In simple terms, this means that the quality provided must suit what the situation requires.

Let us consider two examples. The thermal resistance of building materials is a basic characteristic that needs to be known in order to be able to evaluate such factors as the level of energy use in a building. Testing of thermal conductivity is carried out regularly both by industry and large national laboratories. Insulating materials constitute a major item of trade, both within and between countries. There are therefore two requirements to be considered: one is to be able to state a sufficiently accurate value of thermal conductivity to enable energy use in buildings to be calculated, and the other is to provide the necessary information required for the trade in, and marketing of, different insulating materials. The first of these requirements is often the less demanding in terms of accuracy of test results: there are many other factors that play an important part in determining the level of energy use in a building. However, the second requirement, which reflects regulations in building codes and so on, can result in a need for greater accuracy of measurement if manufacturers push their marketing claims to the last decimal place.

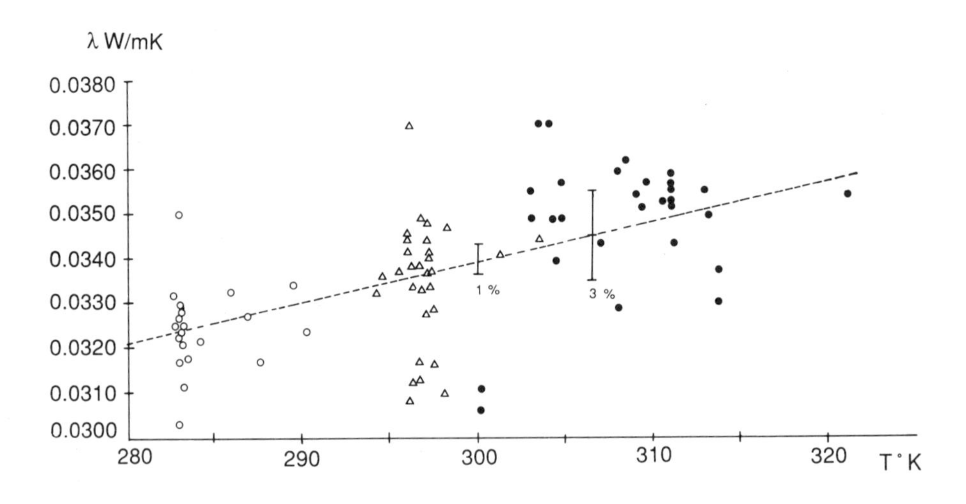


Figure 1. Comparative tests of thermal insulation material.

In practice, comparative measurements at international level between different laboratories have shown that even well-defined methods can give a spread in results. In the case of thermal resistance of thermal insulating materials, this can sometimes amount to as much as 10%. At the same time, good instrumentation for quality control can have an accuracy of about 3%. The best measurements can have an accuracy better than 1%. The maximum spread in these figures is acceptable for the purpose of assessing energy use in a building, while the best accuracy approaches the requirements of the manufacturers for their marketing purposes or product classification.

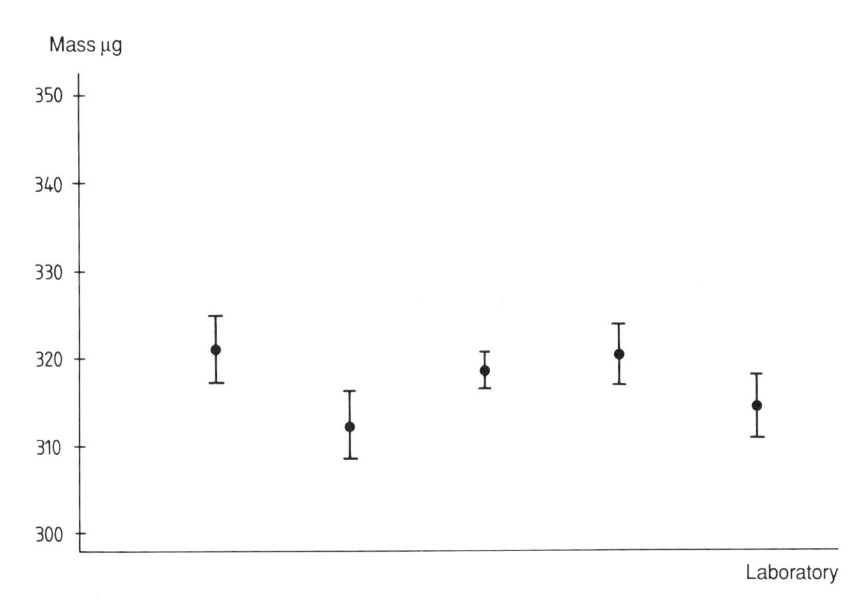


Figure 2. The mass values (above 50 g) of the 50 g transfer standard as measured by national laboratories

Our second example comes from an area of technology that has been of great significance for industry and trade since time immemmorial: the measurement of mass, and its international traceability through primary standards at national and international level and secondary standards and working standards at national level and in industry. Trade and industry today make full use of the best accuracy that can be achieved, which, for the kilogram, represents 1 x $10^{-8}$ (±10 μg). In order to achieve this, special laboratories with expensive equipment and highly trained personnel are required. One reason for this high accuracy is the fact that every step in the traceability chain results in a certain loss of accuracy. Another reason is to be found in the continuous development of production technology, which means in practice that the results of present-day research in measurement technology are already being demanded by industry for everyday application.

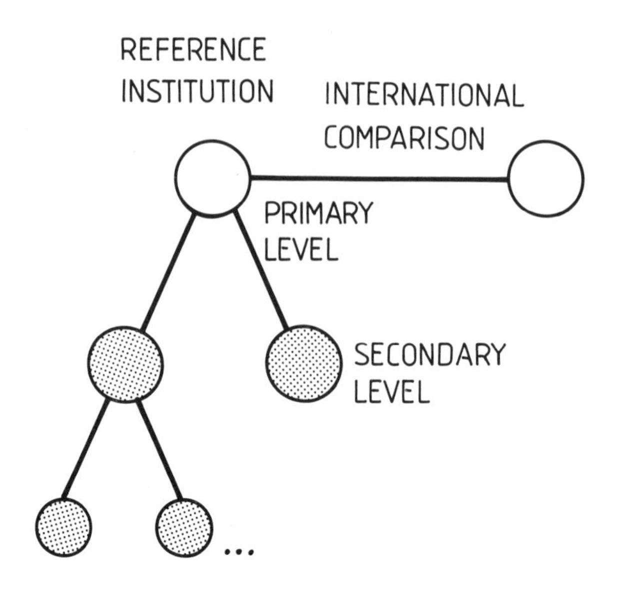


Figure 3. The principle of traceability.

It can be seen from the foregoing that, in one context, the right quality in testing results may be expressed in percentage points, while in another it requires parts per million. Several factors influence the conditions governing achievement of the right quality. Experience from industry and authorities of how they evaluate testing services reveals the following important points:

- The laboratory must be able to choose and apply a method that efficiently provides a technically correct answer to the problem involved.
- The laboratory must be able to provide reliable results, of an accuracy level appropriate to needs.
- Confidentiality and impartiality are important in many cases.
- Reports and presentation of results must be clear and complete, and the recipient must be able to understand them and apply them as needed.
- The actual laboratory work must also be performed efficiently, so that delivery times can be met and costs and charges minimised.
- The laboratory should have the necessary expertise to be able to assist in evaluation of the test results and to provide other relevant technical advisory services.

In general terms, these points can be regarded as the quality criteria of the laboratory.

## 3 How do we provide the quality that we need?

Testing is influenced by a number of elements, all of which influence the quality to a greater or lesser degree and which can therefore be used to direct the results towards the correct, desired quality level. The most important of these elements are as follows:

- A personnel structure having the necessary fundamental competence. This includes personnel training appropriate to the requirements imposed by technical development.
- Knowledge of testing and measurement technology. Many methods need to be interpreted if they are to provide reproducible and correct results in the particular application. Knowledge can be consolidated through participation in standardization work, in research and in method development.
- Properly maintained and calibrated equipment that can be employed in a competent manner, and which is also reasonably modern. It has been found in many comparison measurements that shortcomings in the results can be traced to handling, reading of test instruments and so on, rather than to fundamental errors or inadequacies of calibration. In a complicated test, it is the interaction between man and equipment that results in precision.
- A simple organisation, in which the distribution of responsibility and authority is clearly understood. A smoothly operating organisation is more a question of team spirit, attitudes and open communication than of comprehensive documentation, although a certain amount of documentation is, of course, necessary.
- A clear and simple administrative system that embraces such aspects as handling of materials, production of reports, filing, confidentiality and so on.
- The right technical environment, working environment and general 'culture' in the laboratory or organisation are often decisive for the quality of the work, yet at the same time can be difficult to define in detail.

We are now seeing a trend towards attempting to control these elements to a certain extent through the introduction of quality assurance systems. General rules for such systems are set out in, for example, ISO Guide 25 and in the European standards in the EN 45000 series. These rules are often general and formal in their nature, tending to be concerned largely with documentation associated with the various factors described above. This is illustrated in the guidelines included in the EN standards. The advantages of such systems are that system evaluation creates an apparent openness to all laboratories. At the same time, it is difficult for the standards to make sufficient allowance for variations in different sectors of technology or industry, or for the actual situations and methods of working of different laboratories. Some of these drawbacks can be curtailed by concentrating quality assurance work in terms of system monitoring on the performance of the most important elements, rather than on their actual form. However, the most serious drawback is the fact that the systems do not provide any guarantee that the results obtained will be of the right quality.

In reality, such measures must include such widely differing elements as those concerned with purely technical investigations and the competence and working conditions of the laboratory staff. To some extent, formal rules can be of value in creating the right quality.

## 4 How do we know that we have the necessary test quality?

The simple answer to this question is, of course, provided by satisfied customers. The next question is then how we assure ourselves of this?

Depending on the particular needs of the application, the answer may be found in the simple advice: "Be sure to visit any laboratory you consider using. You do not need to be an expert to spot laxity in house-keeping." It is relatively easy for a customer to recognise order and tidiness in a laboratory. On the other hand, detailed information on the accuracy of results may often be required. This can be provided only by comparative testing and traceability of results, in the same way as we have used our reference standards back into the distant past.

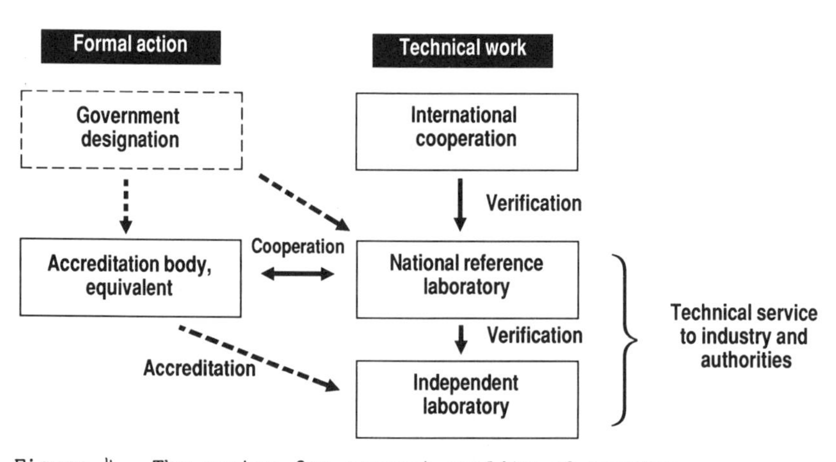


Figure 4. The system for correct quality of testing.

In order to achieve the right quality of testing, we need to make specific efforts. At the same time, testing is an activity that requires a high level of cost awareness and efficiency, not least in comparison with other parts of the process of development, marketing etc. This means that verification of the fact that we are producing the results that are needed must be of such a level as to ensure that we are 'reasonably certain' of the results when set against the possible consequences of an error. There must be a clear cost/value relationship (in terms of the value of the results) that encourages quality assurance work. In certain respects, this can be based entirely on technical and economic factors, i.e. through the avoidance of costs or generation of higher revenue by achieving results within the relevant limits, presented in a manner that is acceptable to the user. In some cases, acceptance can be related to more formal systems that provide the formal right of acceptance of the results. However, it should be noted that such systems do not lessen the liability of the laboratory for its results, their relevance or their applicability. It is particularly unfortunate if formal procedures are developed that are both expensive and technically irrelevant. In such a situation, the cost/value relationship is only initial. It will be followed by a reduction in the credibility of, or trust between, laboratories as 'inexplicable' differences reveal themselves.

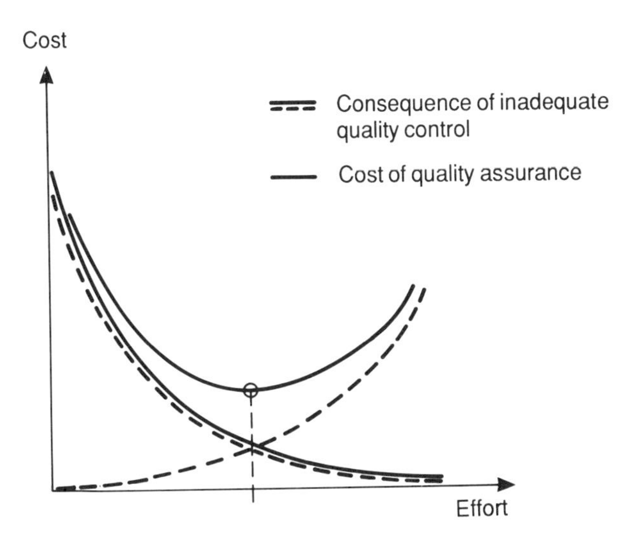


Figure 5. The cost/value effect in quality assurance.

The content of the above is that we should establish and participate in networks of reference laboratories, charged with the task of, and having authority for, creating international traceability in certain given sectors. Responsibility would be vested in the laboratory management, controlling work by such means as a quality assurance system involving rational application of, say, EN 45000 documentation. This is a form of working that already exists within such areas as electrical safety and also in certain parts of the building sector. The national reference laboratories are required to support other laboratories in such a way that results can be demonstrated as according, within reasonable values depending on needs, with the national reference level. When necessary, these forms of working can be complemented by formal verification systems such as accreditation. This will ensure that the available expertise is used as well as possible in order to achieve the required technical results. The basic idea should be that the laboratory is regarded as operating well until any signs to the contrary become apparent, and that it should be open for different technical comparisons.

At present, the best-operating forms of acceptance of results between one country and another are based on technical cooperation between laboratories having the highest national and international competence. In itself, this cooperation defines the state of the art within a particular test sector, in approximately the same way as does joint research in other sectors at national and international level. Formal systems for general openness are limited in their application, and can be regarded as prime contenders mainly in countries in which the technical infrastructure is highly diversified, and in situations typical of the work done by 'routine testing laboratories' carrying out similar, repetitive tests of a clearly defined nature, so that a strict working specification and personnel performance can be documented and checked. This 'robot-like' situation is not typical of test and research institutions operating over a wider field. For them, the quality of evaluation depends on an integral whole of expertise and responsibility, which is not presupposed in normal quality assurance systems. If the work done by the laboratories is complicated, and needs to be carried out by skilled personnel, the whole concept of quality needs to be expanded. Guiding and developing this quality is the responsibility of management, and requires active participation in the work of the laboratory. Quality control is becoming increasingly dependent on communication and on competent planning for good working conditions in the laboratory.

Let us consider the example of full-scale tests in the field of fire behaviour and prevention. For this application, teamwork and planning of the tests are as important as calibration of the instruments, without even considering such factors as a genuine understanding of how heat is transported and temperature measured. Another example is to be found in the field of seismic testing by means of computer-controlled shaker tables and extensive computer processing of accelerometer values. In such a situation, calibration of the accelerometers is clearly an integral part of the chain of measurements. Other links in the chain include questions such as the means of securing sensors to the mechanical system, computer control and so on. In more advanced testing, much of the quality assurance work must be integrated into the actual test procedure and be delegated to those carrying out the work. In such situations, over-all inspection and quality assurance must be organised in a similar way, becoming increasingly comparable to the relationship between professor and students in university research and development work. Methods such as fault analyses and comparative tests of traceability hierarchy can form the basis of quality assurance in well-defined test situations such as calibration of electrical instruments. A simple, easily available form of quality control can be provided by management reading test reports and discussing their contents with those performing the tests. This requires suitable expertise at management level. In this context, attention should also be drawn to an element that is often disregarded in constructing quality assurance systems; that of the production of manuals and documentation of routines. The ability to carry out a task correctly is dependent more on the natural working rules than on the extent of documentation. A well-documented but complicated procedure will result in a poorer level of quality control than a number of simple rules associated with clear and natural instructions that can be taken seriously and maintained fresh through frequent use. Management must make it easy for those performing test work to achieve the right quality.

Against this background, there is reason to question whether the first generation of quality assurance systems for laboratories is not ready for modification or replacement. Formal and often administratively top-heavy forms of quality assurance should be replaced by functional, efficient forms of working, better suited to the skilled personnel who nowadays work in laboratories. This would be equivalent to replacing the mechanistic view of workers on production lines that was current during the early years of the century by more modern forms of work organisation in present-day industry, in which the individual's capabilities, skills and development are regarded as important elements in achieving the right quality.

## 5 Summary

Many different factors influence the level of right quality and right results in testing. Some of these factors are more important than others. Some can be checked. Some can be documented, but not all of them. To a great extent, modern testing procedures are very dependent on the skill of those applying them and on their attitudes to the work. We could attempt to summarise the most important factors as follows:

- Understand the need. This is a prerequisite to be able to define the right quality.
- Concentrate on establishment and maintenance of competence and skills. This applies particularly to personnel, although also to other resources.
- Work logically and tidily. To a reasonable extent, this is essential for the performance and credibility of the organisation.
- Develop traceability within those sectors in which real reliability and accuracy of results are needed.
- Appreciate that the technical and working environments constitute decisive factors in the quality of the work, but are difficult to document and equally difficult to define.

If all the preceding factors apply, conditions should be right for the correct quality of technical investigations, testing and measurement. This applies not only to efficient work, but also to the reliability and utility of the results. Conditions will also then be right for the employment of relatively modest means to incorporate the work in a quality assurance system. The emphasis of such a system would be on the functional elements, and its formal superstructure would be as little as necessary for administrative purposes and to ensure formal acceptance of the results.

# 2 THE NEED FOR MUTUAL RECOGNITION OF TEST LABORATORIES

J.E. WARE
BSI Quality Assurance, Milton Keynes, UK

Abstract
The paper discusses the development of the Single European Market and the consequential need for mutual recognition arrangements between test laboratories. It traces the background which leads to this need and the role and work of international organizations in providing a basic framework to enable such recognition to take place under harmonized and controlled conditions. It highlights particular guides and standards which are relevant.

## 1 Introduction

First of all it seems to me that we need to establish the situation in which mutual recognition of test laboratories becomes important or necessary. After all, one could merely adopt a position in which it could be argued that a test laboratory has a customer who requires a test report for a specific purpose and having provided this the laboratory has discharged its contract with the customer. We therefore have to look wider and explore the reasons for the customer requiring the test report. He presumably wants it to assure his customer, or a third party certification body or a regulatory body of this fact.

It is in the latter two needs that mutual recognition becomes important, particularly where the product is to be marketed in other countries where the national requirement is for certification by a national body or where the mandatory area requires testing by a national body.

An essential component of any attempt at mutual recognition is confidence in the competence of the testing laboratory. This also becomes important in a situation such as has emerged in the concept of the Single European Market, where in order that products may cross frontiers there has to be mutual confidence in the competence of notified testing laboratories.

By its very name therefore, mutual recognition implies third party involvement of one sort or another.

## 2 ISO/CASCO

Before going any further, I would like to introduce you to ISO/CASCO and its work. Formed in 1970, the ISO Council committee on conformity assessment (CASCO) brings together 44 member countries as participating members, 22 countries as correspondent members and a number of important international organizations. Its objectives are:-

- To study means of assessing the conformity of products, processes, services and quality systems to appropriate standards or other technical specifications.
- To prepare international guides relating to the testing, inspection and certification of products, processes and services, and to the assessment of quality systems, testing laboratories, inspection bodies, certification bodies and their operation and acceptance.
- To promote mutual recognition and acceptance of national and regional conformity assessment systems, and the appropriate use of International Standards for testing, inspection, certification, assessment and related purposes.

Although operated by the ISO Central Secretariat it is supported by IEC whose members participate fully and it provides the mechanism through which documents relating to testing matters produced through ILAC task force activities are given wide exposure for national comment leading to publication as ISO/IEC guides.

Over the years a task performed by CASCO has been to create and publish a number of ISO/IEC guides, the intention of which has been to provide a framework for the harmonized development of testing and certification practices so as to assist the recognition process and to ease world trading problems.

It is now widely recognized that trade is hampered by the existence of 'technical barriers', one of which is undoubtedly the varying national requirements for testing and certification. One of the features of the concept of the Single European Market is to free manufacturers of the need to go through multiple testing and certification in their various market areas. 'One stop testing or certification' is demanded and must be achieved.

Equally, in the worldwide sense, manufacturers are seeking the same facility. It is not easy because of the need to address the legal responsibilities and conditions but if, technically we can develop the confidence necessary to accept each others' work, then the legal problems can often be addressed through suitable contract arrangements.

CASCO produced, in 1972, a booklet entitled 'Certification - Principles and Practice' which was published jointly by ISO and the International Trade Centre (ITC). Although now somewhat outdated and currently in the course of review, the identification of eight basic systems by which product conformity can be demonstrated are still valid. These are:

1 Type testing.

2 Type testing followed by subsequent surveillance through audit testing of factory samples purchased on the open market.

3 Type testing followed by subsequent surveillance through audit testing of factory samples.

4 Type testing followed by subsequent surveillance through audit testing of samples from both the open market and the factory.

5 Type testing and assessment of factory quality control and its acceptance followed by surveillance that takes into account the audit of factory quality control and the testing of samples from the factory and the open market.

6 Factory quality control assessment and its acceptance only.

7 Batch testing.

8 100% testing.

It is interesting to note that, with the exception of system 6, testing has a direct role. In system 6, testing is important because although it concentrates on the manufacturers control through a structured quality system, such a system has to have a means of verification of conformity which usually can only be achieved through competent testing.

Whilst the simple type test is a method of indicating conformity it can only be regarded as showing that the individual item was in conformity. It cannot, of itself give any assurance about subsequent products other than by inference. Thus the audit and quality system methods have been developed.

It was clear to CASCO in the mid 70's that a key feature for effective third party certification was competent testing. Thus an early guide, Guide 25, addressed the subject and formed the basis for developments in ILAC, CASCO and worldwide. Gradually other needs surfaced and there are now a number of ISO/IEC guides relating to testing, namely:-

ISO/IEC Guide 25:1982 'General requirements for the technical competence of testing laboratories' (Second edition).
ISO/IEC Guide 38:1983 'General requirements for the acceptance of testing laboratories'.
ISO/IEC Guide 43:1984 'Development and operation of laboratory proficiency testing'.
ISO/IEC Guide 49:1986 'Guidelines for development of a Quality Manual for a testing laboratory'.
ISO/IEC Guide 54:1988 'Testing laboratory accreditation systems - General recommendations for the acceptance of accreditation bodies'.

ISO/IEC Guide 55:1988 'Testing laboratory accreditation systems - General recommendations for operation'.

Many of these guides have been adopted nationally. In Europe they formed the basis for developing the standards required to achieve mutual recognition.

### 3 Mutual recognition

Thus we have in place the mechanisms by which mutual recognition becomes a practical possibility.

I have mentioned that it is generally in the third party or regulatory area where benefits resulting from mutual recognition can arise.

The Single European Market can be used as an example of how the standards relating to testing are an important ingredient.

The development of harmonized certification schemes, particularly in the electrotechnical area and in Europe, required, in the absence of any other mechanism that the testing laboratories participating should have some form of evaluation, commonly through assessment by their peers. In some cases where this was not present, difficulties did arise as it was not possible, because of legal liability questions, for test reports to be accepted without some degree of check testing.

Similarly, Directives emerged in which there were requirements covering both third party certification and independent testing. It was recognized that in these areas, it was left for member states to initiate competent Testing and Certification Bodies to discharge their obligations under Directives. It was equally recognized that again there was no common criteria against which the competence of such Certification Bodies and Testing Bodies could be evaluated. Thus, over this period several Guides have been developed covering third party certification and assessment activities.

Thus, in the light of 1992, the Commission established two Working Groups in 1986, one to look at the development of criteria for Certification Bodies; the other to develop similar criteria for test laboratories.

The Commission had recognized that harmonization of testing and certification can create the environment for reciprocal recognition and facilitate trade between Member States. It was therefore seen that the development of these common procedures and criteria would provide the basis for:

(a) recognition of the competence of Certification Bodies and Test Laboratories;
(b) agreements between Certification Bodies and Test Laboratories;
(c) agreements between National Bodies responsible for recognizing Certification Bodies and Test Laboratories;

(d) the nomination of Certification Bodies and Test Laboratories for regulatory purposes by Member Governments.

The work of these Groups eventually saw the light of day in CEC documents Certif 87/15 and Certif 87/16. These documents were based, as far as possible, on existing ISO/IEC Guides and drafts moving through that system. Through the member Governments, they should have received exposure nationally to ascertain national views and commentary on these documents.

The result of this work was that towards the latter part of 1987, the two documents were issued as final drafts having gone through a consultation process by national Governments.

The next stage was that the Commission mandated CEN/CENELEC, the joint European Standards Body, to produce the two documents as European Standards. A Working Group was formed and first met in June this year and their first task was to review the comments that had arisen as a result of the CEN/CENELEC consultation process through member bodies. This increased exposure in fact resulted in 100 pages of comments for each of the documents. The Working Group took the view that rather than two standards as had been proposed, the documents should be separated into a number of standards, each dealing with the separate chapter subjects and that initially those chapters dealing with criteria would be published. As a result, there are now seven proposed standards, of which three are for testing as follows:-

EN 45001 'General criteria for the operation of test laboratories'.
EN 45002 'General criteria for the assessment of test laboratories'.
EN 45003 'General criteria for laboratory accreditation bodies'.

## 4 Standards for testing

The standards of interest in the testing field are EN 45001, 45002 and 45003. We can now look at these and try to highlight some of the significant aspects.

First of all I think that we need to be very clear on the meaning of accreditation. Unfortunately, due to the way in which the use of this word has developed, it is now used with two rather different meanings. Initially, accreditation became used as a result of the establishment of ILAC. In this sense, accreditation in my view is a form of certification in that a Body is carrying out an assessment and certification of a test laboratory in exactly the same way that a Certification Body carries out this exercise with a supplier of goods or services.

Accreditation of Certification Bodies on the other hand creates another level through which the Certification Bodies carrying out the certification are themselves assessed and certified.

This tends to cause some confusion but should be borne in mind when considering the titles of these European Standards. It is seen most clearly in considering EN 45003 and EN 45011. EN 45003 deals with the general criteria for Laboratory Accreditation Bodies i.e. Bodies carrying out the certification of laboratories; EN 45011 deals with the general criteria for Certification Bodies operating product certification i.e. those that certify manufacturers and their products.

**4.1 EN 45001**

One can see from its field of application that it specifies general criteria for the technical competence of a testing laboratory, including calibration laboratories, irrespective of the sector involved. It deals with all those aspects which it is necessary to consider in the operation of a competent laboratory.

As one might expect, it also addresses the question of impartiality, independence and integrity. Under this heading it requires that the laboratory and its personnel shall be free from any commercial, financial and other pressures which might influence their technical judgement.

It deals with the management and organization of the laboratory; its use of qualified personnel; the way in which it maintains and controls its equipment. It requires that there is a laboratory quality system with detailed working procedures and addresses the question of test reports.

This item is of utmost importance in that a test report has to be presented clearly and unambiguously, with the test results and all other information. Two important considerations arise:

- firstly, it requires that the test report carries a statement to the effect that the test results relate only to the items tested;
- secondly, it requires that a test report should not include any advice or recommendation arising from the test results.

Perhaps we should consider the reasons why.

In the first instance, it is important that when considering the question of third party certification or product conformity, the user is left in no doubt as to whether the report deals only with a single item as tested or whether it covers any follow up action.

As far as the second is concerned, it ensures that the test report is a factual account of the tests carried and the results of those tests, leaving any judgement or opinion in the hands of those people qualified to so do. This of course does not preclude a test laboratory from issuing an attestation.

**4.2 EN 45002**

This Standard outlines the guidance for the way in which a laboratory should be assessed in terms of compliance with the requirements of EN 45001. It is important that this guidance be prepared in the context of 1992 and the desire for mutual recognition.

**4.3 EN 45003**

In dealing with the requirements for those Bodies who carry laboratory accreditation, this Standard provides the same opportunity for co-ordination.

## 5 Conclusion

I have attempted to indicate how the mechanisms have been constructed over the years to enable mutual recognition to be achieved. I think we are well on the way.

# 3 L'IMPORTANCE DES ESSAIS DANS LE SECTEUR DU BÂTIMENT (The importance of testing in the building sector)

P. CHEMILLIER
Centre Scientifique et Technique du Bâtiment, Paris, France

Les essais jouent un rôle essentiel dans le bâtiment. On y fait appel en maintes circonstances qui peuvent être regroupées en trois familles :

- Les essais de recherche concernant les phénomènes et les ouvrages.

- Les essais de caractérisation des produits de construction.

- Les essais de contrôle sur ouvrages terminés.

Nous allons examiner ces trois familles et en souligner l'importance.

## 1 Essais de recherche concernant les phénomènes et les ouvrages

Une des difficultés du bâtiment, avec la multiplication de produits nouveaux et le raffinement sans cesse croissant des besoins et exigences des usagers, est de connaître le comportement des ouvrages et du bâtiment dans son ensemble sous l'action de phénomènes divers, souvent simultanés.

Cette connaissance est en effet indispensable pour adapter les ouvrages et trouver ainsi de bonnes réponses aux exigences.

Or s'agissant le plus souvent de phénomènes nouveaux, sinon dans leur principe du moins dans leurs modalités, on ne dispose guère des enseignements de l'expérience accumulée au cours des ans.

Il s'agit donc d'imaginer des essais reproduisant les phénomènes et d'étudier le comportement des ouvrages au cours de ces essais.

C'est ainsi qu'on a mis au point des essais pour étudier les phénomènes suivants : propagation du son dans les structures ou à travers les parois, transferts de chaleur et d'humidité à travers les parois, effets du vent sur les structures et sur les façades, naissance et propagation du feu et de la fumée, vieillissement en atmosphère saline, etc...

Il est généralement difficile de concevoir et de réaliser des essais vraiment représentatifs de la réalité car la réalité est complexe et met en jeu des paramètres qu'on ne sait pas toujours cerner.

Ces essais se font sur modèle réduit lorsque l'on est en mesure de procéder pour tous les paramètres à la similitude voulue. Mais il apparaît de plus en plus souvent nécessaire de faire les essais sur des éléments en vraie grandeur pour deux raisons : la première est qu'on ne sait pas toujours représenter à échelle réduite certaines composantes du phénomène étudié, la seconde est que l'étude porte sur des éléments d'ouvrage qu'on ne sait pas représenter valablement à échelle réduite (c'est le cas de joints de façade par exemple). Tout cela exige des équipements de grande taille.

A partir de ces essais on met généralement au point des modèles théoriques qui permettent ensuite dans l'étude de projets de se dispenser de faire des essais que d'ailleurs on ne pourrait pas faire puisqu'à ce stade le bâtiment n'est pas encore construit et qu'on imagine mal de faire un bâtiment prototype comme cela se pratique couramment dans l'industrie automobile.

Ces modèles se présentent en général sous la forme de méthodes de calcul, parfois sous la forme de tableaux de valeurs.

Ils permettent de déterminer les caractéristiques à donner aux ouvrages.

On voit immédiatement d'après ce qui vient être dit toute l'importance que revêt la qualité de ces essais puisque c'est sur eux que reposent les méthodes de calcul qui seront ensuite utilisées dans la pratique.

Ils exigent de solides connaissances scientifiques en physique, en mécanique, en chimie pour analyser et schématiser les phénomènes à étudier et de solides compétences en métrologie pour capter les informations nécessaires en cours d'essai. De nouveaux capteurs sont souvent à inventer.

## 2 Essais de caractérisation des produits de construction

Les produits de construction ont des caractéristiques (propriétés physiques et performances) qui vont déterminer les performances de l'ouvrage dans lequel ils sont incorporés, et auxquelles il vient d'être fait allusion.

S'il s'agit de propriétés physiques les essais qui permettent de les déterminer sont généralement classiques mais s'il s'agit de performances les essais sont souvent à inventer et à mettre au point.

Prenons le cas d'une fenêtre : parmi ses performances il y a la perméabilité à l'air et la perméabilité à l'eau. Il n'est pas évident a priori de définir ces deux performances de façon claire et d'imaginer les essais qui permettront de les mesurer par une grandeur ou, à défaut, de les situer sur une échelle conventionnelle de repères (par exemple des classes ou des niveaux).

Mais il y a des cas plus difficiles lorsque les phénomènes en jeu sont mal connus. On en trouve un exemple avec la durabilité de la transparence d'un double vitrage, performance essentielle pour un tel produit.

Il s'agit en l'occurrence d'essais de recherche tels que décrits plus haut à propos des ouvrages. Là encore on cherche à analyser les phénomènes et le comportement du produit devant ces phénomènes.

Lorsque les essais portent sur des produits innovants on ne sort pas complètement de la phase de recherche pour caractériser les produits et apprécier leur aptitude à l'emploi. Les essais de caractérisation comportent encore une part de recherche et ne peuvent donc pas être effectués par des laboratoires ordinaires même bien équipés. Il faut l'intervention d'experts ayant de bonnes connaissances scientifiques pour interpréter les résultats des essais. Plus le produit est innovant plus la part d'interprétation est grande.

C'est ce qui se passe en France dans la procédure de l'Avis technique qu'on assimile à tort à un essai de type conforme à un protocole d'essai bien établi. Il en ira de même pour l'Agrément technique européen si, comme prévu, il est utilisé pour les produits innovants.

Mais à supposer que la phase de la recherche soit achevée et que les essais soient au point, soient même normalisés, d'importants problèmes se posent car les essais s'appliquent à des produits et ces produits sont mis sur le marché, sont commercialisés, sont en concurrence avec d'autres produits. Les fabricants ou leurs mandataires mettent en avant autant qu'ils le peuvent, ce qui est normal, les caractéristiques de leurs produits. Pour ce faire ils se fondent le plus souvent sur les essais et font même parfois appel à des organismes tiers qui certifient les caractéristiques en question ou qui certifient la conformité à une spécification technique de référence (une norme par exemple) qui fournit les caractéristiques.

Les résultats d'essais deviennent ainsi un argument commercial.

Il est donc essentiel pour une information honnête des utilisateurs des produits et pour une concurrence loyale entre produits que les essais soient de qualité.

Il est également important que sur l'ensemble du marché les mêmes produits soient partout caractérisés par les mêmes essais ce qui n'est pas le cas aujourd'hui dans les pays de la Communauté européenne.

Mais même si l'essai est le même il est essentiel que les résultats soient exprimés de la même façon, avec les mêmes unités.

On doit être conscient des enjeux industriels qui se trouvent derrière les valeurs annoncées par les uns et les autres. Par exemple, de très légers écarts entre les valeurs de la conductivité thermique de deux matériaux isolants peuvent déclencher une véritable guerre commerciale et des contestations.

C'est pour cette raison que les fabricants les plus sérieux, ceux qui pratiquent dans leur usine une politique de la qualité, sont favorables à la certification des caractéristiques de leurs produits par un organisme tiers dont les essais ne soient pas contestables, plutôt qu'à une simple déclaration du fabricant basée sur des essais effectués par lui.

On observe ce phénomène en France dans plusieurs secteurs où les industriels ont été favorables à la mise en place de certifications de produits, basées sur des règlements techniques très stricts faisant appel à des essais très élaborés. Ces industriels ont d'ailleurs constaté que la pratique de ces essais les conduisait à faire progresser la qualité de leurs produits et qu'ainsi les essais étaient, au-delà d'un moyen de contrôle et de promotion commerciale des produits, un instrument de progrès technologique. Le surcoût que représentent pour le fabricant des essais sérieux et complets est largement compensé par les progrès dans la production. Améliorer la qualité des essais fait intervenir les mêmes ingrédients qu'améliorer la productivité de fabrication.

### 3 Essais de contrôle sur ouvrages terminés

Il existe des essais qui sont pratiqués dans les bâtiments terminés au titre du contrôle. C'est le cas lorsque la satisfaction d'une exigence dépend beaucoup de la qualité d'exécution des travaux. Cela se rencontre en acoustique lorsqu'il s'agit d'évaluer l'isolement entre locaux aux bruits aériens ou aux bruits d'impact.

Certaines réglementations, comme la réglementation française pour l'acoustique en matière de logements, imposent un tel contrôle dans les bâtiments achevés. D'autres comme la réglementation anglaise pour l'acoustique des logements offrent plusieurs moyens de prouver la conformité aux objectifs imposés et parmi ces moyens il y a celui qui consiste à faire des essais dans des bâtiments existants dans lesquels ont été utilisées les mêmes techniques que celles qui sont envisagées dans le projet à juger.

En thermique il existe aussi certains essais qui permettent de déceler des ponts thermiques dans des façades de bâtiments réalisés.

Tous ces essais in situ sont généralement délicats car la maîtrise des divers paramètres est évidemment moins aisée qu'en laboratoire. Mais ils correspondent à une voie qui mériterait d'être explorée. En effet ce qui compte pour l'usager ce ne sont pas les performances d'un bâtiment évaluées par le calcul sur la base des plans, mais ce sont les performances réellement fournies par le bâtiment une fois réalisé. Un gros effort devrait être fait pour développer des essais permettant in situ de mesurer ces performances.

Le contrôle prendrait alors toute sa dimension et serait authentiquement fondé sur une démarche performancielle, puisqu'on jugerait sur le résultat final.

**En conclusion** il convient d'insister sur le rôle des essais à tous les stades du déroulement d'un projet de construction.

Des essais appropriés et réalisés par des laboratoires qui pratiquent pour eux-mêmes une politique de la qualité sont de nature à favoriser l'amélioration de la qualité de la construction, en même temps qu'ils assainiraient les rapports entre partenaires et concurrents.

Dans le marché unique européen où en principe les mesures protectrices d'ordre réglementaire disparaîtront et ne pourront plus s'opposer à la pénétration de produits étrangers dans un Etat, il est probable que les essais joueront un rôle essentiel. C'est sur eux que repose la crédibilité de ce qu'on appelle la "nouvelle approche" visant à la libre circulation des produits. Si les résultats d'essais ne sont pas crédibles, les utilisateurs, les acheteurs et les Etats reconstitueront de nouvelles barrières pour se protéger contre la pénétration de produits douteux.

Pour que le dispositif fonctionne il faut donc :

- des laboratoires compétents, neutres à l'égard des intérêts en cause

- des méthodes d'essai harmonisées

- des équipements d'essai régulièrement contrôlés dans le cadre de politiques d'assurance-qualité des laboratoires.

# 4 LES BESOINS SPÉCIFIQUES DU GÉNIE CIVIL
## (The specific needs in civil engineering)

J.F. COSTE
Laboratoire Central des Ponts et Chaussées, Paris, France

**Abrégé**

Il convient de tenir compte des besoins spécifiques du Génie Civil pour l'assurance qualité des laboratoires d'essais et la relation de celle-ci avec la qualité de service de l'ouvrage terminé.
La spécificité des essais pour le Génie Civil est abordée en trois points successifs :
- *les essais de matériaux* : s'agissant de matériaux composites et complexes, on a recours non seulement à des essais sur échantillons en laboratoire, mais aussi à des essais in situ caractéristiques du milieu à tester,
- *les essais de contrôle* : la qualité de service de l'ouvrage, en général prototype unique, est très dépendante de la mise en oeuvre des matériaux et des essais de contrôle doivent être faits tout au long de la construction ; ces essais portent tant sur les résultats performanciels que sur la manière de mettre en oeuvre,
- *les essais de conformité* : l'ouvrage ne fait pas l'objet d'une réception unique une fois achevé, mais est réceptionné en général par étapes successives par le client qui exige des essais de conformité pour chaque partie d'ouvrage.

En conclusion, la qualité des prestations d'un laboratoire d'essais de Génie Civil est très dépendante du contexte extérieur au laboratoire dont il convient de tenir compte dans l'établissement d'un Plan Qualité.

Keywords : Qualité des essais des matériaux, contrôle de mise en oeuvre, conformité et durabilité des ouvrages.

# 1 Introduction

Les essais de laboratoire interviennent en Génie Civil depuis la conception de l'ouvrage jusqu'au moment où il est livré au gestionnaire de l'ouvrage. Ainsi peut-on distinguer :
- *les essais préalables des matériaux* constitutifs de l'ouvrage au cours de l'élaboration du projet,
- *les essais de contrôle* au cours de la construction de l'ouvrage,
- *les essais de conformité* au moment de sa réception.

Ces types d'essais successifs qui portent sur des matériaux constitutifs variés participent à la garantie de la qualité de service de l'ouvrage et possèdent leurs spécificités propres au Génie Civil et aux usages de la Profession. Ils interviennent en complément d'autres essais "in situ", souvent non destructifs, au cours des différentes étapes de la construction de l'ouvrage.

Ils ont par eux-mêmes une incidence économique importante : la résistance d'un béton n'est-elle pas conforme au cahier des charges et c'est tout l'ouvrage - ou une partie de l'ouvrage- qui risque d'être refusé par le Maître d'oeuvre.

La qualité des essais de laboratoire a donc une importance tant du point de vue technique qu'économique. C'est dans ce double contexte que nous allons examiner les particularités des essais de Génie Civil au cours des trois étapes précitées.

° °
°

# 2 Les essais préalables de matériaux

Les ouvrages de Génie Civil étant implantés dans la nature ont recours en général à la fois à des matériaux fabriqués à partir de produits de base (ciments, sable, ...) et à des matériaux en place (sol de fondation par exemple).

## 2.1 Cas des matériaux fabriqués à partir de produits de base

Les essais préalables de matériaux en laboratoire ont deux objectifs essentiels :
- caractériser la méthode optimum de fabrication du matériau constitutif de l'ouvrage ou d'une partie de cet ouvrage,
- vérifier que le matériau constitutif testé conduit bien aux performances souhaitées.

**Le premier objectif** implique que l'on tienne compte des conditions réelles de mise en oeuvre du matériau sur le chantier : à quoi bon proposer un béton de très haute résistance si ce dernier est trop sec et ne peut être pompé ? De même, les conditions de soudabilité d'un acier peuvent le rendre inapte à son utilisation au soudage sur chantier.

L'essai préalable apparaît donc, en premier lieu, comme un essai d'interface entre la méthode de fabrication du matériau et sa mise en oeuvre dans la structure.

Il s'ensuit que le matériau retenu résultera d'un compromis entre ses qualités performancielles (résistance, déformation, ...) et ses qualités de mise en oeuvre.

**Le second objectif** implique que les performances puissent être caractérisées par des grandeurs mesurables et faire l'objet d'essais présentant une répétitivité suffisante pour donner lieu à une exploitation statistique significative, assurant ainsi une bonne représentativité du matériau mis en oeuvre.

Or les matériaux de Génie Civil sont très souvent des géomatériaux, matériaux composites complexes, constitués d'un mélange de produits industriels (aciers, ciments) de produits naturels précalibrés (sable, granulats) ou même naturels (graves par exemple).

Il s'agit souvent à l'origine de sous-produits de l'industrie (cas de la fumée de silice) qui ont des caractéristiques complexes, variant suivant la provenance ou le mode de fabrication sans que l'on en maîtrise les paramètres autrement que de façon empirique, à l'exemple des bitumes routiers.

Le matériau constitué de ces produits de base devra également tenir compte du processus de fabrication réel de chantier qui est parfois difficilement reproductible en laboratoire, notamment par suite de l'effet d'échelle entre les méthodes de production sur le chantier et celles de fabrication en laboratoire.

Ce matériau est donc difficile à modéliser et à caractériser par des grandeurs physiques, compte tenu du grand nombre de facteurs d'influence qui peuvent intervenir.

Le plus souvent, on devra se contenter de mesurer des grandeurs d'usage, d'après des essais qui simulent la manière dont le matériau travaillera dans l'ouvrage ou sont censés le faire : les matériaux à base de produits bitumineux tels que les enrobés bitumineux en sont un exemple flagrant où chaque caractéristique fait l'objet d'un essai spécifique et empirique : résistance à l'écrasement, fluage, vieillissement, fragilité au froid, orniérage...

Il en résulte deux inconvénients majeurs :

1) **Le premier inconvénient** est une grande sensibilité des résultats d'essais aux variations du mode opératoire, sans que l'on puisse vraiment savoir avec précision si la variation des résultats obtenus affecte avec la même sensibilité la qualité de l'ouvrage construit avec ce matériau.

Cette sensibilité est d'autant plus grande que les performances du matériau croissent : 5 % de baisse de résistance à la compression d'un béton ordinaire ayant une résistance moyenne de 35 MPa représente 1,75 MPa. Pour un béton à haute performance ayant une résistance moyenne de 90 MPa, cette baisse représente 4,5 MPa !

La tendance des normes d'essai est alors à un contrôle parfois excessif de tous les paramètres pouvant intervenir dans les résultats (diamètre des moules, températures, humidité...) qui exigent une qualité des essais en laboratoire très pointue.

Afin de mieux cerner le comportement d'un matériau, on a souvent recours à une batterie d'essais dont il faut assurer la cohérence.

Pour ce faire, les derniers sont regroupés dans des programmes d'essais qui comportent :

- le choix des méthodes d'essais pertinentes (choix des essais, répétitions, séquences...),
- les conditions d'échantillonnage et de préparation,
- l'exécution des essais, par un laboratoire accrédité,
- l'interprétation des résultats,
- la présentation de ces résultats.

Les Grands Equipements de Recherche en Génie Civil participent également à une meilleure approche du comportement des matériaux en permettant de simuler leur comportement en vraie grandeur ou du moins à des échelles et des conditions proches du contexte réel de travail. Grâce à une instrumentation fine, ils fournissent des mesures qui peuvent être mises en corrélation avec les résultats sur éprouvettes du même matériau en laboratoire.

Ainsi se sont développés :

- dans le domaine routier : les manèges de fatigue pour expérimenter le comportement des revêtements de chaussées,
- dans le domaine des structures : les essais sur assemblages, ou les composants eux-mêmes, les bancs d'essai des câbles, les tables de vibration ...

2) **Le second inconvénient** est que les méthodes d'essais et les normes de matériau varient la plupart du temps d'un pays à l'autre sans que l'on puisse recouper les résultats autrement que par des corrélations statistiques, compliquant par ailleurs les essais d'intercomparaison entre laboratoires. Cet inconvénient devrait cependant être progressivement atténué avec le développement de normes internationales (ISO ou EN).

*La qualité des essais de laboratoire aurait donc tout à gagner d'une meilleure modélisation du comportement des matériaux de Génie Civil, mettant en relation dans des modèles les lois de comportement du matériau décrites par des grandeurs physiques avec ses propriétés d'usage qui sont mises en évidence par les essais courants.*

## 2.2 Essais préalables des matériaux sur place

Tout ouvrage de Génie Civil repose sur le sol dont il convient de connaître les caractéristiques avant de définir le type de fondation, qu'il s'agisse d'une route ou d'un ouvrage d'art. Les caractéristiques de ce même sol peuvent du reste influer fortement sur le type d'ouvrage qui sera projeté.

La connaissance des caractéristiques du sol en place à l'aide d'essais de laboratoire se heurte alors à une difficulté supplémentaire : celle d'obtenir des échantillons du matériau en place sans en pertuber les caractéristiques au cours de son extraction : c'est-à-dire celle d'obtenir des "échantillons dits intacts".

Bien souvent les résultats de laboratoire devront être recoupés par des essais in situ avec toutes les difficultés de relier les caractéristiques mesurées sur éprouvette en laboratoire et celles repérées sur place.

# 3 Les essais de contrôle de mise en oeuvre

Un ouvrage de Génie Civil est en général un prototype ; il appartient au mieux à une série limitée (cas des ponts-types d'autoroute par exemple) dont chaque exemplaire doit s'adapter au contexte local : d'où la nécessité de contrôler la mise en oeuvre des matériaux.

Même si les opérations de fabrication des matériaux et des éléments de base sont devenues très mécanisées, leur mise en oeuvre est encore soumise à une recherche de productivité croissante, sans pour autant être le fait d'une main-d'oeuvre spécialisée.

Le "contrôle intégré" de la part de l'Entreprise permet de s'assurer de la qualité de la fabrication des matériaux, et de contribuer à la qualité de leur mise en oeuvre par exemple en garantissant un certain nombre de relevés de données concernant celle-ci (coffrages, ferraillage, vibration du béton...).

Il convient de souligner à cette occasion que la qualité de la mise en oeuvre ne se traduit pas uniquement en termes de performances techniques ; en effet, la qualité esthétique de l'ouvrage et son insertion dans l'environnement sont des éléments qui prennent de plus en plus d'importance et le contrôle de la mise en oeuvre aura à s'exercer également sur ces points : aspect des parements en béton, réglage des talus, finition des équipements de voirie...

Les essais de laboratoire sur des échantillons prélevés sur place ne viendront donc qu'en complément des contrôles sur place.

Leur qualité sera elle-même très dépendante de la manière dont le matériau a été prélevé sur place et les éprouvettes réalisées et protégées au cours de leur transport au laboratoire : combien de fois la baisse de résistance d'un béton observée sur un chantier reflète en réalité non pas la baisse de qualité des performances du béton en place, mais celle des prélèvements !

Cependant la qualité des essais en laboratoire reste essentielle du double point de vue technique et économique :

- technique, car les performances du matériau n'étant souvent obtenues qu'au bout d'une certaine période (cas de bétons, des graves-ciment,...), il faut connaître avec précision si les caractéristiques du matériau -comme sa résistance par exemple- sont atteintes avant de lancer une nouvelle phase de construction,
- économique, car si les résultats livrés par le laboratoire montrent que le matériau n'a pas les caractéristiques requises, c'est toute la partie de l'ouvrage correspondante qui est mise en cause, avec des enjeux financiers importants à la clé.

## 4 Les essais de conformité

### 4.1 Les essais de conformité

Les Entreprises ont une obligation de conformité de l'ouvrage aux spécifications du marché. La vérification de celle-ci passe par la réalisation des essais de conformité qui doivent tenir compte de deux caractéristiques des ouvrages de Génie Civil :

- chaque ouvrage constitue en général un exemplaire unique, ou fait partie d'une petite série, de sorte qu'il n'est pas possible de contrôler par échantillonnage la qualité de l'ouvrage,
- chaque ouvrage est implanté sur son site définitif, ce qui implique que le laboratoire procède au moins à une partie des essais sur place en instrumentant le dit ouvrage ; ces essais devront donc être exécutés dans des conditions climatiques et matérielles non contrôlables et au mieux peu contrôlables.

Ainsi, la conformité de l'ouvrage de Génie Civil sera examinée à partir d'essais effectués sur l'ouvrage terminé ou sur des parties d'ouvrage au cours des phases de construction. Ces essais se résument en général à vérifier la conformité de la géométrie de l'ouvrage à vide et sa déformabilité sous charge maximum, ainsi que son étanchéité si l'ouvrage doit remplir cette fonction.

Lorsque l'ouvrage présente un linéaire important (cas d'une chaussée par exemple), ces essais sont réalisés à l'aide d'appareils à grand rendement.

## 4.2 Conformité et durabilité

Contrairement aux essais de type qui permettent de caractériser le comportement à l'usage d'une production industrielle en service, les essais de conformité ne garantissent que de façon approchée le comportement de l'ouvrage de Génie Civil dans le temps et permettent mal de prévoir sa durabilité, c'est-à-dire sa résistance aux agents atmosphériques ainsi que sa résistance à la fatigue vis-à-vis des actions de service.

Par ailleurs, les essais en laboratoire n'apportent aujourd'hui que des garanties sur certains de ces éléments constitutifs : corrosion des aciers, résistance des peintures aux agressions atmosphériques...

Or, la durabilité des ouvrages devient, pour le gestionnaire des infrastructures, un objectif primordial, car ce dernier doit pouvoir garantir aux usagers un service constant alors que le patrimoine à entretenir s'accroît.

L'amélioration de la prédiction de la durée de service d'un ouvrage nécessite de progresser dans trois directions :
- meilleure connaissance statistique des sollicitations naturelles ou de service,
- meilleure connaissance de l'état de l'ouvrage à un instant donné car le comportement du matériau dans une structure dépend tout autant de son évolution physico-chimique (micro-fissuration, altération de ses composants...) que des sollicitations extérieures,

- enfin, meilleure connaissance du comportement des matériaux constitutifs dans le temps sous les sollicitations naturelles et de service.

Pour les deux premiers points, il faut faire appel à des mesures sur les ouvrages eux-mêmes et par la suite à leur instrumentation par le laboratoire qui assurera le relevé des données et leur suivi. C'est l'un des objectifs des Projets Nationaux développés dans le cadre du Programme de Recherche en Génie Civil (PROGEC).

Une fois mieux connues les sollicitations et le comportement du matériau au sein des structures, il deviendra possible de répondre au troisième point grâce à des essais accélérés en laboratoire : à partir des données statistiques des sollicitations (données climatiques, charges de service...) et des données relatives à l'état du matériau interne à la structure (humidité, fissuration...), le comportement de l'ouvrage dans le temps pourrait être mieux prédit à l'aide de modèles de simulation sur ordinateur, calés sur les essais précédents. (Cette approche a été décrite dans l'article de MM. Niels PLUM, Joern JESSING et Per BREDSDORF paru en Mars 1966 dans le Bulletin n°30 de la RILEM).

## 5 Conclusion

La prise en compte de la qualité des essais de Génie Civil exécutés en laboratoire, par référence au Manuel Qualité, a été un progrès décisif de ces dernières années.

Cependant, elle ne constitue qu'une première approche de la qualité de l'ouvrage en service, car l'objectif final est bien l'approche performancielle de la qualité de l'ouvrage.

Tout progrès dans la qualité des essais de Génie Civil passe désormais moins par un respect méticuleux des conditions d'essais que par une amélioration de la connaissance des fondements scientifiques et techniques de ces essais au regard du comportement des matériaux dans les structures soumises aux sollicitations diverses.

PART TWO

# TECHNICAL COMPONENTS OF TEST QUALITY

# 5 CALIBRATION AND MAINTENANCE OF TESTING EQUIPMENT

P.J.H.A.M. VAN DE LEEMPUT
Nederlandse Kalibratie Organisatie, Delft, The Netherlands

Abstract
In testing it is of importance that the testing equipment functions properly. Regular maintenance and periodic calibrations are inevitable. This paper describes what calibration and traceability are. Moreover information is given on recalibration intervals and the European infrastructure which is set up to provide traceability for measurement and for testing laboratories.
Keywords: Maintenance, calibration, traceability, calibration intervals, Western European Calibration Cooperation.

## 1 Introduction

Certification of quality systems is becoming more and more important. When setting up such quality systems, one of the main aspects is testing and, as a consequence, calibration. The goal of testing is to obtain correct results, together with a realistic statement of the actual measurement uncertainty. To obtain correct results it is necessary that the measuring equipment functions properly. Apart from regular maintenance, this means that tests with different instruments, intended to measure the same property, give the same results (good reproducibility). In this paper the need for calibration is described and the way to obtain traceable results.

There are two factors which mainly influence the accuracy of test results:

The way in which the test is performed.
The accuracy of the testing equipment used.

The accuracy of testing equipment is checked by calibration. Apart from the results of periodical calibrations the history and maintenance of the equipment can give a lot of information which results in more confidence in the performance of the equipment. The way in which tests are performed is the subject of other contributions to this symposium.

## 2 The purchase and maintenance of testing equipment

The accuracy of test results is mainly influenced by the type of equipment used. This does not mean that, when planning the purchase of new equipment, one must buy the most advanced and most modern equipment. More advanced technology often means more possibilities of deterioration of the equipment. It is very important to formulate the requirements and the specifications in advance; factors such as required measurement uncertainty, frequency of use and the price should be taken into account.

To have a good insight into the possibilities of measuring and testing equipment it is required to build up a history of the equipment. An adequate registration must be kept for that purpose. These records include:

The name of the item of equipment.
The name of the manufacturer, the type identification and serial number or other unique identification.
The date the item is received and the date on which it is placed in service.
The location where the item is place (where appropriate).
The condition when it was received; e.g. new, used, reconditioned.
A copy of the instructions of the manufacturer.
The dates and the results of the calibrations.
The recalibration interval or the date of the next calibration.
Details of maintenance carried out, and planned for the future.
Details of any damage, modification, repair or malfunction.

To ensure that the equipment is always in the proper condition, a maintenance program should be developed, and maintenance procedures must be laid down.

## 3 Calibration and traceability

Maintenance alone is not sufficient. Calibration is required, calibration in a way that the measurements are traceable. The Vocabulaire International de Métrologie defines calibration as [1]:

> The set of operations which establish, under specified conditions, the relationship between values indicated by a measuring instrument or measuring system, or values represented by a material measure, and the corresponding known values of a measurand.

The result of a calibration permits the estimation of errors of indication of the measuring instrument, measuring system or material measure, or the assignment of values to marks on arbitrary scales. Key words of the definition are: "known values of a measurand". Normally these known values have also an assigned measurement uncertainty (better than the device which is calibrated), obtained from another calibration.

Going back with calibrations to the level of national or international standards, we reach traceability:

> the property of a result of a measurement whereby it can be related to appropriate measurement standards, generally international or international standards, through an unbroken chain of comparisons [1].

National standards are developed, maintained and disseminated by national standards laboratories, e.g. National Physical Laboratory (NPL, United Kingdom), Physikalisch Technische Bundesanstalt (PTB, Federal Republic of Germany), Van Swinden Laboratory (VSL, Netherlands) and National Institute for Standards and Technology (NIST, USA).

International cooperation and comparisons by the national standards laboratories ensures that the results of calibrations by these laboratories are of equivalence and reliable.

**4 Recalibration intervals**

The results of calibrations are normally recorded in a calibration certificate or a calibration report. These results obtained on the moment of calibration are no guarantee that the same results will be obtained after a period of time, and that he initial calibration results are still reliable. This means that the equipment must be recalibrated after a certain period of time.

It is impossible to define these recalibration intervals in general, as many factors can influence these periods. Amongst these factors are:

Type of the equipment.
Recommendation of the manufacturer.
Trend data obtained from previous calibrations.
History of maintenance and servicing.
Extent and severity of use.
Tendency to wear and drift.
Frequency of cross-checking against other reference standards.
Frequency and quality of in-house check calibrations.
Environmental conditions (temperature, humidity, vibration, etc.).
Accuracy of measurement sought.

Task Force E of the International Laboratory Accreditation Conference (ILAC) prepared in 1983 a report on recalibration intervals of measuring equipment used in testing laboratories [2]. In 1984 this report was published by the Organisation International de Métrologie Légale (OIML) as International Document no 10 [3].
The first part of this document gives guidelines for the determination of recalibration intervals, whereas part 2 contains examples of initial recalibration intervals.
To illustrate the complexity of the subject some examples:
For analog voltmeters (reference standards, not very often used and

in general well cared for) the Netherlands and Finland recommend a recalibration interval of 12 months, whereas New Zealand accepts 60 months. For the voltmeters which are used frequently all these countries recommend 12 months.
The recommended recalibration intervals for roundness standards vary from 6 months (Mexico) to 60 months (South Africa).

Having bought a measuring instrument one of the tasks is to choose the initial recalibration period. First of all the risk of an instrument going out of tolerance must be as small as possible; but second, laboratories wish to keep the annual calibration costs to a minimum.

Although intuition and general knowledge of the engineer are normally the basis of the initial recalibration interval, the most important factors to take into account are: the manufacturer's recommendation, the expected extent and severity of use, the influence of the environment and the accuracy of measurement sought. Having built up a certain history of the equipment it can be necessary to review the recalibration interval, taking into account also the other factors mentioned in the beginning of this chapter. For some techniques how to do this I refer to references [2] and [3].

## 5 Organizations which provide traceable calibrations

Most of the countries have national standards laboratories which calibrate the measurement standards of laboratories and industries. Some of the national laboratories also carry out traceable calibrations on a lower accuracy level.

Laboratories and industries carry out traceable calibrations for internal purposes and for external purposes, e.g. as part of servicing the supplied measurement instruments.

It is self-evident that there are differences between the quality of the calibration laboratories. Some laboratories take this task very seriously and deliver good quality and traceability.
In order to distinguish these laboratories from others and in order to provide an infrastructure of traceability, a lot of national standards laboratories have set up a laboratory accreditation scheme for calibration (and measurement) laboratories. A list of the western European calibration services (both EC and EFTA countries) is given in table 1.

These organizations accredit laboratories against criteria which are based on EN 45001 [4]. Special attention is paid to the traceability, including recalibration intervals and demonstrated best measurement capabilities. These additional criteria are laid down in WECC Doc. 17 (see Annex 1). Measurement audits are part of the assessment procedure and surveillance scheme. In these respects the laboratory accreditation schemes for calibration and measurement laboratories differ from the schemes for the accreditation of testing laboratories. In general however they apply the same rules [5] and are organized in a similar way [6].

In the EC and EFTA (European Free Trade Association) the national calibration services mentioned in table 1 (and representatives of Greece, where a calibration service will be set up in the near

Table 1. Western European calibration services

Austria: Österreichischer Kalibrierdienst (ÖKD)
Belgium: Belgische Kalibratie Organisatie (BKO,
Organisation belge d'Etalonnage (OBE)
Denmark: Statens Tekniske Provenaevn (STP)
Federal Republic of Germany: Deutscher Kalbrierdienst (DKD)
Finland: Mittauspalvenu (MSF)
France: Bureau National de Métrologie (BNM)
Ireland: Irish Laboratory Accreditation Board (ILAB)
Italy: Servizio di Taratura in Italia (SIT)
Netherlands: Nederlandse Kalibratie Organisatie (NKO)
Norway: Norsk Kalibreringstjeneste NKT)
Portugal: Instituto Português da Qualidade (IPQ)
Spain: Sistema de Calibración Industrial (SCI)
Sweden: Svensk Mätplatsorganisation (SMO)
Switserland: Swiss Calibration Service (SCS)
United Kingdom: National Measurement Accreditation Service (NAMAS)

future) cooperate in the Western European Calibration Cooperation (WECC). WECC started informally in 1975; on 9 June 1989 a Memorandum of Understanding was signed by all participants. The main goal of WECC is to reach mutual acceptance of each others calibration certificates. Initially this was achieved on a bilateral basis. On 1 December 1989, however, a multilateral agreement was signed by the Federal Republic of Germany, Finland, France, Italy, the Netherlands, Sweden, Switzerland and the United Kingdom.

This means that many hundreds of laboratories have demonstrated that they have a quality system which ensures that the laboratories provide traceable calibrations (and measurements) with an associated measurement uncertainty.

Testing laboratories can make use of this European infrastructure for metrology to improve their quality system by having calibrated (whenever possible) the equipment by a laboratory which is accredited for that purpose.

A list of WECC Documents is given in Annex 1.

## 6 Traceability is not available; what to do between calibrations

In a number of cases it is not possible to be traceable. This can be the case e.g. when there are no accredited laboratories or even no national standards laboratories to provide traceability. Solutions should be found in other techniques to obtain confidence in the performance of the equipment.

Intercomparisons with other laboratories or even with other equipment in your own laboratory is a solution. The periodic check with certified reference materials is another technique to obtain reliable test results. Interlaboratory testing and reference materials are subject of other contributions so I restrict myself to some remarks.

### 6.1 Intercomparisons

Both in metrology and in testing intercomparisons are organized. In general these intercomparisons have at least the following three purposes:

- To verify whether accredited laboratories achieve the best measurement capability they claim.
- To establish, by comparing the results with the results of a standards laboratory, acceptable traceability for accreditation.
- On a voluntary basis to verify the performance of equipment and of laboratories.

Guidelines for intercomparisons are given in [7] and [8].

### 6.2 Reference materials

Certified reference materials play an important role in establishing the performance of a measurement process. For a guide on the uses of reference materials we can refer to ISO/IEC Guide 33 [9]. Apart from taking over the role of calibration of the equipment they are also inevitable for in-between checks. For this purpose 'working' reference materials can be prepared which are traceable to the certified ones.
In the chemical field it is obvious that certified reference materials should be traceable to national or international standards (e.g. mass). Due to the initiative of some leading laboratories in Europe EURACHEM has been established with the main task to discuss the problem of traceability in analytical chemistry.

## 7 Concluding remarks

Testing can only give good results when the performance of the equipment is well known. Maintenance, traceability and calibration are the basic ingredients for this performance. In western Europe an infrastructure has been set up to provide traceability to both calibration and testing laboratories. The problem of traceability in analytical chemistry is in discussion.
ISO TC 176 is working on an International Standard on measuring and testing equipment [10].

Annex 1

List of WECC Documents

01-1990 List of WECC Documents
02-1990 Information Brochure
03-1990 Cooperating Services
04-1990 Circulation of Audits
05-1990 Plenary Meetings
06-1990 Expert Meetings
07-1990 Reciprocal Declarations of Equivalence
08-1990 A Memorandum of Understanding between National Calibration Services
09-1990 Multilateral Agreement between the National Calibration Services of Finland, France, the Federal Republic of Germany, Italy, The Netherlands, Sweden, Switzerland and the United Kingdom
11-1987 Desirable Features of a National Calibration Service
12-1988 Requirements and Recommendations Concerning Certificates Issued by Accredited Laboratories
13-1987 Establishment and Maintenance of Confidence Between Calibration Services
14-1987 Monitoring the Performance of Calibration Services
15-1987 WECC International Measurement Services
16-1988 The Evaluation of Calibration Services
17-1988 Additional General Requirements for the Accreditation of Calibration Laboratories
18-1989 Guidelines for the Expert Groups of WECC
19-1990 Guidelines for the Expression of the Uncertainty of Measurement in Calibrations

WECC(89)1 Requirements for the accreditation of calibration laboratories by national calibration services
WECC(89)2 Recommended environmental conditions

Annex 2

## References


1 BIPM, IEC, ISO, OIML, International Vocabulary of Basic and General Terms in Metrology. Geneva, International Organization of Standardization, 1984.

2 Report of Task Force E to ILAC 83.

3 OIML International Document no 10 - Guidelines for the determination of recalibration intervals of measuring equipment used in testing laboratories
Part 1: The Guidelines
Part 2: Examples of initial re-calibration intervals

4 EN 45001 - General criteria for the operation of testing laboratories

5 EN 45002 - General criteria for the assessment of testing laboratories

6 EN 45003 - General criteria for laboratory accreditation bodies

7 ISO 5725 - Precision of test methods - Determination of repeatability and reproducibility by inter-laboratory tests (in revision).

8 ISO/IEC Guide 43 - Development and operation of laboratory proficiency testing.

9 ISO/IEC Guide 33 - Uses of certified reference materials.

10 ISO DP 10012-1 - Quality assurance - Requirements for the control of measuring and testing equipment.

# 6 REPEATABILITY, REPRODUCIBILITY AND SOME PARAMETERS INFLUENCING THE PRECISION OF TEST METHODS

H. SOMMER
Forschungsinstitut der Vereinigung der Österreichischen Zementindustrie, Vienna, Austria

**Abstract**
Today a high precision of testing is more important than in former years: Production has become more uniform, the actual product quality should exceed requirements only as much as necessary for economical reasons, but because of the new legal regulations concerning product liability, it should never fail to pass. On the other side, with the increasing international trade more laboratories and different countries are concerned, making it more difficult than ever to keep the precision of testing at a high level.

High quality equipment, good staff, and an excellent quality assurance manual alone do not guarantee satisfactory precision. Frequent exercise and continuous comparisons are also a pre-requisite for attaining and maintaining a high quality in testing. This is demonstrated using the experiences reported at the Rilem-Workshop "Evaluation of Cement and Concrete Laboratory Performance" 1989 in Tel Aviv.

When testing cement for strength the deviations between well equipped laboratories may be twice that of laboratories which test very frequently and are connected by frequent comparative tests.

## 1 Introduction

So far as building materials are concerned the interest in the precision of test methods often was a rather theoretical one in former years: The actual quality was considerably above requirements, the materials were not exported to other countries and they were tested always by the same laboratories well known to the customers.

Today, in some cases, production has become so uniform, that precision of testing is no longer better than uniformity of production. On the other side the product should exceed requirements only as much as necessary for economical reasons; but because of the new legal regulations concerning product realiability, it should never fail to meet the requirements.

The increasing international trade does not make things easier: The more laboratories and different countries are concerned, the more difficult is it to keep the precision of testing at a high level.

The precision of a method of test is by no means a constant, even if all the laboratories have excellent staff and equipment and work to the same quality assurance manual: Exercise and continuous comparisons with other laboratories are a prerequisite for maintaining test quality at a high level.

This is demonstrated using the experiences reported at the Rilem-Workshop "Evaluation of Cement and Concrete Laboratory Performance" 1989 in Tel-Aviv /1/.

## 2 Definitions and procedures

Repeat tests in one laboratory using the same equipment represent repeatability conditions. The standard deviation is smallest when the same operator repeats the tests the same day; it is greater when the tests are repeated by different operators at longer intervals of time.

Identical samples tested at different laboratories represent reproducibility conditions.

The procedures for determining repeatability and reproducibility are described in ISO 5725 /2/ "Precision of test methods - Determination of repeatability and reproducibility for a standard test method by inter-laboratory tests".

## 3 Components of standard deviation (fig. 1)

To determine compressive strength - the most important property of cement and concrete - concrete or mortar mixtures are made

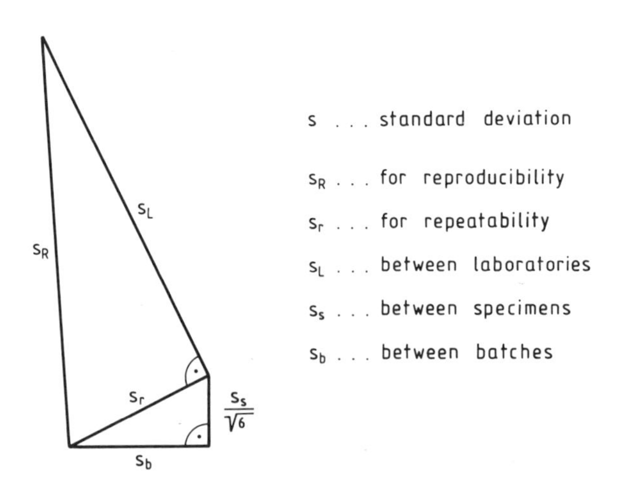


Fig. 1. Components of standard deviation.

and from these 3 specimens are made. From the differences between the 3 or 6 single values (3 mortar prisms give 2 compressive strengths each) the standard deviation $s_s$ between specimens of one batch can be calculated. Its influence is reduced to $s_s/\sqrt{3}$ or $s_s/\sqrt{6}$ by using the mean of the 3 or 6 single values.

When cement is tested for standard strength there is also a standard deviation $s_b$ between different batches and the standard deviation $s_r$ for repeat tests is obtained only by geometrically adding $s_b$ und $s_s$.

When the standard deviation under reproducibility conditions with $s_R$ determined by inter-laboratory tests, the standard deviation for repeat tests (as an avarage for different laboratories) has to be substracted geometrically to find the standard deviation $s_L$ between different laboratories.

## 4 Repeatability

### 4.1 Influence of practice (fig. 2)

The cement research institute in Vienna is doing the third party testing for all the cements produced in Austria, using the methods of test described in the relevant Austrian standard.

In 1982 the CEN method (EN 196 part 1) was introduced as an additional method. To determine repeatability the same cement was tested twice by 5 operators /3/. The results are plotted in fig. 2: in 1982 the relative standard deviation was 2,8 % but in 1984 and the following years the values were between 1,0 and 1,5 %.

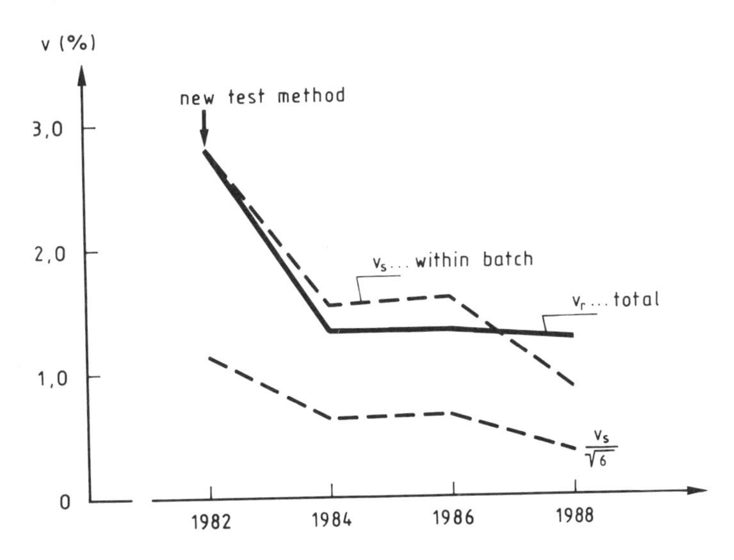


Fig. 2. Testing cement for standard strength - repeatability and influence of practice /3/.

This illustrates the importance of practice - even if the laboratory is well experienced with a very similar method of test and only details (w/c-ratio 0,50 instead of 0,60, different sand and compaction by machine instead of by hand) are changed.

**4.2 Influence of operators** (fig. 3)
In fig. 2 also the relative standard variation $v_s$ between the six compressive strengths found on the specimens of the same batch are plotted.

Though $v_s$ (because of $s_r^2 = s_b^2 + s_s^2/6$) adds only little to the total variation, it appears to be a good indicator for repeatability and operator performance.

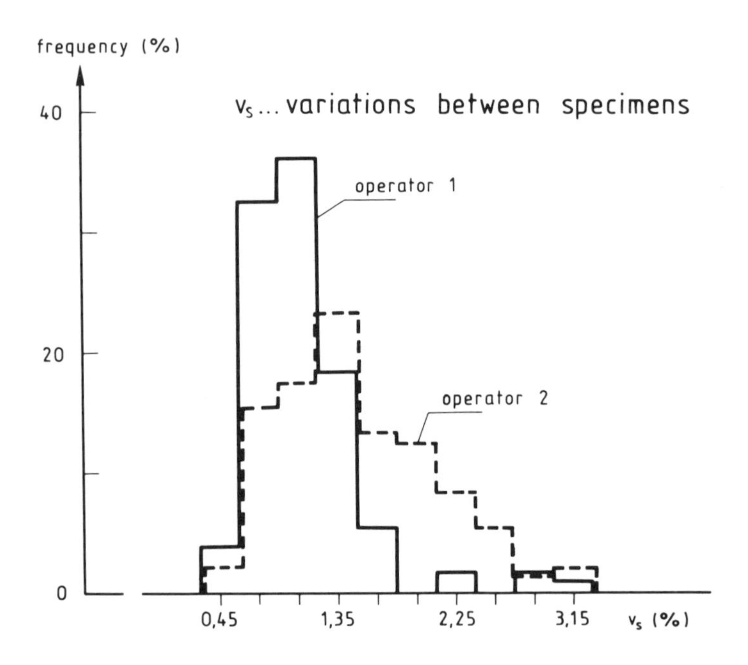


Fig. 3. Testing cement for standard strength - coefficients of variation between specimens of the same batch, plotted for 2 operators /3/.

Fig. 3 shows the relative standard deviations $v_s$ between specimens calculated from the results (6 single values each) of some hundreds of standard strength determinations. For operator 1 the values for $v_s$ are smaller and scatter less than for operator 2 who had joined the laboratory only recently and had less experience.

But the opposite may happen as well: Operators who have done the same, rather monotonous work for many years may develop less uniformity and may need to be transferred to another place of work.

### 4.3 Recommended minimum test frequency

Statistical evaluations as shown in fig. 3 should be made every 2 or 3 months in order to prevent changes of the season or changes in the condition of the equipment used from overshadowing the picture.

If 2 operators are considered necessary (to have one available in case of illness or holiday) and if 30 tests for each operator are considered desirable for statistical reasons this would mean 240 to 360 tests per year.

Of course, test frequency can be relatively low with a direct method of test where there are fewer parameters influencing precision or - as with the Blaine-test f.i. - if it is easy to calibrate the results against a reference material.

### 4.4 Other time-dependant influences

To ensure that no other changes (for instance in curing conditions) have taken place, repeat tests at regular intervals using identical samples are useful.

## 5 Reproducibility

### 5.1 Outliers

When evaluating the results of inter-laboratory tests sometimes up to 25 % of the participating laboratories are eliminated as outliers. Of course, this results in a precision possible only in theory, and does not reflect what happens under practical conditions.

Therefore, results should be only eliminated if they significantly deviate from the other values (compare fig. 4), that means if the Cochran-Test (described in /2/) shows them to be outliers with a high statistical probability of 99 % - and if the cause can be found and removed or the laboratory in question can be excluded from tests of that kind in the future.

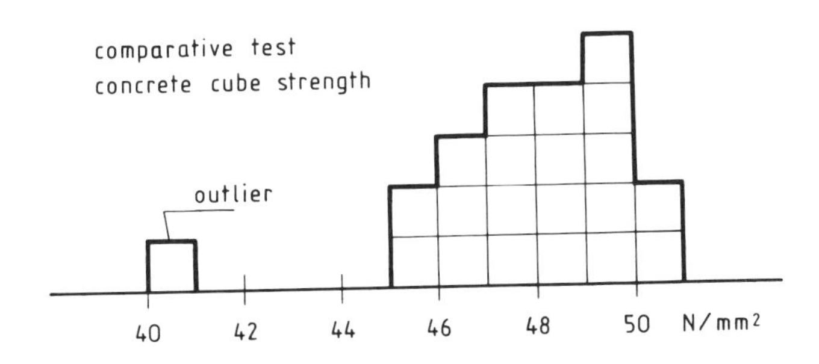


Fig. 4. Results of a comparative test for concrete cube strength.

## 5.2 Influences of curing

Many organisations organize comparative tests at short intervals of time in order to be able to identify and correct disagreements between the laboratories at an early stage.

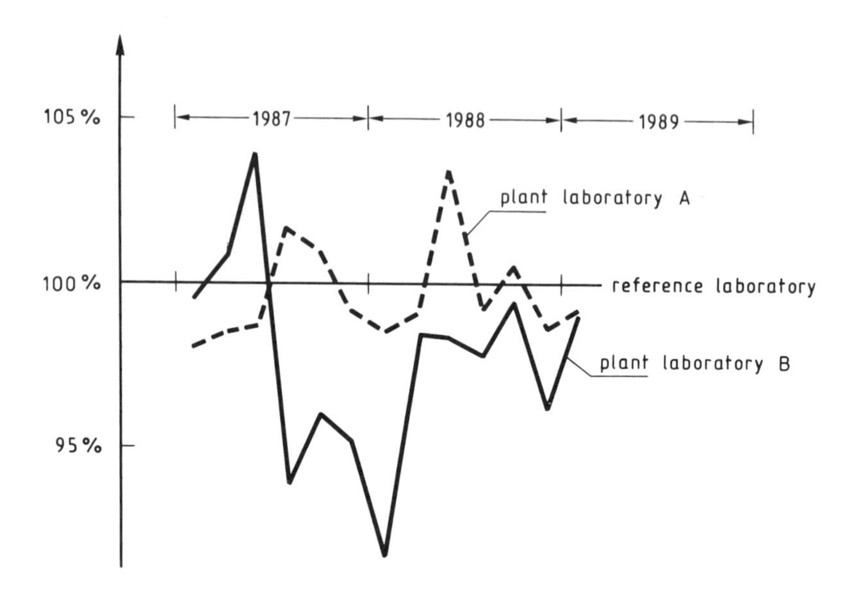


Fig. 5. Differences in the standard strengths found on split samples by a reference laboratory and 2 plant laboratories /4/.

Fig. 5 is an example for comparative tests between a cement reference laboratory and plant laboratories /4/. In winter 1988/89 plant laboratory B found - in contrast to the time before - exceptionally lower strengths. The differences between the strengths found by the reference laboratory and the strengths reported by 15 plant laboratories are plotted in fig. 6. For the time in question plant laboratory B was the only one to differ so much from the reference laboratory.

Inspection of the plant laboratory revealed the cause for the lower strengths: The moist cabinets used for curing the specimens were built into a wall and the adjoining room had not been heated in wintertime; though the thermometer showed the air temperature to be within the range specified, the backside of the moist-cabinet was cooler than formerly.

An additional insulation glued to the backside of the wall was a sufficient remedy.

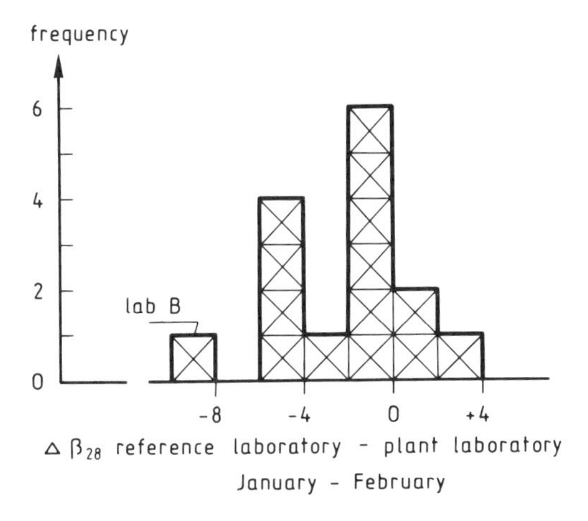


Fig. 6. Differences in the standard strengths found on split samples by a reference laboratory and 15 plant laboratories /4/.

### 5.3 Influence of testing machines

Fig. 7 summarizes the results of the comparative tests organized for many years by the research institute of the German cement industry /5/. Amongst other factors,testing machines with different speeds of loading contributed significantly to the scatter of results. By advising and training the laboratories the standard deviations were reduced from values between 1,8 and 3,2 N/mm$^2$ to values between 1,1 and 2,0 N/ mm$^2$.the scatter of results. By advising and training the laboratories the standard deviations were reduced from values between 1,8 and 3,2 N/mm$^2$ to values between 1,1 and 2,0 N/mm$^2$.

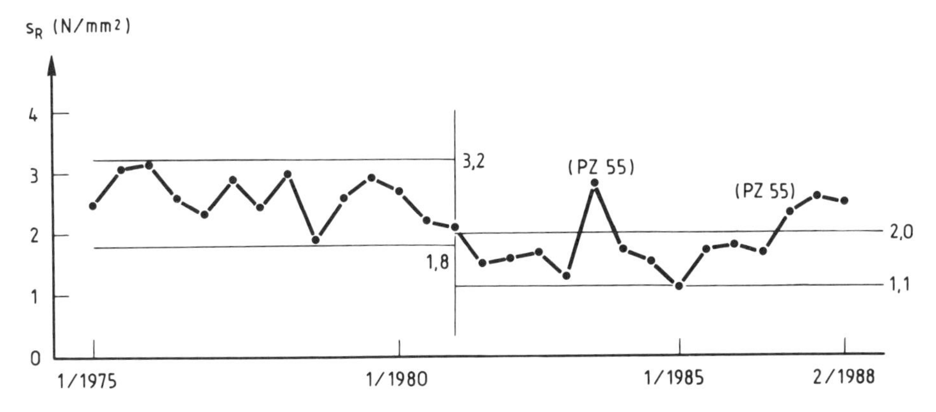


Fig. 7. Testing cement for strength - results of comparative tests between 20 accredited laboratories /5/.

When high strength cements (PZ 55) were tested the standard deviations remained high because "soft" and "stiff" machines still performed differently.

**5.4 Influence of exercise and of frequent comparative tests**
The results of a comparative test are plotted in fig. 8 /4/. Samples of the same cement were sent to 14 plant laboratories and to 14 accredited laboratories. The accredited laboratories had a standard deviation of 3,8 N/mm² - twice that of the plant laboratories.

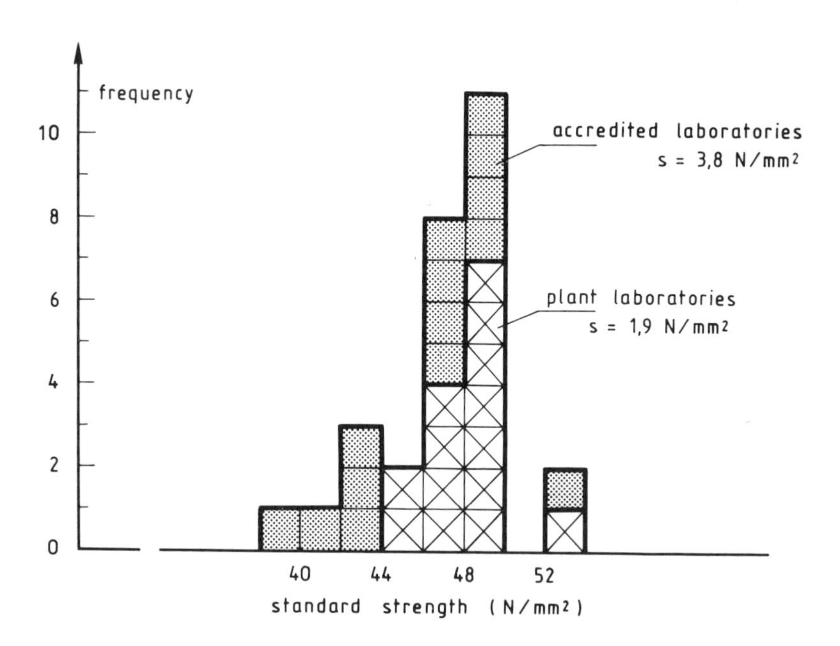


Fig. 8. Results of a comparative test on cement standard strength /4/.

The plant laboratories test split samples of each sample tested by the cement reference laboratory, which evaluates the results of these frequent comparisons and suggests improvements to the test methods. Some of the accredited laboratories test cement samples only by the dozens a year and take part in comparative tests only every one or two years.

This proves that official recognition, good staff, high quality equipment, and quality assurance systems alone do not guarantee optimum results. Frequency of testing, regular comparative tests and continuous efforts to improve the test methods are necessary, too.

**Literature**

/1/ Evaluation of Cement and Concrete Laboratories. Proceedings of the workshop of Rilem Committee TC 91-CRL 1989, Tel-Aviv. Published by United States Department of Commerce, NIST, Gaithersburg/Maryland, USA.
/2/ ISO 5725: Precision of test methods - Determination of repeatability and reproducibility for a standard test method by inter-laboratory tests. Second edition 1986, ISO, Geneve, CH.
/3/ Schütz, P., and Sommer, H.: Test Frequeny and Precision of Results. Contribution to /1/.
/4/ Sommer, H.: Experience with a Cement Reference Laboratory in Austria. Contribution to /1/.
/5/ Wischers, G.: Comparative Tests between 20 accredited Laboratories. Contribution to /1/.

# 7 MATÉRIAUX DE RÉFÉRENCE: UTILISATIONS ET DÉVELOPPEMENTS (Reference materials: uses and developments)

A. MARSCHAL
Laboratoire National d'Essais, Paris, France; REMCO, Paris, France

## A/ LA NECESSITE DES MATERIAUX DE REFERENCE POUR L'ETALONNAGE

Une analyse chimique est une mesure, et "à la base de toute mesure, il existe une référence, un étalon".

Jusqu'à la première moitié de ce siècle, la référence pour le chimiste était essentiellement la boîte de masses permettant le raccordement des dosages gravimétriques et titrimétriques *(Schémas 1 et 2)*. Certaines procédures complémentaires pouvaient être mises en oeuvre pour des techniques particulières *(Schéma 3)*. A partir des années 50, l'évolution des techniques analytiques vers les méthodes instrumentales - relatives ou comparatives - a, dans la pratique, rendu ce type de raccordement caduc.

Les méthodes relatives ont pu, en première approximation faire croire que le corps pur associé à la balance résolvait tous les problèmes *(Schéma 4)*, mais il est vite apparu que tel n'était pas le cas lorsque des analyses exactes étaient nécessaires, et que le personnel compétent était trop coûteux ou trop rare pour être occupé à la préparation d'étalons internes complexes.

La qualité d'une référence ne doit pas, en effet, être appréciée uniquement de manière théorique, en prenant en compte sa seule incertitude intrinsèque. Elle doit aussi être appréciée à la toise de l'effet qu'elle induit sur l'exactitude de la mesure finale qu'elle permet d'effectuer...

Les méthodes modernes d'analyses et d'essais des matériaux étant généralement sensibles aux effets de matrices de diverses natures, le développement des références devait prendre en compte la réalité de cette situation, ce qui a conduit aux développements de MR dans de nombreux pays industrialisés

On peut, comme base de réflexion, considérer que la qualité métrologique d'une référence est la résultante de :

- son incertitude intrinsèque,
- sa capacité à assurer un étalonnage approprié dans les conditions pratiques d'usage.

*RACCORDEMENT DE METHODES ABSOLUES*

*- Gravimétrie -*

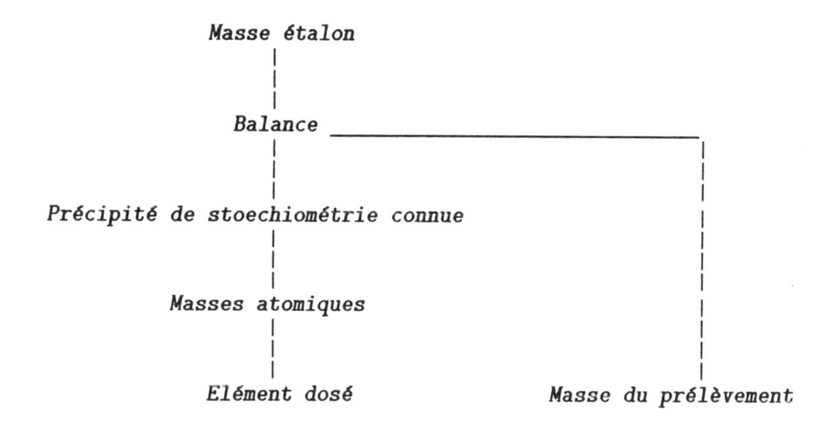


Schéma 1

*RACCORDEMENT DE METHODES ABSOLUES*

*- Titrimétrie -*

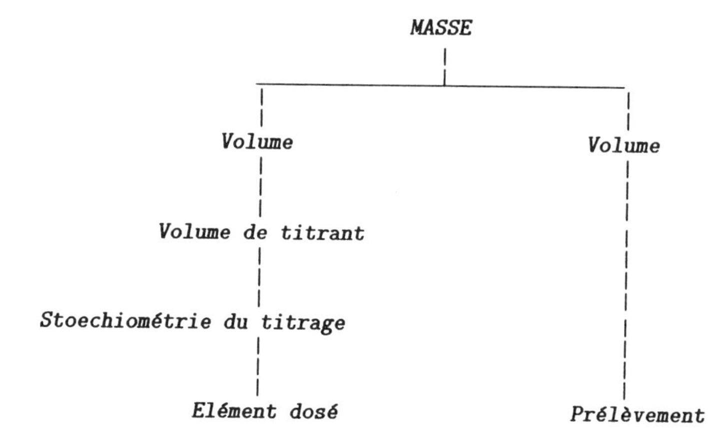


Schéma 2

## RACCORDEMENT DE METHODES ABSOLUES

- Coulométrie -

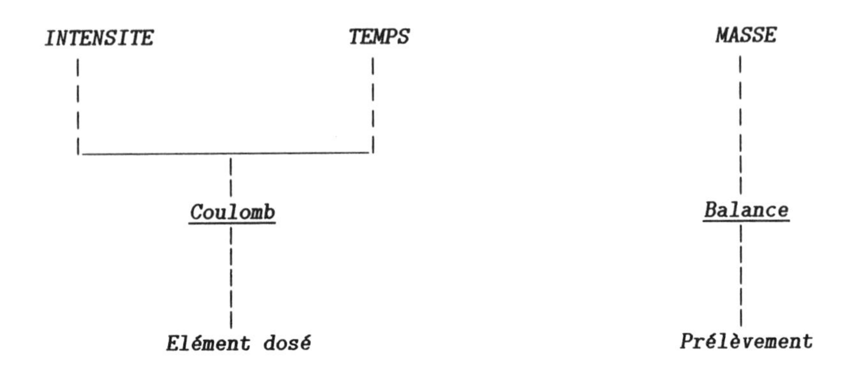


Schéma 3

## RACCORDEMENT DE METHODES RELATIVES

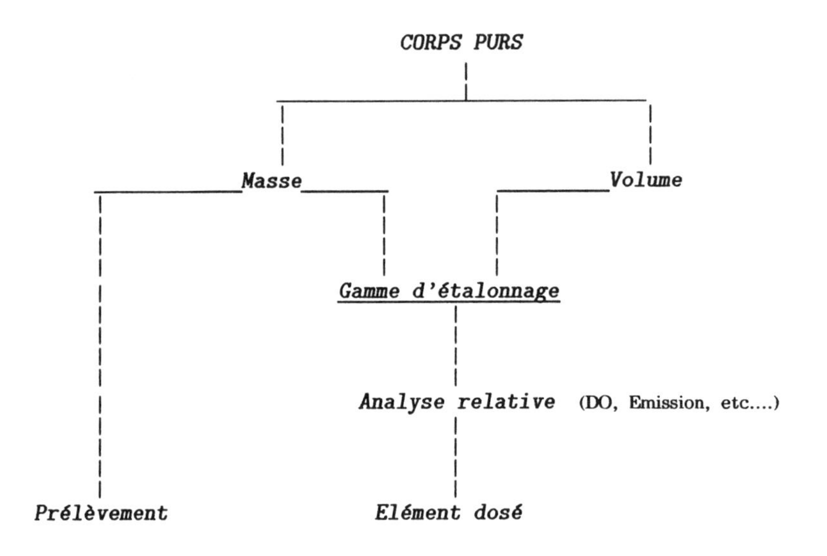


Schéma 4

Si on retient de combiner ces deux composantes quadratiquement, on peut représenter des situations typiques de la manière suivante *(schéma 5)*

Dans le cas où la composante "erreur d'adéquation" n'est plus chiffrable, la conclusion est que le raccordement n'est plus véritablement assuré, puisque son incertitude n'est pas chiffrable.

Le schéma général de raccordement des analyses chimiques peut aujourd'hui être représenté comme suit *(schéma 6)*.

On y voit deux bras :

1) Le développement des MR internes

Ils doivent assurer :

- Certification interne techniquement valide
- traçabilité démontrée - Incertitude estimée

Ils ne peuvent donc pas être des échantillons mal connus, récoltés au hasard de quelque troc, ou dont la valeur est estimée par une procédure de justesse inférieure à l'analyse qu'elle doit étalonner.

2) L'usage de MRC

° Lot de MR certifiés à l'aide d'une procédure techniquement valide, grâce à la collaboration des meilleurs laboratoires, ce type de MR apporte la meilleure garantie d'exactitude.

° Ils bénéficient d'une reconnaissance internationale.

° Ils présentent un meilleur rapport performance / prix, compte tenu du nombre d'échantillons sur lequel le prix des analyses est réparti.

La certification des MRC peut être assurée par 4 types de campagnes :

- **valeur de consensus** (plus de 50 laboratoires "de bonne volonté")
- **campagne interlaboratoires** (environ 20 : de traçabilité démontrée et d'exactitude courante)
- **métrologie analytique** (2 x 2 : de traçabilité démontrée et d'exactitude optimisée)
- **méthode définitive** (1 seule méthode dont l'exactitude est très supérieure au besoin)

Le choix et les bonnes pratiques de ces procédures pourront être approfondis à la lecture des documents édités par l'ISO-REMCO, et discutés avec les organismes responsables des Matériaux de Référence.

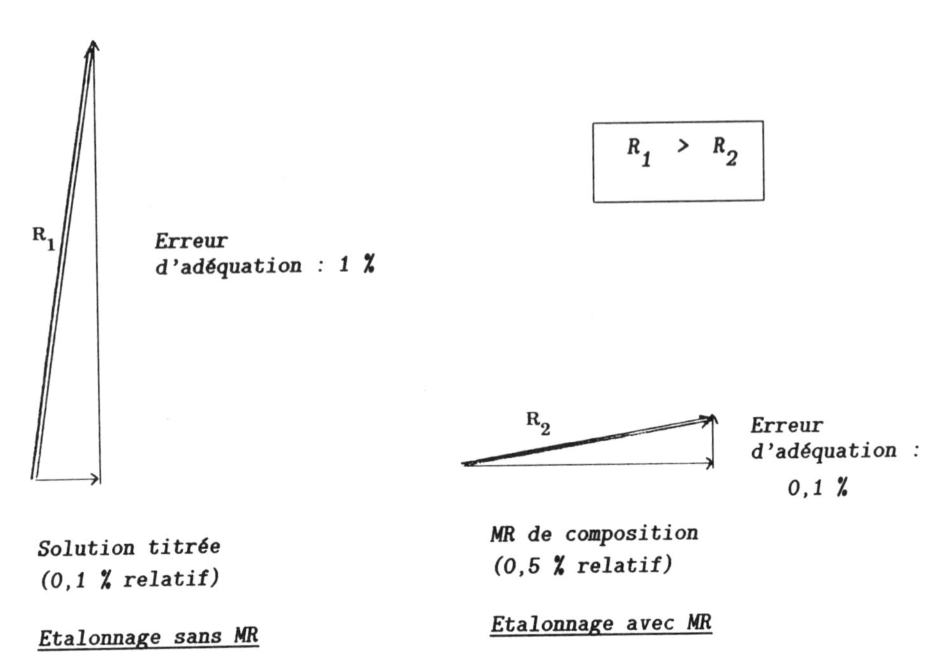

Schéma 5

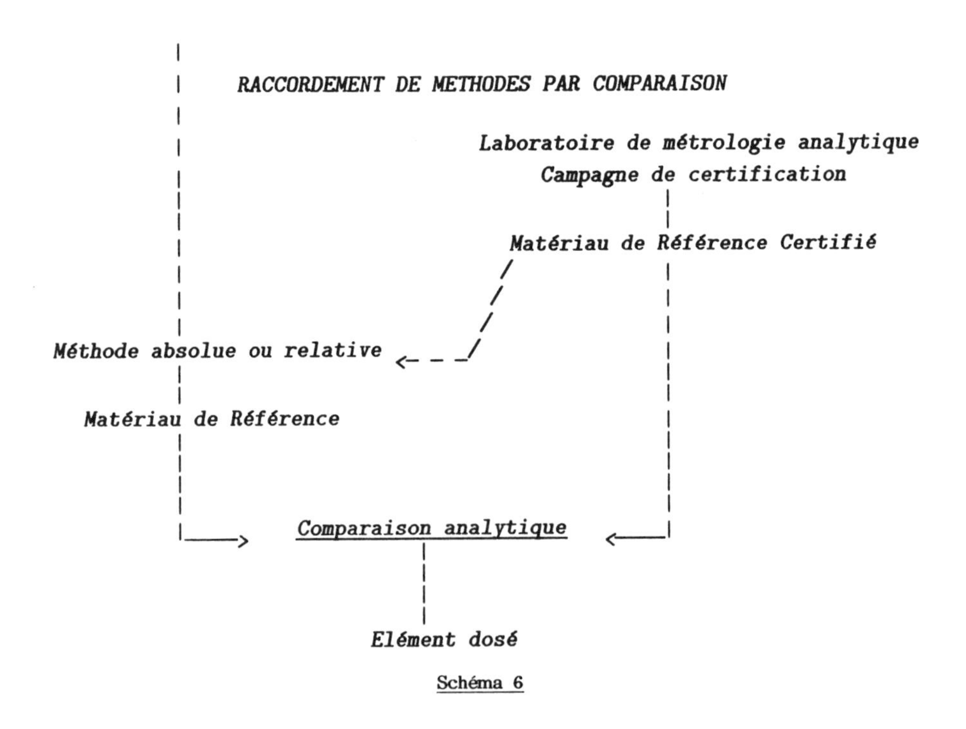

Schéma 6

B/ ASSURANCE QUALITE DES ANALYSES ET ESSAIS

Un des modes de preuves les plus efficaces et les plus crédibles de l'assurance qualité des analyses est la participation à des actions d'intercomparaisons.

La participation à des campagnes d'intercomparaisons classiques est une façon d'atteindre cet objectif, mais sa mise en oeuvre implique que de telles campagnes soient effectivement organisées au moment opportun. En outre, les enseignements que peuvent en retirer les laboratoires risquent de se révéler décevants si la quantité ou la qualité des résultats rassemblés est inférieure aux prévisions.

La comparaison à des MRC présente divers avantages :

. Elle peut être mise en oeuvre à tout moment

. Elle permet aux laboratoires de comparer leurs résultats à des valeurs validées, obtenues par des laboratoires sélectionnés

. Elle assure en général la possibilité d'intercomparaison à une plus vaste échelle, limitant les risques de divergences systématiques entre pays ou continents.

C/ DEVELOPPEMENT DES MATERIAUX DE REFERENCE CERTIFIES (MRC)

En règle générale, le développement de MR est toujours dicté par l'identification d'un besoin concret. Cet état de fait confère aux utilisateurs de MRC une part de responsabilité qu'il leur appartient d'assumer en connaissance de cause.

1) Duplication de MR existants

Le développement de MRC est le plus souvent une opération complexe, longue et coûteuse, et il est souhaitable - chaque fois que cela est possible - d'éviter d'initier des duplications d'efforts non justifiées. Confrontés à un besoin nouveau, les utilisateurs devront donc avant tout s'informer sur l'existence dans le monde entier de MRC susceptibles de répondre à leur attente. La mise en place récente de COMAR, Banque de données internationale sur les MRC, leur apportera une aide certaine dans cette démarche.

S'il apparaît qu'une étude d'aptitude à l'emploi est nécessaire avant de prescrire l'usage d'un MRC donné pour une application donnée, il est recommandable de l'effectuer de façon coordonnée et d'en faire connaître les conclusions, en particulier au producteur dont les remarques pourront être utiles pour l'ensemble des utilisateurs.

La simple duplication d'un MRC existant dans un autre pays est le plus souvent une perte d'efficacité, voire une erreur, en particulier si la motivation est l'espoir d'obtenir le MR à un prix inférieur. Par contre, l'établissement de collaborations dans le cadre de programmes joints ou associés devrait être encouragé et recherché.

2) Développement de MRC d'usage multisectoriel

Un certain nombre de MRC d'intérêt général sont destinés à être utilisés par plusieurs milieux scientifiques et professionnels. C'est par exemple le cas pour les MRC chimiques, les produits purs, les solutions titrées, les MRC d'étalonnage instrumentaux - ou pour les étalons de viscosité, densité optique, masse volumique, etc.., pour les propriétés physiques -.

Une profession confrontée à ce type de besoin devra avant tout l'exprimer avec détermination auprès des organismes nationaux responsables des MR. Elle pourrait ensuite, à juste titre, s'associer à toute action visant à rassembler les moyens intellectuels, techniques et financiers nécessaires aux développements de ce type de MR.

Il est par contre peu recommandé, sauf cas d'espèce, d'initier le développement de tels MRC ou de succédanés, dans le cadre limité d'un seul milieu industriel. Cette approche conduit à une sous-exploitation des investissements et des bénéfices ; elle génère trop souvent des MR de qualité métrologique douteuse sous l'aspect exactitude et traçabilité.

3) Développement de MRC d'usage sectoriel

Le développement de MRC destinés à un secteur professionnel particulier implique la mise en oeuvre de connaissances et de moyens complémentaires relevant, les uns de la profession, les autres de la métrologie. On constate que, selon les pays et les circonstances, la maîtrise d'oeuvre de tels projets est assurée soit par la profession concernée, soit par l'organisme Métrologie - Matériaux de Référence.

Chacune de ces deux approches présente des avantages et inconvénients dont l'analyse dépasse le cadre de cet exposé. Par contre, il est essentiel de garder en mémoire que chacune de ces composante est vitale au bon développement du MRC considéré.

Il faut donc éviter que :

- la profession ne s'en remette passivement à la collectivité ou à la providence

- les métrologues se retirent dans leur laboratoire par manque de moyens ou manque d'intérêt pour les vrais besoins du monde industriel et scientifique

- et surtout que les deux parties, arguant de quelque motivation occulte, n'agissent comme s'ils n'avaient rien à recevoir et à donner à l'autre.

Un tour d'horizon préalable du problème, réalisé en concertation, permettra de mettre en évidence, au cas par cas, la meilleure contribution aux différentes tâches.

4) <u>Développement de MRC d'usage spécifique</u>

On constate encore et toujours des demandes pour des MRC correspondant à des besoins d'applications spécifiques ou occasionnelles. Les utilisateurs doivent être conscients du fait qu'un MRC dont la demande annuelle prévisible sur plusieurs années n'atteint pas plusieurs dizaines d'unités ne sera vraisemblablement pas développé selon la procédure générale, à l'initiative du seul producteur.

Il convient donc que, devant ce type de situation, les utilisateurs n'attendent pas l'apparition sur le marché d'hypothétiques MR, et qu'ils considèrent les possibilités suivantes :

* Créer et subvenir à la constitution d'un club spécifique d'utilisateurs s'attachant à ce problème

* Contribuer à la constitution du capital nécessaire à un investissement non rentabilisable

* Développer leurs propres MRC, en respectant les règles de cet exercice

* Recherche d'autres procédures de travail n'imposant pas l'usage de MRC.

## CONCLUSION

Les analyses chimiques et les essais de matériaux sont vraisemblablement une des variétés de mesures dont l'usage est le plus général dans les différentes activités scientifiques et industrielles. Elles contribuent à la qualité des prestations et des produits dans tous ces domaines ; leur contribution sera en rapport avec leurs qualités propres.., et les Matériaux de Référence sont un élément déterminant de cette qualité.

# 8 PROFICIENCY TESTING AS A COMPONENT OF QUALITY ASSURANCE IN CONSTRUCTION MATERIALS LABORATORIES

J.H. PIELERT
Center for Building Technology, US National Institute of Standards and Technology, Gaithersburg, USA

Abstract
Proficiency testing is a procedure for using results generated in interlaboratory test comparisons for the purpose of assessing the technical competence of participating testing laboratories. This gives users of laboratory services confidence that a testing laboratory is capable of obtaining reliable results. Interlaboratory testing involves the organization, performance and evaluation of tests on the same or similar materials by two or more different laboratories in accordance with predetermined conditions. Interlaboratory testing may also be used for checking the individual performance of laboratory staff, evaluating the effectiveness of a test method, and determining characteristics of a material or product. Programs in the United States which use proficiency testing to evaluate laboratory performance are discussed along with international programs which have been identified by RILEM Technical Committee 91-Cement Reference Laboratories. The U.S. Construction Materials Reference Laboratories which distributes over 6000 samples of 13 different construction materials is highlighted, along with ASTM standardization activities related to proficiency and interlaboratory testing.
Keywords: cement, concrete, construction, interlaboratory, laboratory, precision, proficiency, testing

## 1 Introduction

ISO Guide 43 defines proficiency testing as a "method of checking laboratory testing performance by means of interlaboratory tests" [ISO, 1984]. Interlaboratory testing is the organization, performance and evaluation of tests on the same or similar materials by two or more different laboratories in accordance with pre-determined conditions. Proficiency testing methods vary depending on the type of material or product being tested, the method being used, and the number of laboratories participating. The comparison of test results obtained by one testing laboratory with those obtained by one or more other testing laboratories is a common feature of these methods.

ISO Guide 43 lists three types of proficiency testing programs:

Type A- Item or material to be tested is circulated successively from one participating laboratory to the next.

Type B- Randomly selected sub-samples from a source of a suitable degree of homogeneity are distributed simultaneously to participating testing laboratories.

Type 3- Samples of a program or a material are divided into two or more parts (split samples) with each participating laboratory testing one part of each sample.

Locke has described these types of proficiency sample programs and other variations [Locke, 1984].

This paper discusses standardization activities of the American Society for Testing and Materials (ASTM) related to proficiency and interlaboratory testing, programs in the United States for evaluating and accrediting testing laboratories, and the work of RILEM Technical Committee 91-Cement Reference Laboratories (TC91-CRL). TC91-CRL which is concerned with the evaluation of cement and concrete testing laboratory performance has shown interlaboratory testing to be an important component of many national programs. TC91-CRL held a workshop in Tel Aviv in 1989 to collect information on these programs. There were 21 programs identified in 16 countries, several of which will be discussed in this paper.

## 2 Standards for Proficiency and Interlaboratory Testing in the United States

ASTM has taken a lead role in integrating statistical principles into the standards development process in the United States [Ullman, 1985; ASTM, 1963]. All ASTM test methods must include statements on precision and bias, and the statements must, when possible, be developed through an interlaboratory test program [ASTM, 1986]. ASTM defines precision as "the closeness of agreement between selected individual measurements or test results," and bias as "a systematic error that contributes to the difference between a population mean of the measurements or test results and an accepted reference or true value." If a standard test method is to be considered reliable, then each committee must demonstrate the quality of the results the method will provide. The data obtained in the interlaboratory study and the detailed analyses of the data shall be on file with ASTM. ASTM procedures recognize that situations exist where it is not possible to provide definitive precision and bias statements, such as when interlaboratory data are not available. However, standards committees are expected to work toward obtaining such data.

ASTM has prepared the following standards related to interlaboratory testing [ASTM, 1990].

E177 Practice for Use of the Terms Precision and Bias in ASTM Methods

E456 Terminology for Statistical Methods

C1067 Standard Practice for Conducting a Ruggedness or Screening Program for Test Methods for Construction Materials

E691 Practice for Conducting an Interlaboratory Study to Determine the Precision of a Test Method

C802 Practice for Conducting an Interlaboratory Test Program to Determine the Precision of Test Methods for Construction

Materials

C670 Practice for Preparing Precision Statements for Test Methods for Construction Materials

It is important that test methods be well developed before being subjected to interlaboratory testing. ASTM Standards E177 and E456 provide definitions of important terms. ASTM Standard C1067 provides procedures for detecting and reducing sources of variation in test methods for construction materials early in their development and prior to an interlaboratory study. The standard requires that a study be conducted with a small number of laboratories to determine if the test method is well written and good enough to justify a larger interlaboratory study. ASTM Standard C1067 outlines the theory of ruggedness or screening analysis and includes an example of its application to an asphalt material. Ruggedness is defined as "the characteristic of a test method that produces test results that are not influenced by small differences in the testing procedure or environment." Screening is defined as "the detection of significant sources of variation as compared to chance variation."

ASTM Standard E691 describes general techniques for planning, conducting, analyzing, and interpreting results of an interlaboratory study conducted to evaluate a test method. ASTM Standard C802 provides a standard practice for conducting interlaboratory tests on construction materials. A requirement for using this standard is the existence of a valid and well-written test method developed in competent laboratories, including the provisions for a ruggedness or screening procedure. ASTM Standard C670 offers guidance in preparing precision and bias statements for ASTM methods pertaining to certain construction materials.

## 3 Proficiency Testing Programs in the U.S.

Laboratory evaluation and accreditation programs operating in the United States use proficiency testing as a method of evaluating laboratory performance. An additional benefit of proficiency testing is the availability of data which can be used in the preparation of estimates of precision for incorporation into test method standards. The major programs in the U.S. concerned with the evaluation of construction materials testing laboratories will be discussed including the use of proficiency testing.

### 3.1 Construction Materials Reference Laboratories

The Construction Materials Reference Laboratories (CMRL) located at the National Institute of Standards and Technology (NIST) has the goal of improving the quality of testing of construction materials. The CMRL consists of the Cement and Concrete Reference Laboratory (CCRL) and the AASHTO Materials Reference Laboratory (AMRL). These NIST Research Associate Programs are sponsored by ASTM and the American Association of State Highway and Transportation Officials (AASHTO), respectively.

The major functions of the CMRL are the inspection of testing laboratories and the distribution of proficiency test samples [Pielert, 1989]. Figure 1 shows the scope of current programs which are utilized by almost 800 laboratories located in Australia, Canada,

Denmark, France, Greece, Mexico, Norway, South Africa, Spain, Sweden, United Kingdom and the United States. The levels of laboratory participation in the proficiency sample programs are shown in Figures 2 and 3. Laboratories are not rated, certified or accredited by the AMRL; however, existing laboratory accreditation programs do use certain of these programs in the evaluation of laboratory performance as will be discussed later.

CMRL proficiency sample programs provide comparisons of results of standard tests within or among laboratories for the purpose of aiding in the recognition of, and correction of, deficiencies. All proficiency sample programs are operated in a similar way. At intervals of either six or twelve months, quantities of two slightly different lots of a given material are procured, carefully homogenized, and divided into two groups of individual test samples.

<u>Laboratory Inspection</u>

CCRL

- o cements
- o concrete
- o aggregates
- o reinforcing steel
- o pozzolans

AMRL

- o soils
- o aggregates
- o asphalt cement
- o bituminous concrete
- o plastic pipe

<u>Proficiency Sample</u>

CCRL

- o portland cement
- o blended cement
- o masonry cement
- o portland cement concrete
- o fly ash

AMRL

- o soils
- o fine aggregates
- o coarse aggregates
- o asphalt cement
- o cut-back asphalt
- o bituminous concrete
- o emulsified asphalt
- o paint

Fig. 1. CMRL Laboratory Inspection and Proficiency Sample Programs

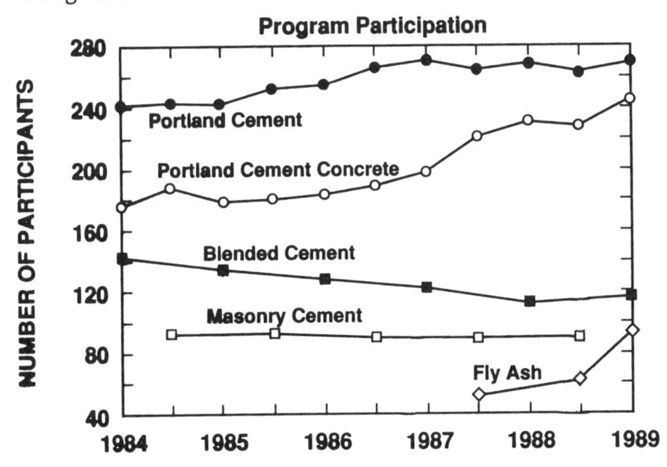


Fig. 2. Participation Levels in the CCRL Proficiency Sample Programs

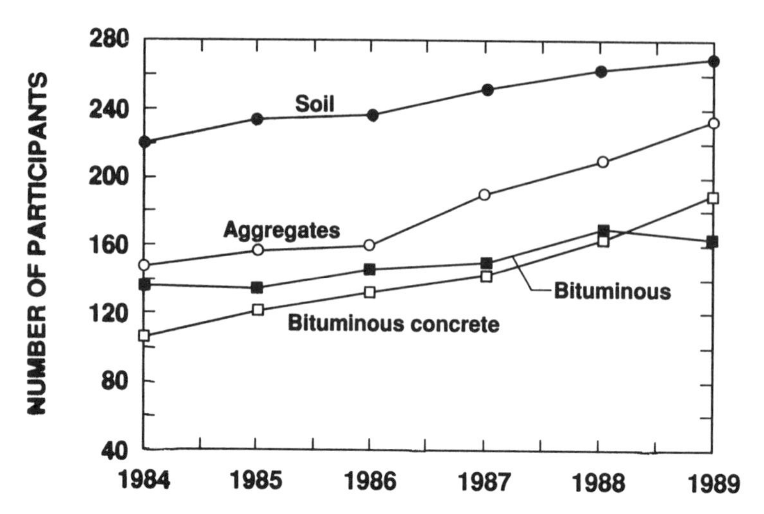


Fig. 3 - Participation Levels in the AMRL Proficiency Sample Programs

Each participating laboratory receives a pair of samples (one from each group), performs the specified tests on each and returns the results to the CMRL. Within approximately two months after sample distribution, a final report is distributed to all participants. The report contains average values, standard deviations, scatter diagrams and other statistical information obtained using the procedures set forth in papers by Youden [1959], and by Crandall and Blaine [1959]. Figure 4 shows a typical statistical presentation of results. Each laboratory is given a code number so that it may distinguish its test results, but not the results of other laboratories.

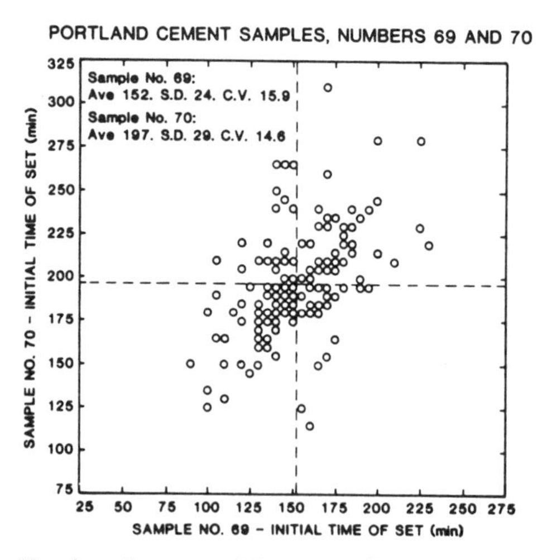


Fig. 4. Youden Scatter Diagrams for Initial Time of Set - Gilmore Needles

Summaries of the results obtained by a particular laboratory for specific proficiency testing programs are issued periodically in the form of performance charts to provide a clear picture of the laboratory's overall performance for the past ten pairs of samples (Figure 5).

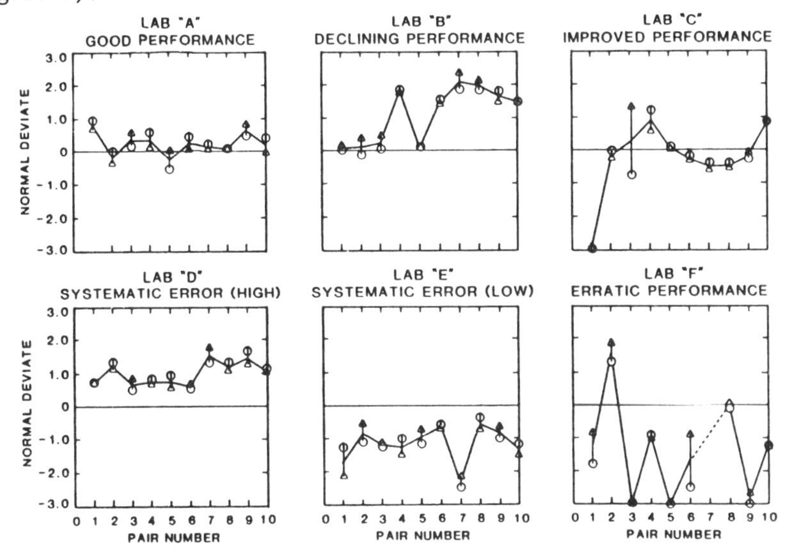


Fig. 5. Performance Charts Showing Levels of Laboratory Performance Over Time

Data from CCRL programs may also be used by standards committees in assessing the adequacy of current test methods or determining the impact of revisions to standards. Final reports from the various proficiency sample programs are routinely distributed to the appropriate ASTM committees. Committee C1 on Cement and C9 on Concrete and Concrete Aggregates have found proficiency sample data from the CCRL programs very useful in assessing the adequacy of existing test methods, in determining the impact of revisions to standards, and in the development of precision statements.

ASTM Standard C670 provides a method for using CCRL proficiency sample results for formulation of precision statements. Estimates of precision for selected ASTM cement test methods have been prepared [Pielert, Haverfield, and Spellerberg, 1985]. Other applications of CCRL data in the development of standards are discussed in a paper by Pielert and Spring [1987].

### 3.2 AASHTO Accreditation Program

AASHTO started the AASHTO Accreditation Program (AAP) in 1988 for materials testing laboratories [AASHTO, 1989]. The AAP certifies the competency of testing laboratories in carrying out specific tests on soils, asphalt cements, cut-back asphalts, emulsified asphalts, bituminous mixtures, bituminous concrete aggregates, and portland cement concrete and aggregates. The laboratory inspection and proficiency sample programs of CCRL and AMRL are used to evaluate the performance of laboratory testing. AMRL provides technical support to AASHTO in the administration of the AAP. There are 46 laboratories in the U.S. accredited by AAP as of April 1990.

**3.3 National Voluntary Laboratory Accreditation Program**

The U.S. National Voluntary Laboratory Accreditation Program (NVLAP) is administered by NIST. NVLAP's function is to accredit public and private testing laboratories based on evaluation of their technical qualifications and competence for conducting specific test methods in specific fields of testing. NVLAP accreditation in the construction testing services is available for methods of test for concrete, aggregates, cement, admixtures, geotextiles, soil and rock, and bituminous materials [Gladhill, 1989].

Proficiency testing is an integral part of the NVLAP accreditation process. CMRL proficiency sample programs for cements, concrete and aggregates are utilized for laboratories seeking accreditation in these fields of test. A within-laboratory proficiency program for concrete cylinder load testing is also required by NVLAP.

**3.4 American Association for Laboratory Accreditation**

The American Association for Laboratory Accreditation (A2LA) was formed in 1978 as a nonprofit, scientific, membership organization dedicated to the formal recognition of testing organizations which have been shown to be competent [Locke, 1990]. A2LA has granted accreditation in the following fields of testing: biology, chemistry, construction materials, geotechnical, electrical, mechanical, and nondestructive testing. Most construction materials are included in the construction materials field of testing. A2LA requires its participating laboratories to participate in the applicable proficiency sample programs of CCRL and AMRL.

**4 International Activities**

RILEM TC91-CRL held a Workshop on the Evaluation of Cement and Concrete Laboratory Performance in Tel Aviv in September 1989 to consider: (1) quality assurance systems in laboratories; (2) assessment of the technical performance of laboratories; and (3) the role of reference laboratories. The purpose of the workshop was to collect information on national systems and to prepare model guidelines for the evaluation of cement and concrete laboratory testing. The workshop attracted 30 participants from Austria, Finland, France, Germany F.R., Israel, Spain, Sweden, Switzerland, United Kingdom and the United States. Workshop proceedings are in preparation [Pielert, 1990].

The workshop resulted in the following conclusions:

1. While achieving recognition as an accredited laboratory may be expensive and time consuming, the laboratory has much to gain.
2. The following factors should be taken into consideration in defining good laboratories
   - tests are conducted every day;
   - equipment is according to the appropriate specification and is operating satisfactorily;

- o staff is motivated and well trained; and
- o procedures exist for checking the validity and reproducibility of test results.

3. Interlaboratory comparative tests are an excellent method for assessing the technical performance of laboratories, and if run regularly, they will improve agreement between laboratories.
4. The frequency of testing in a laboratory has a direct impact on the quality of testing.
5. Care must be taken when talking about "accredited" laboratories since the term varies between countries.
6. There is need for properly defining the term "reference laboratory." It was suggested that a laboratory should only be so designated if it has:
   - o a high frequency of testing;
   - o a testing machine that is frequently calibrated and reserved for reference laboratory testing only;
   - o a testing machine that is more accurate than that of a commercial laboratory and has been checked for stiffness and alignment;
   - o well documented and established test procedures; and
   - o thoroughly trained staff and tight control over sample preparation, storage, handling, and testing.
7. The performance of "reference laboratories" should be monitored by encouraging more comparisons between the reference laboratories in different countries using a common test method.

The following information on interlaboratory testing programs of several nations is summarized from the Tel Aviv Workshop proceedings.

### 4.1 Austria [Sommer, 1990]

The Austrian Cement Research Institute in Vienna does the third party testing for all the cements produced in Austria. This includes monitoring the test results obtained by the local laboratories at the cement plants and requiring that each sample taken and tested by the Institute be also tested by the local laboratory. This frequent comparative testing assures good agreement between the laboratories when required for quality control. This system of combining laboratory evaluation and product quality control is thought to be both very effective and economical.

### 4.2 France [Hawthorn, 1990]

The French cement interlaboratory testing program is organized by the Association Technique de l'Industrie des Liants Hydrauliques (ATILH). Annually, each participant receives a quantity of cement and sand to run specified tests. Statistical interpretation of the resulting data are made and each laboratory is provided all results and an identification of its data. Laboratories are able to compare their results with others in the program. This often leads to improvement in test procedures and the elimination of poor performing equipment. It is believed that the quality of testing in French cement plants has reached a high level of competence because of this program.

### 4.3 German F.R. [Wischers, 1990]

The Verein Deutcher Zementwerke (VCZ), or the Technical Association of German Cement Works, is an accredited certification body for cement established in 1877. It has operated its own accredited testing laboratory, Forschungsinstitut der Zementindustrie (FIZ), since 1902. There are currently 20 national and foreign testing laboratories accredited for supervising and/or certifying cement.

Since 1971, all accredited testing laboratories for cement are required to participate in a comparative testing program twice a year, with FIZ acting as a reference laboratory. FIZ offers to check equipment or provide technician training when a laboratory has a test result markedly different from the average. In most cases, laboratory performance is improved.

### 4.4 Israel [Even and Koster, 1990]

Uniform samples of portland cement are prepared by the Standards Institution of Israel or the National Building Research Institute and sent to four laboratories for testing according to an Israeli standard. This is an attempt to reach a consensus as to the strength of the cement. The testing procedure includes a specified sampling procedure and simultaneous testing in the four laboratories. The testing program also includes a study of the relationship between various concrete mixes prepared with the cement. The system provides a more accurate analysis of cement, and produces important data on the various types of cement manufactured and used in Israel.

### 4.5 Finland [Vaittinen, 1990]

The Technical Research Centre of Finland (VTT) has for the last 15 years systematically arranged comparative cube tests between ten officially approved testing laboratories. This, along with annual inspection visits to the laboratory, provides the basis for certification of the laboratories. VTT prepares the 150 mm cube specimens and randomly distributes them to participating laboratories. All tests are performed according to the Finnish standard at the same time in every laboratory. The compressive strength and density are determined for each specimen. Test results are sent to each participating laboratory and compared to the reference testing machine at VTT.

### 4.6 United Kingdom [Atkinson, 1990]

The British Cement Association (BCA) introduced the British Comparative Cube Testing Program in 1961. The program was developed in response to large discrepancies which had been discovered between the values obtained from machines which conformed to British Standard 1610 load verification measurement. The comparative cube test program allows a machine user to check alignment of platens and ball seating together with other aspects of machine performance. In 1980 BCA gained approval from the National Measurement Accreditation Service (NAMAS) for the program which is shortly to be covered by a new British Standard. Over 200 tests are carried out each year with BCA analyzing the results. The program is valuable in monitoring the performance of testing in cement and concrete testing laboratories.

## 5 Conclusions

Proficiency testing is an important mechanism for evaluating the performance of testing laboratories. The results of such testing may provide a basis for accrediting laboratories, evaluating test methods, and preparing standards. There is increasing interest in the subject on the international level as trade between nations grows in importance.

## 6 References


AASHTO, (1989) Procedures Manual - AASHTO Accreditation Program, Washington, D.C.

ASTM, (1963) **Manual for Conducting an Interlaboratory Study of a Test Method**, Philadelphia, PA.

ASTM, (1986) **Form and Style for ASTM Standards**, Philadelphia, PA.

ASTM, (1990) **Annual Book of ASTM Standards**, Philadelphia, PA.

Atkinson, C., (1990) British Comparative Cube Testing Program, **Proceedings of the Workshop on the Evaluation of Cement and Concrete Laboratory Performance**, NIST Special Publication, Gaithersburg, MD (in-press).

Crandall, J.R. and Blaine, R.L., (1959) **Statistical Evaluation of Interlaboratory Cement Tests**, Proceedings, American Society for Testing and Materials, Vol. 59, p. 1129.

Even, J. K. and Koster, G. (1990), Evaluation of Cement Strength Testing by Four Laboratories, **Proceedings of the Workshop on the Evaluation of Cement and Concrete Laboratory Performance**, NIST Special Publication, Gaithersburg, MD (in-press).

Gladhill, R.L., (1989) **Program Handbook-Construction Testing Services**, NISTIR 89-4039, National Institute of Standards and Technology, Gaithersburg, MD.

Hawthorn, F., (1990) Experience with Interlaboratory Testing in France, **Proceedings of the Workshop on the Evaluation of Cement and Concrete Laboratory Performance**, NIST Special Publication, Gaithersburg, MD (in-press).

ISO, (1984) Guide 43 **Development and Operation of Laboratory Proficiency Testing**, International Organization for Standardization.

Locke, J.W., (1984) **Proficiency Testing**, ILAC Conference, October 23, 1984, London.

Locke, J. W. (1990) Quality Assurance in the Construction Materials Laboratory, **Proceedings of the Workshop on the Evaluation of Cement and Concrete Laboratory Performance**, NIST Special Publication, Gaithersburg, MD (in-press).

Pielert, J.H., Haverfield, J.W. and Spellerberg, P.A., (1985) Application of CCRL Data in the Formulation of Precision Statements for Selected Cement Standards, **Cement, Concrete and Aggregates**, ASTM, Vol. 7, No. 1, pp. 37-42.

Pielert, J. and Spring, C., (1987) Application of CCRL Data in the Development of Cement Standards, **ASTM STP 961**, E. Farkas and P. Klieger, Ed., ASTM, Philadelphia, PA.

Pielert, J.H., (1989) Construction Materials Reference Laboratories at NIST- Promoting Quality in Laboratory Testing, **ASTM Standardization**

**News**, December 1989, pp 40-44.
Pielert, J. H. (1990), Proceeding of the Workshop on Evaluation of Cement and Concrete Laboratory Performance, Tel Aviv, Israel, June 1989, **NIST Special Publication**, Gaithersburg, MD (in-press)
Sommer, H., (1990) Experience With a Cement Reference Laboratory in Austria, **Proceedings of the Workshop on the Evaluation of Cement and Concrete Laboratory Performance**, NIST Special Publication, Gaithersburg, MD (in-press).
Ullman, N.R., (1985) The Use of Statistics in Standards Development, **ASTM Standardization News**, July 1985, pp 32-36.
Vaittinen, K., (1990) Comparative Compression Cube Tests Between Approved Testing Laboratories in Finland, **Proceedings of the Workshop on the Evaluation of Cement and Concrete Laboratory Performance**, NIST Special Publication, Gaithersburg, MD (in-press).
Wischers, G., (1990) Comparative Tests Between 20 Accredited Laboratories, **Proceedings of the Workshop on the Evaluation of Cement and Concrete Laboratory Performance**, NIST Special Publication, Gaithersburg, MD (in-press).
Youden, W.J., (1959) **Statistical Aspects of the Cement Testing Program**, Proceedings, American Society for Testing and Materials, Vol. 59, p. 1120.

# 9 INTERLABORATORY PROGRAMMES, ROUND-ROBIN TESTS

M. DELORT
Association Technique de l'Industrie des Liants Hydrauliques, Paris, France

Abstract
Interlaboratory programmes including round-robin tests are commonly organized in the french and european cement industry for 30 years with benefit for the participants.

They have been useful to improve the quality of testing of the laboratories of cement producers, users and inspection bodies. They allowed setting-up of certification schemes for cements with third parties and reliable autocontrol by the factories.

They will take place in a single certification scheme for european countries. Through the ATILH's experience which is described, the essential parameters to be followed are given in this paper.
Keywords : Interlaboratory programmes, round-robin tests, test procedure, cement, standardization, certification.

## 1 Introduction

Cement is commonly produced in most countries of the world, but this single word covers a very wide range of types, compositions and qualities. As a matter of fact, cements are manufactured according to national or even local conditions of production and use.

However, all cements are analysed and tested for production and quality control purposes. Their chemical, physical and mechanical characteristics are controlled according to standardized or commonly used methods.

This paper will show the interest of round-robin tests for comparing results and methods ; it will also present the international experience of CERILH-ATILH interlaboratory testing programme.

## 2 Round-robin tests, a need for the cement industry

### 2.1 Historical interest

Evolution of techniques in the fields of construction and civil works, combined with availability of raw materials and national or local traditions, led the cement industry to produce many different types of cements. However, production and quality control of different products such as a pure Portland cement for high strength concrete and a masonry cement, are mostly monitored by means of the same chemical, physical and mechanical characteristics. Among others, parameters like chemical composition (free lime, $SO_3$, MgO...), volumic mass, fineness, initial setting time, 28-day strength, are controlled by all manufacturers and many users. Doing so, a classification of all products named "cement" is possible. Unfortunately, the national standards use generally different methods for testing the same parameter and the values cannot be compared.

As cements are defined by a set of characteristic values, which allows the products to be compared, it is particularly useful to compare also the results from different laboratories. This comparison of results must be carried out, using a well defined procedure in order to avoid scattering due to the product itself.

From their origin (1960 for CERILH - France), round-robin tests programmes provided benefits to the participants.

The main interest of round-robin tests is to make the participants able to compare their own results with those of other laboratories. Many laboratories analyse and test a restricted number of well known products ; some others specially those of users, only analyse products from time to time. Participation to round-robin tests gives to them an opportunity of testing their results on an unknown product and to estimate among other factors, the calibration of equipment and the skillness of technicians.

As a matter of fact, once or twice a year, participation to such tests obliges the laboratories to a strict application of methods and gives an opportunity of detecting possible drifts.

Standardized reference methods, specially in wet chemistry, are not commonly used for production and quality control. Participation to round-robin tests is a good training for the technicians and, besides, enables the laboratories to compare results obtained with reference methods to those obtained with alternative methods.

In addition, round-robin tests allow a statistical evaluation of the different methods as far as it is possible and help laboratories in developing new methods and equipment. For example, the systematic introduction of X-ray fluorescence in 1974 has been definitely helped by round-robin tests which provided comparison with results of other methods : gravimety and complexometry.

**2.2 Further interest : evolution of standards, certification schemes**

A 20-year experience in participating to interlaboratory testing programmes brought up the french laboratories to a high level of competence and reliability. This unabled them to take place into the national scheme of certification of products complying with NF P 15-301 standard.

French certification scheme is carried out by an approved body (AFNOR) which delegates inspection and testing to Laboratoire de la Ville de Paris. It is based upon autocontrol by the factory laboratory and, 12 times per annum, control by the inspection body.

This scheme imposes the use of repetable and reproductible methods and a high level of skillness for technicians. Both factors are controlled, as we have seen before, when laboratories participate to round-robin tests. Today, results of the french plant laboratories are not more dispersed than those of the "pilot laboratories".

An interlaboratory testing programme on standardized and other commonly measured characteristics, provides to the organizer many data which unfortunately cannot be entirely exploited. For instance, there is no procedure for comparing results of tests done according to AFNOR standards to others, done according to other standards. However the most important parameters of cement testing were indentified and this has been useful in establishing the european standards for "methods of testing cement" ie EN 196 serie of standards.

As a general matter, it can be said that interlaboratory testing programmes are useful for modifying existing standardized methods.

When a new method is to be standardized, it is very useful to settle a round-robin test, as it differentiates scattering due to the method itself (results of pilot laboratories) and scattering due to the laboratories. Among other examples, CEN methods were tested in ATILH's interlaboratory testing programme before coming into force.

## 3 Organization of an interlaboratory testing programme - ATILH's experience

### 3.1 Procedure

Once a year, generally in december, ATILH sends a 10 kg sample of cement to the participants with preprinted sheets for tabulating the results.

A sufficient amount of cement (around 2 tonnes) is sampled in a factory and homogenized in a laboratory. Quartering is then carried out to obtain 10 kg samples representative of the original product. Furthermore, participants are requested to homogenize their sample when receiving it, in order to eliminate a possible segregation during transportation.

Laboratories are encouraged to use AFNOR standardized methods, but other methods are welcomed, as comparison of methods is a secondary purpose of the programme. A set of AFNOR sand bags is sent to foreign laboratories.

Participants are classified in four populations :

- . French pilot laboratories,
- . European pilot laboratories,
- . French laboratories,
- . Foreign laboratories.

Pilot laboratories are asked to double analyses and tests with different technicians in order to estimate the repetability of used methods. For that purpose, they receive 2 or more samples.

When ATILH receives results, moreoften by telefax, they are computerized and selected according to the following criteria :

- . type of test,
- . method used,
- . population of laboratories.

Mean values and standard deviations are calculated.

Student test at 99 % level with a risk of error of 5 % is used to determine upper and lower limits and aberrant values are eliminated. Afterwards, means and standard deviations are recalculated for the remaining values.

Repetability is checked according to Cochran test on values obtained by pilot laboratories (2 for each) ; reproductibility within the pilot laboratory population is checked according to Dixon test (application of ISO 5725).

Results are plotted in histograms and given in tables where each participant is able to find its own results through a code number. It can compare them with the mean values and the results of pilot laboratories.

All the results are given in a report which is sent to

the participants with a confidential cover letter. This letter details, for each laboratory, its aberrant and suspected values. Even values close to limit values, calculated according to student test, are pointed out.

Besides, as far as it is possible, the report gives comments on the methods, in terms of accuracy and comparison.

### 3.2 Evolution of participating laboratories

In France, CERILH, the former technical centre of the cement industry, carried out an interlaboratory testing programme since 1960. Starting with 25 participants, all french, participation reached 145 laboratories in 1978 after the programme was opened to foreign laboratories. Since 1984, the number of participants lowered to 120 for three reasons which are detailed below.

Crisis of petroleum in 1974 involved a reduction of cement consumption in France and in Europe. Several old-fashioned kilns were shut down definitively and associated laboratories were closed.

Concentration of the production in the hands of a restricted number of bigger companies also involved a reduction of production works.

Some companies, the major ones, with 10 laboratories or more, have realized the interest of interlaboratory testing and decided to organize once or twice a year their own interlaboratory programme. Consequently, the central laboratory only participates to ATILH's programme to calibrate its results with those of other major laboratories.

However, the participation seems to be now stabilized with around 120 laboratories from 24 different countries which are classified as follows :

. French pilot laboratories : 5
. European pilot laboratories : 6
. Other french laboratories : 55
. Foreign laboratories : 51

. Total (1989) : 117

## 4 Essential parameters for a good interlaboratory programme

Someone could find obvious that parameters which are listed below have to be strictly followed, however success of

ATILH's interlaboratory programme is most probably due to their strict application.

First of all, homogeneization and quartering of the original sample must be perfectly done, otherwise drifts occur in the data interpretation.

The origin of sample must remain unknown for all participants. It is only specified whether cement is a pure Portland (CPA) or a blended cement (CPJ).

Interpretation must be done independantly of the origin of results.

All results are available for all participants but with a code number as confidentiality must be preserved. Round-robin test is not a competition between laboratories. There are neither winners nor losers. Even laboratories getting poor results have the opportunity to improve.

## 5 Future of interlaboratory programmes-Tomorrow, the European certification

Although a version of ENV 197, the european pre-standard for cements, has been rejected in 1989, on-going discussions will lead CEN/TC 51 to achieve drafting of a european standard within a couple of years. Besides, the european directive on construction products (89/106/EEC) specifies that all standardized products shall be certified.

We assume that certification will be carried out by laboratories notified to the EEC by national governments. The certification scheme will include the principles of EN 29 002 on quality organization and insurance but will, in addition, take into account the control of products.

This notion of certification is new for several european countries which will have to adapt themselves to such scheme.

On the other hand, the notified laboratories will have to calibrate their results together, as they will deliver the same "EC conformity mark".

Interlaboratory programmes will surely take place in the european certification setting up :

- in helping laboratories which do not already practise autocontrol ;
- in testing the results of notified laboratories to verify their capacity for delivering the EC mark.

Several round-robin tests are presently organized in european countries and many laboratories participate to two or more of these. In the near future, it could be profitable to undertake discussions among the organizers with the aim to propose a single organization for all european countries, but open to other participants. This would allow to send several (2 or 3) samples per annum to perhaps 250 or 300 laboratories. By increasing both the frequency of tests and number of particpants, the quality of the statistical interpretation should normally improve.

## 6 Conclusion

Organization of an interlaboratory programme requires careful application of a restricted number of criteria : homogeneity of samples, objective statistical interpretation, confidential expression of results.

Benefit for the laboratories in participating to such test programmes has changed. Starting 30 years ago with the single purpose of calibration, it moved to testing of new methods and then to setting up of certification schemes.

Today, laboratories of the cement industry have reached a high level of skillness and reliability but the job never ends : the future european standardization and certification of cements justifies a development of interlaboratory programmes which will have to be adapted to the european context.

# 10 THE UNCERTAINTY OF MEASUREMENTS

G. BRUNSCHWIG
PIARC, Paris, France

Abstract
Every measurement is fatally prone to errors, the nature and causes of which are very numerous : the "true value" of the measured quantity is never accessible, and moreover, it is not possible to compute the total error on the result of measurement. Nevertheless one can estimate the different errors and combine them in order to obtain the uncertainty of the measurement and its components. The expressed result must always mention the uncertainty with which it is known. Some work is nowadays in progress at ISO, and it would lead to standards or guides, aiming at helping testing or metrological laboratories in that delicate question, especially in the framework of recommendations issued by the "Bureau International des Poids et Mesures".
Keywords: Communication, Measurement, Error, Uncertainty, Standardisation, Metrology.

## 1 Introduction

The only measurement procedure which can lead to a certain result, if one pays attention and does not make gross mistakes, is the counting of items. In every other case, whatever the caution taken, the result of a measurement, either simple or complex, never gives the certainty of having determined the "true value" of the measured quantity, so that some people consider the latter expression as incongruous and are unwilling to use it.

On the contrary everybody agrees, to quote the first words of the French Standard X 06-044, that "the expression of a measurement result is not complete if it is not accompanied by data related to the uncertainty affecting it". About the way of determination of these data, the agreement is far from being unanimous, leaving its actuality, in 1990, to the following statement from a 1939 paper by Raymond.T.Birge : "The question of what constitutes the most reliable value to be assigned as the uncertainty of any given result is one that has been discussed for many decades and, presumably, will continue to be discussed. It is a question that involves many considerations and by its very nature has no unique answer."

## 2 The nature of measurement uncertainty

The present definition of the uncertainty of measurement, which may be updated some day, is given by VIM (International Vocabulary of Metrology) as follows : " An estimate characterizing the range within which the true value of a measurand lies". To determine this estimation, one must analyse the different errors that are made during the measurement.

They are of several natures, and may come from the measuring instrument and its calibration, the operator, the method itself, the rounding of results, the "imported" errors, if one uses values which are measured elsewhere ; on the other hand, the measurand may be submitted to influencing factors, such as temperature, hygrometry, pressure, etc. ; if the variation of such factors is neglected, some new errors may be introduced.

It is about the classification of errors that serious differences occur.

## 3 The "orthodox" school

Up to about the 1980's, the most general way was to classify the components of measurement errors into random and systematic. The former ones caused no problem for their estimation : they were treated by statistical methods applied to repeated measurements. The latter ones, which affect all measurements in a similar manner, could be "estimated" only as upper limits, through the knowledge and information of the experimenters. What was important was that the error propagation had to be considered separately, allowing for their type : by addition of variances for the random errors, and linearly for the other ones. The overall uncertainty was obtained by adding a confidence interval of the overall random error and the algebraic sum of the systematic errors.

On this base a number of standards have been established.

## 4 The new approach

Facing a number of criticisms opposed to the classical method, particularly on the conditions of repeatability of the measurements and on their mutual independence, facing also some difficulties encountered in international comparisons of standards or units, the BIPM (Bureau International des Poids et Mesures) launched in 1978 a questionnaire among some thirty national laboratories. They were unanimous to consider that the random part of error must be characterized by a standard deviation. But different opinions were issued about its systematic part and the composition of errors allowing to express the overall uncertainty.

Subsequently, the BIPM published recommendations aiming at distinguishing the two kinds of uncertainties, not by the nature of the errors that they estimate, but by the way through which the estimation is made ; the words "random" and "systematic" were to be avoided and replaced by "type A" (accessible by statistical methods) and "type B" (accessible by other methods). But it was also recommended that type B uncertainties were to be characterized by quantities considered as approximations of the corresponding variances, the existence of which is assumed.The composition of both types must be done by addition of variances ; finally, it is indicated that one must take into account covariances, if they exist.

A working group, from ISO/TAG 4 (metrological matters) is in charge of preparing, on these bases, a "Guide to the expression of uncertainty in physical measurements". Work is in progress.

It happens that, at the same time, another working group, from ISO/TC 69 (statistical methods) began to prepare a Committee Draft on the same topic, though in a less general framework (measurement of a unique quantity, not measurement of a complex quantity, needing a series of "ancillary" measurements), but without any reference to the BIPM recommendations.

It has been thought necessary to try avoiding any inconsistencies between the future publications, and a joint group has been constituted to this effect.

## 5 What must be thought of the quarrel "Ancients" vs. "Moderns"?

The word quarrel is not excessive : each school of thought defends its point of view, with arguments that are often serious, sometimes less, and recent papers have been rather polemic...

It is clear that the new approach, if standardized, cannot be followed for the unique reason that it conforms to the BIPM recommendations : such an argument is far too formal, and instead of that the many users of the classical method have to be convinced that the new approach, which counters their habits, is based on solid grounds. The approach by the variances-covariances matrix seems to be a good argument : it allows to start from the relationship between the measurand and all the real quantities, relationship which may be either a simple or a complex one.Its derivation gives an excellent basis for analysing the errors.

The series of French standards X 06-044 to 06-047, published between 1984 and 1987, do not refer explicitly to the BIPM recommendations, but one may think that their point of view conforms to them. Although mentioning the existence of systematic errors, they recommend that corresponding corrections must be applied to the results, and give detailed indications on the methods to be used for their evaluation. Afterwards, it is assumed that residual errors are random ones. They are combined through addition of known and estimated variances, and one can express the overall uncertainty by a standard deviation, or an interval.

The draft standard presently prepared by ISO/TC 69/SC 6/WG 3 is partly inspired by the classical approach, recommending that the overall uncertainty must be expressed as the addition of a confidence interval and credible limits of the systematic errors. But it introduces, among systematic errors, an internal classification, distinguishing between "strictly systematic" and "locally systematic", the latter being some "frozen" random errors in a given measurement operation.

However it does not seem impossible that, through some reflection on the context within which lies the problem of uncertainty, the two points of view might be, not completely reconciled, but given each its own place.

## 6 The context

Some of the comments received on the first drafts of the standard drew attention to a point which had not been evoked, either in the standard or in the drafted guide.

Mr. Harry H.Ku, from the Statistical Engineering Division of NIST (National Institute of Statistics and Technology) seems to me as having made the best statement on this subject.

Having stated that the "orthodox" method, which requires linear addition of all systematic errors, is perhaps too conservative, and that the new approach, based on addition of variances, could be too liberal, Mr. Ku underlines the fact that one might take into account the use the uncertainty statement is aimed at.

There are on one hand measurement situations where "a series of results are continuously monitored for an established measurement process, such as a calibration system for a standard".

On the other hand, international comparisons of standards lead to wish that all national laboratories are to be put on the same basis, and to look for differences between them. The linear addition of systematic errors can lead to an uncertainty which is important enough to camouflage some differences, and conclude that results are comparable, which in fact are not.

Thus the first context is relevant of the "orthodox" approach, and the second one of the new approach.

## 7 Conclusion

At the time when the present note is written, it is too early to presume how ideas and practices will evolve : the texts are still under study.

But one may be inclined to think that considering the context within which is to be used the expression of a measurement uncertainty would diminish the differences, and one can pay homage to the perspicacity of Raymond T. Birge, who foresaw in 1939 that no unique answer could be brought to the question.

# 11 LA MESURE DES CARACTÉRISTIQUES DU RÉSEAU DE BULLES D'AIR CONTENUES DANS LE BÉTON DURCI PAR EXAMEN MICROSCOPIQUE: INFLUENCE DE L'OPÉRATEUR ET PRÉCISION STATISTIQUE (Characteristic measurements of air void spacing in hardened concrete by microscopic examination: operator influence and precision)

M. PIGEON and R. PLEAU
Département de Génie Civil, Université Laval, Sainte-Foy, Québec, Canada

Résumé
Parmi les paramètres qui influencent la précision de la détermination des caractéristiques des vides d'air par analyse microscopique (ASTM C457), la subjectivité de l'opérateur est souvent décrite comme une source majeure d'imprécision. En comparant les résultats obtenus par différents opérateurs sur les mêmes bétons, nous avons observé que l'écart moyen entre deux opérateurs est de l'ordre de 10%. Cet écart est appréciable mais n'est pas excessif si on le compare à la variabilité de la mesure dans son ensemble. Selon le paramètre étudié, cette variabilité se situe entre 15% et 30% lorsque les exigences minimales de la norme sont respectées.
Mots-clés: bulles d'air, analyse microscopique, précision, opérateur, variabilité statistique, durabilité, gel-dégel.

## 1 Introduction

La mesure des caractéristiques du réseau de vides d'air par analyse microscopique de sections polies (selon la norme ASTM C 457) est souvent utilisée pour évaluer la durabilité des bétons exposés à des cycles de gel-dégel. Malheureusement, la précision de cette mesure est influencée par plusieurs paramètres (Pleau et al., 1990) et, parmi ceux-ci, la subjectivité de l'opérateur est souvent décrite comme étant une source majeure d'imprécision. A l'université Laval, nous disposons d'un grand nombre de mélanges de béton pour lesquels les caractéristiques du réseau de bulles d'air ont été mesurées par différents opérateurs. En particulier, quatre opérateurs ont examiné 30 éprouvettes du même béton. L'analyse statistique des résultats obtenus permet d'évaluer l'erreur qui est directement imputable à l'opérateur.

## 2 Description de la mesure

La norme ASTM C 457 (1989) propose deux méthodes pour déterminer les caractéristiques du réseau de bulles d'air dans le béton durci. Nous avons opté pour la *modified point count method* parce qu'elle est moins fastidieuse et surtout beaucoup plus rapide que la *linear traverse method*, tout en donnant d'aussi bons résultats (Sommer, 1979). Pour chaque béton étudié, deux sections polies de 100x100mm ont été examinées au microscope (100X) et, sur chacune de ces sections, nous avons marqué 1500 points d'arrêt répartis sur une longueur de 1125 mm et une surface minimale de 50 $cm^2$. Pour le total des deux sections, nous avons donc parcouru un minimum de 2250 mm et effectué 3000 points d'arrêt répartis sur une surface d'au moins 100 $cm^2$ ce qui respecte les exigences de la norme ASTM C 457 (i.e. un minimum de 1350 points répartis sur 2286 mm et 71 $cm^2$). A chaque point d'arrêt, l'opérateur doit décider si la croisée du réticule du microscope est située au-dessus d'un granulat, d'une bulle d'air ou de la pâte de ciment. Ces résultats sont enregistrés et compilés de telle sorte qu'à la fin de la mesure on connaît le nombre total de points d'arrêt ($S_t$), le nombre d'arrêts sur une bulle d'air ($S_v$) et sur la pâte de ciment ($S_p$), de même que le nombre de bulles d'air interceptées par la ligne de traverse (N). Sur la base des lois de la stéréologie on peut alors, en admettant certaines hypothèses simplificatrices, calculer les principales caractéristiques du réseau de bulles d'air comme la teneur en air (A), la surface volumique ($\alpha$) et le facteur d'espacement des bulles d'air ($\bar{L}$) (Powers, 1949).

## 3 Précision de la mesure

La précision des résultats obtenus de l'analyse microscopique est fonction de plusieurs facteurs comme, bien sûr, les caractéristiques de la procédure utilisée (nombre de points d'arrêt, longueur parcourue et surface examinée); mais aussi l'hétérogénéité du matériau, le grossissement du microscope, la qualité du polissage et la subjectivité de l'opérateur. Une étude récente, basée sur l'analyse statistique des résultats obtenus sur plus de 600 bétons à air entraîné ($\bar{L} \leq 400$ µm), a permis d'évaluer la précision de la mesure pour la procédure utilisée dans notre laboratoire (i.e. 1500 points, 1125 mm et au moins 50 $cm^2$ sur chaque éprouvette) (Pleau et al., 1990). Le tableau 1, qui est tiré de cette étude, montre l'erreur maximale (i.e. l'écart maximal entre la valeur mesurée et celle que l'on aurait obtenue si le même opérateur avait examiné un très grand nombre d'éprouvettes provenant du même béton) en fonction du nombre d'éprouvettes examinées pour les trois principales caractéristiques du réseau de bulles d'air. La dernière ligne du tableau fournit également la valeur estimée de la précision obtenue lorsque l'on s'en tient strictement aux exigences minimales de la norme ASTM C 457. On constate que la mesure du facteur d'espacement est plus précise que celle de la surface volumique, qui est elle-même plus précise que celle de la teneur en air. Cela s'explique par le fait que la teneur en air est uniquement fonction de $S_v$ alors que la surface volumique est fonction de N et $S_v$ et que le facteur d'espacement est fonction de N, $S_v$ et $S_p$. Evidemment, plus on augmente le nombre de variables, plus la probabilité d'obtenir simultanément une erreur maximale sur chacune de ces variables devient faible, d'où la précision accrue.

Tableau 1 — Précision statistique de la mesure pour un même opérateur. (associée à une probabilité de dépassement de 5%)

| nombre d'éprouvettes | erreur maximale (%) | | |
|---|---|---|---|
| | teneur en air | surface volumique | facteur d'espacement |
| 1 | 28 | 25 | 19 |
| 2 | 20 | 18 | 14 |
| 3 | 16 | 14 | 11 |
| 4 | 14 | 12 | 10 |
| **ASTM C 457** | **≈ 28** | **≈ 20** | **≈ 16** |

## 4 Influence de l'opérateur

Pour étudier l'influence de l'opérateur, nous avons fabriqué 30 éprouvettes du même béton qui ont été examinées au microscope par quatre opérateurs expérimentés. Les résultats obtenus sont présentés au tableau 2 qui indique, pour chaque paramètre étudié (N, $S_v$, $S_p$, A, $\alpha$ et $\overline{L}$), la moyenne ($\overline{x}$) et l'écart-type (s) des valeurs mesurées par chacun des opérateurs (désignés par les lettres A, B, C et D). Pour chacune des six paires d'opérateurs (A-B, A-C, A-D, B-C, B-D et C-D), nous avons effectué un test statistique (student-fisher) afin de déterminer, avec une probabilité d'erreur de 5%, si les moyennes obtenues par les deux opérateurs étaient significativement différentes (lorsque le test est négatif on suppose que l'écart entre les deux moyennes peut être simplement attribuable à la variabilité de la mesure). Le tableau 3 montre l'écart maximal ($\Delta$x max.) et l'écart moyen ($\Delta$x moy.) des valeurs obtenues par deux opérateurs pour les cas où cet écart est significatif. La dernière ligne du tableau ($H_0$) indique sur combien de paires (sur un total de 6) le test statistique conclut à une différence significative attribuable à l'opérateur. Ces résultats indiquent que la différence moyenne entre les opérateurs est de l'ordre de 10% et que, à une exception près, l'écart maximal n'excède pas 14%.

Tableau 2 — Résultats obtenus par quatre opérateurs pour l'examen microscopique de 30 éprouvettes provenant du même béton.

| paramètre étudié | | opérateur | | | | | | | | total | |
|---|---|---|---|---|---|---|---|---|---|---|---|
| | | A | | B | | C | | D | | | |
| | | $\overline{x}$ | s | $\overline{x}$ | s | $\overline{x}$ | s | $\overline{x}$ | s | $\overline{x}$ | s |
| mesures | N | 492 | 9,9 | 450 | 9,1 | 533 | 10,5 | 446 | 10,8 | 480 | 8,0 |
| | Sv | 118 | 11,2 | 109 | 10,6 | 123 | 14,9 | 111 | 12,6 | 115 | 8,7 |
| | Sp | 356 | 7,5 | 370 | 7,9 | 380 | 8,4 | 334 | 7,5 | 360 | 5,7 |
| résultats | A | 7,9 | 11,2 | 7,2 | 10,6 | 8,2 | 14,9 | 7,4 | 12,6 | 7,7 | 8,7 |
| | α | 22,3 | 12,0 | 22,2 | 7,8 | 23,5 | 15,1 | 21,7 | 12,3 | 22,4 | 7,8 |
| | $\overline{L}$ | 136 | 8,4 | 155 | 4,4 | 135 | 12,6 | 142 | 11,6 | 142 | 6,2 |

Tableau 3 — Estimation de la différence moyenne entre deux opérateurs pour la série de 30 éprouvettes provenant du même béton.

| | mesures | | | résultats | | |
|---|---|---|---|---|---|---|
| | N | $S_v$ | $S_p$ | A | α | $\overline{L}$ |
| Δx max. (%) | 18 | 12 | 13 | 12 | 8 | 14 |
| Δx moy. (%) | 12 | 9 | 8 | 9 | 8 | 12 |
| $H_0$ | 5/6 | 4/6 | 5/6 | 4/6 | 1/6 | 3/6 |

Toujours dans le but d'évaluer l'influence de l'opérateur, nous avons également répertorié les mélanges de béton qui ont été examinés par plus d'un opérateur. Pour chacun des paramètres étudiés (N, $S_v$, $S_p$, A, α et $\overline{L}$), le tableau 4 donne, pour 8 opérateurs différents (désignés de A à H), le nombre de mélanges examinés (n), la différence moyenne ($\Delta\overline{x}$) entre le résultat obtenu par cet opérateur et ceux obtenus par les opérateurs témoins (les valeurs négatives de $\Delta\overline{x}$ signifient que l'opérateur obtient, en moyenne, un résultat inférieur à celui des autres opérateurs) ainsi que le nombre de fois où l'opérateur obtient un résultat supérieur et inférieur à celui de ses confrères (+/-). On constate, par exemple, que l'opérateur A a mesuré, 29 fois sur 41, une valeur de N inférieure à celle obtenue par ses confrères et que la différence est, en moyenne, de 9%. Lorsqu'un opérateur obtient un nombre sensiblement égal de valeurs supérieures (+) et inférieures (-) à celles obtenues par les autres opérateurs, la valeur de $\Delta x$ n'est pas significative car elle peut être due uniquement à la variabilité de la mesure et non à la subjectivité de l'opérateur. Le tableau 4 montre que pour certains opérateurs, l'opérateur C par exemple, on ne relève aucune différence significative alors que pour d'autres les différences sont importantes et peuvent atteindre 15%. Les zones ombragées du tableau 4 indiquent les cas pour lesquels la différence entre les opérateurs est significative alors que la dernière ligne du tableau donne la valeur moyenne de cette différence. On constate que les différences sont significatives pour environ la moitié des cas et que la moyenne de ces différences varie entre 3 et 10% selon le paramètre étudié.

Tableau 4 — Estimation de l'erreur associée à différents opérateurs.

| | opér. | n | mesures | | | résultats | | |
|---|---|---|---|---|---|---|---|---|
| | | | N | $S_v$ | $S_p$ | A | $\alpha$ | $\bar{L}$ |
| $\Delta\bar{x}$ (%)<br>+/− | A | 41 | **-9**<br>12+/29− | **-4**<br>17+/24− | **+3**<br>28+/13− | **-4**<br>17+/24− | **-4**<br>17+/24− | **+8**<br>25+/16− |
| $\Delta\bar{x}$ (%)<br>+/− | B | 11 | **-15**<br>2+/9− | **-4**<br>2+/9− | **-2**<br>3+/8− | **-4**<br>2+/9− | **-11**<br>3+/8− | **+12**<br>9+/2− |
| $\Delta\bar{x}$ (%)<br>+/− | C | 26 | -4<br>13+/13− | **-6**<br>11+/15− | 0<br>13+/13− | **-6**<br>11+/15− | +1<br>13+/13− | +1<br>12+/14− |
| $\Delta\bar{x}$ (%)<br>+/− | D | 15 | +2<br>7+/8− | +3<br>8+/7− | **-3**<br>3+/12− | +3<br>8+/7− | -1<br>8+/7− | -4<br>6+/9− |
| $\Delta\bar{x}$ (%)<br>+/− | E | 32 | 0<br>17+/15− | +2<br>16+/16− | **-1**<br>13+/19− | +2<br>16+/16− | -1<br>15+/17− | 0<br>15+/17− |
| $\Delta\bar{x}$ (%)<br>+/− | F | 16 | **+13**<br>16+/0− | **+9**<br>14+/2− | -2<br>7+/9− | **9**<br>14+/2− | **+4**<br>13+/3− | **-9**<br>16+/0− |
| $\Delta\bar{x}$ (%)<br>+/− | G | 12 | **-5**<br>4+/8− | 0<br>6+/6− | -1<br>5+/7− | 0<br>6+/6− | **-5**<br>3+/9− | **+5**<br>10+/2− |
| $\Delta\bar{x}$ (%)<br>+/− | H | 19 | 0<br>9+/10− | -1<br>10+/9− | +2<br>63 | -1<br>10+/9− | 1<br>10+/9− | 1<br>10+/9− |
| moyenne des $\Delta\bar{x}$ significatifs (%) | | | **11** | **6** | **2** | **6** | **6** | **9** |

note: les zones ombragées indiquent les cas pour lesquels la valeur de $\Delta x$ est significative.

## 5 Discussion et conclusion

Nos résultat indiquent clairement que, même pour des opérateurs expérimentés, le résultat de la mesure des caractéristiques du réseau de bulles d'air est influencé par la subjectivité de l'opérateur. L'erreur que l'on peut imputer à l'opérateur est évidemment très variable et diffère selon le paramètre étudié. Comme on peut le constater aux tableaux 3 et 4, l'erreur commise sur N est habituellement plus grande que celle commise sur $S_V$ et $S_p$ ce qui s'explique par le fait que, pour N, l'opérateur est plus souvent amené à porter un jugement que pour $S_V$ et $S_p$ où la décision est généralement plus facile à prendre. De la même façon, on remarque que l'erreur commise sur le facteur d'espacement est environ 50% plus élevée que celle commise sur la teneur en air et la surface volumique. Cela vient du fait que, en général, les opérateurs qui obtiennent une teneur en air plus élevée obtiennent aussi une surface volumique plus élevée et que ces différences ont toutes deux pour effet de conduire à l'obtention d'un facteur d'espacement moins élevé.

Pour tenir compte de l'influence de l'opérateur sur la précision de la mesure, nous recommandons d'ajouter 3% (pour A et $\alpha$) et 5% (pour $\overline{L}$) aux valeurs qui sont données au tableau 1. Les nouvelles valeurs ainsi obtenues reflètent la précision d'ensemble de la mesure lorsque celle-ci est effectuée par un opérateur expérimenté. Sans être négligeable, l'erreur associée à la subjectivité de l'opérateur n'est cependant pas excessive si on la compare à l'ensemble de la mesure. Notre expérience démontre cependant que, pour des opérateurs peu expérimentés, la différence peut être beaucoup plus élevée et qu'elle excède parfois 50%. Cela démontre bien l'importance de la formation des opérateurs et on ne saurait trop insister sur l'importance d'exercer des contrôles périodiques et de procéder à des comparaisons inter-laboratoire afin de s'assurer de la fiabilité de la mesure.

## Références


ASTM (1989) Recommanded Practice for the Microscopical Determination of Air Content and Parameters of the Air-Void System in Hardened Concrete, ASTM C 457-82a, **Annual Book of ASTM Standards,** vol. 04.02.

Pleau, R., Plante, P., Gagné R. et Pigeon, M. (1990) Practical Considerations Pertaining to the Microscopical Determination of Air-Void Characteristics of Hardened Concrete (ASTM C 457 Standard), **Cement, Concrete, and Aggregates,** vol. 12, no. 1, pp. 5-14.

Sommer, H. (1979) The Precision of the Microscopical Determination of the Air-Void System in Hardened Concrete, **Cement, Concrete, and Agg gates,** vol. 1, no. 2, pp. 49-55.

# 12 LA VÉRIFICATION DES MACHINES D'ESSAI EN BELGIQUE
## (The verification of testing machines in Belgium)

J.-P. ELINCK
Université Libre de Bruxelles, Génie Civil, Bruxelles, Belgium

Résumé
A la demande de son Ministère des Travaux Publics, la Belgique a établi une norme pour la vérification des machines d'essai de matériaux et plus particulièrement pour celles utilisées pour le contrôle des produits en béton, et ceci afin d'uniformiser les procédures en usage dans les laboratoires de contrôle.
Lors des discussions au sein de la commission mise en place par l'Institut Belge de Normalisation, il est apparu que, dans les entreprises industrielles, les machines étaient souvent mal installées ou mal entretenues.
Dès lors, la norme décrit tous les critères qui doivent être rencontrés lors de la vérification annuelle obligatoire des machines d'essai.
En plus de l'imposition des critères d'exactitude (justesse, fidélité, réversibilité,...) du système de mesure de la force appliquée par la machine, la norme attire l'attention sur les conditions d'installation , de conservation et d'entretien des machines d'essai.
Mots - clés : Etalonnage, Tarage, Justesse, Fidélité.

### 1 Introduction

Lorsque les laboratoires exécutent des essais de contrôle à la demande d'un client, le résultat produit qualifie le matériau fourni et ce résultat est souvent un élément d'une transaction de nature commerciale.

En effet, si une fourniture de béton n'a pas à un certain âge la résistance à la compression prescrite par le cahier des charges, soit une amende est imposée à l'entrepreneur, soit il est nécessaire de recommencer une partie des travaux, ce qui de toute manière provoque une perte de temps et d'argent souvent importante.

La responsabilité du laboratoire d'essai peut donc être considérable.

D'ailleurs, la première réaction est souvent de se précipiter au laboratoire pour "voir" si la machine ayant servi à l'essai est bien "juste".

Mais qu'est-ce-qu'une machine juste?

Pour répondre à cette importante question, l'Institut Belge de Normalisation (IBN) a institué, à la demande du Ministère des Travaux Publics, une commission composée de personnes appartenant à tous les milieux intéressés par le problème, notamment des représentants des administrations publiques, des organisations professionnelles, des laboratoires d'essai privés, des établissements d'enseignement technique et des Universités. Cette commission s'est réunie pour la première fois le 11 octobre 1982 sous la présidence de l'Inspecteur Général des Ponts et Chaussées du Service de l'Infrastructure du Ministère des Travaux Publics.

## 2 Normalisation

Au moment de cette première réunion, existaient la norme belge NBN 11-401 de 1980 traitant de l'étalonnage direct des machines d'essai de traction statique des matériaux métalliques, un projet d'Euronorm (Eu 157 de 1980), des projets de normes français (AFNOR) et allemands (DIN) et un projet ISO (ISO/TC 164/SC 1 N 131 F).

Ces différents projets comportaient de nombreux points communs, mais aussi quelques disparités.

Dès lors, les procédures d'étalonnage suivies et les instruments utilisés variaient d'un organisme de contrôle ou laboratoire à l'autre.

Comme les machines utilisées en Belgique sont toutes de provenance étrangère, la commission décida d'élaborer une norme s'inspirant des diverses prescriptions étrangères et, en perspective du futur grand marché européen, il apparut qu'il serait sage de s'inspirer le plus largement possible des documents internationaux déjà en projet, à savoir le projet d'Euronorm et le projet ISO. Ainsi, lors de l'introduction d'une norme européenne, de bonnes habitudes seraient déjà bien ancrées.

Après 9 réunions réparties sur quelque 3 ans, et après enquête publique, la première édition de la norme est diffusée par l'IBN en janvier 1987, sous le titre "Classification et étalonnage des machines et systèmes (p.ex : portique + vérin + capteurs de force) d'essai des matériaux - Machines de compression, de flexion et de traction - NBN X07 - 001".

## 3 Les qualités d'une machine d'essai

La norme a pour but de préciser les caractéristiques communes auxquelles les machines d'essai et d'une façon plus étendue les systèmes d'essai doivent répondre lors de la mise en charge d'un matériau ou d'un élément de structure.

La détermination de ces caractéristiques se fait par l'étalonnage et porte sur la justesse, la fidélité et, le cas échéant, sur la réversibilité de la machine.

Donc, pour que la machine soit "fiable", ou "précise", encore que ces termes ne soient pas définis dans la norme, il faut qu'elle soit à la fois juste et fidèle.

Supposons que lors d'un étalonnage pour la graduation 100 kN indiquée par la machine, nous obtenions les valeurs 90 - 100 et 110 kN indiquées par le dispositif d'étalonnage.

En moyenne, nous obtenons donc 100 kN et nous pouvons donc conclure que la machine est parfaitement juste, puisque l'écart entre la valeur vraie donnée par le dispositif d'étalonnage et celle indiquée par la machine est de 0%.

Le critère de justesse est insuffisant puisque nous constatons qu'il existe un écart de 20% entre les valeurs extrêmes mesurées. En effet, dans ce premier cas considéré, la propriété de fidélité n'est pas satisfaite. La machine est juste mais n'est pas fidèle.

Considérons maintenant le cas d'une seconde machine qui, pour une valeur indiquée 1000 kN au cadran, mesure en fait 1099 - 1100 et 1101 kN, soit en moyenne 1100 kN.

L'erreur de justesse est dans le cas présent de 10%, alors que l'erreur de fidélité est inférieure à 1 pour mille. La machine n'est donc pas juste, mais elle est fidèle.

Sans doute serait-il plus sûr d'utiliser cette seconde machine pour autant que l'on connaisse l'erreur de justesse.

En fait, aucune des deux machines n'est fiable. Pour qu'elles le soient, il faudrait que les erreurs de justesse et de fidélité ne dépassent pas simultanément une certaine valeur à convenir.

Une troisième erreur relative qui ne peut dépasser une certaine valeur est l'erreur de réversibilité. La norme définit la réversibilité comme étant l'aptitude d'une machine à donner en charge croissante et en charge décroissante la même valeur de la grandeur mesurée.

La fixation des écarts tolérables entre les valeurs lues sur le dispositif de mesure de la force de la machine et sur le dispositif de vérification supposé exact est l'un des objets de la norme.

La grandeur de ces erreurs dépendra du soin avec lequel la machine a été construite et en particulier de la qualité du dispositif de mesure de la force appliquée et définira en outre la classe de cette machine.

La grandeur de ces erreurs dépendra aussi de la qualité de l'installation de la machine. En effet, dans certaines entreprises, l'installation des machines n'est pas toujours aussi soignée qu'il serait souhaitable (absence de socle en béton, de fixation au sol, de mise à niveau ...). Nous avons rencontré par exemple une machine placée "provisoirement", vu l'urgence de sa mise en service, pendant cinq ans, sur deux gîtes de bois.

La grandeur de ces erreurs dépendra encore de la qualité de l'entretien de la machine. Certaines machines anciennes bien entretenues continuent à fournir d'excellents résultats, alors que certaines, beaucoup plus récentes, faute d'entretien, nécessitent une révision urgente.

Ces erreurs dépendront encore de la qualité de la conservation de la machine.

En effet, ces machines sont des équipements de précision et méritent, comme tels, un local approprié pour leur installation. Nous avons rencontré des machines, qui sous prétexte qu'elles servaient uniquement pour le contrôle de fabrication de produits préfabriqués en béton, étaient simplement abritées par un toit et donc exposées à tout vent. Au vu des résultats non satisfaisants des étalonnages successifs, suite à nos observations et suite à la dégradation progressive de son investissement et des frais élevés de réparation, l'industriel a, après trois ans, fermé le local et y a installé le chauffage.

Nous en avons rencontré une autre située dans un local non chauffé et non éclairé, ne comportant pas de prise de courant à laquelle raccorder notre équipement de vérification. Nous avons "étalonné" cette machine un mois de février, par un beau jour ensoleillé, par -5 °C, la porte du "laboratoire" ouverte, pour y voir un peu clair.

Il est donc nécessaire, si l'on veut un contrôle fiable des produits industriels fait au sein même de l'entreprise, que les inspecteurs des organismes de contrôle soient plus sévères quant aux conditions d'installation et d'entretien des machines.

Etant donné que pour des raisons de rapidité et d'économie, il a été décidé d'accorder une marque de qualité à divers produits sur base d'essais effectués au sein de l'entreprise même, avec uniquement un contrôle réduit en laboratoire officiellement reconnu, l'administration des travaux publics a souhaité que ces laboratoires reconnus satisfassent certains critères, dont celui de posséder des machines d'essai étalonnées, manipulées par un personnel qualifié.

En effet, actuellement, pour pouvoir exécuter des essais pour compte des administrations publiques les laboratoires doivent être reconnus selon une procédure instituée par arrêté royal et appliquée par le Ministère des Travaux Publics.

## 4 La procédure d'étalonnage

En son chapitre 2, la norme impose le principe de la procédure. Elle dit : "La machine n'est étalonnée que si elle est en bon état de fonctionnement. Une inspection générale de la machine est faite obligatoirement avant l'étalonnage du système de mesure de la charge (voir annexe A)". Cette annexe impose un examen très complet de la machine et notamment un examen critique de son installation et de sa conservation.

L'étalonnage proprement dit doit en principe porter sur toutes les échelles de la machine et l'influence sur les erreurs relatives des dispositifs accessoires couramment utilisés, tels que aiguille suiveuse, enregistreur, ... doit être vérifiée.

L'étalonnage se fait en comparant les charges lues sur la machine à étalonner à celles correspondantes lues sur le dispositif d'étalonnage. Celles-ci sont donc considérées comme les charges réelles. Ceci implique à son tour que le dispositif d'étalonnage soit juste et fidèle et que les erreurs tolérées pour cet équipement soient d'environ un ordre de grandeur plus petit que celles tolérées pour la machine.

La norme donne en son chapitre 5 le tableau 1 ci-dessous, indiquant les caractéristiques auxquelles le dispositif de vérification doit répondre pour être admis dans une certaine classe. Les différentes erreurs admissibles reprises au tableau 1 seront définies ci-dessous.

Le dispositif d'étalonnage doit être couvert par un certificat d'étalonnage délivré par le Service de la Métrologie du Ministère des Affaires Economiques.

L'étalonnage des dynamomètres jusqu'à 2500 kN, utilisés lors de la vérification des machines d'essai,se fait dans ce service à l'aide d'une machine qui applique sur le capteur des poids morts tarés. Actuellement,selon le règlement d'agrément, cet étalonnage doit avoir lieu tous les deux ans.

Tableau 1.

| Classe du dispositif d'étalonnage | Valeur des erreurs admissibles pour le dispositif d'étalonnage (%) | | | | |
|---|---|---|---|---|---|
| | Erreur relative | | | Faute d'interpolation | Résolution relative |
| | de fidélité b | de réversibilité \|u\| | de zéro fo | | a |
| 0,5 | 0,1 | 0,15 | ±0,02 | ±0,10 | 0,25 |
| 1 | 0,2 | 0,30 | ±0,04 | ±0,20 | 0,50 |
| 2 | 0,4 | 0,60 | ±0,06 | ±0,40 | 0,50 |
| 3 | 0,6 | 0,90 | ±0,10 | ±0,60 | 0,50 |

Le chapitre 6 de la norme traite de l'étalonnage du système de mesure de la charge de la machine et selon le degré de performance atteint, la machine sera classifiée dans une classe bien précise.

Les classes envisagées dans la norme correspondent à celles définies par les normes concernant la construction des machines d'essai.

Le tableau 2 ci-après indique les performances à réaliser selon la classe de la machine.

Tableau 2.

| Classe de la machine | Valeurs maximales admissibles en % | | | | | |
|---|---|---|---|---|---|---|
| | Erreur relative | | | | | Résolution relative |
| | de justesse q | de fidélité b | de réversibilité \|u\| | de zéro fo | due aux accessoires | a |
| 0 | ±0,5 | 0,50 | ±0,75 | ±0,05 | ±0,25 | 0,25 |
| 1 | ±1,0 | 1,00 | ±1,50 | ±0,10 | ±0,50 | 0,50 |
| 2 | ±2,0 | 2,00 | ±3,00 | ±0,20 | ±1,00 | 1,00 |
| 3 | ±3,0 | 3,00 | ±4,50 | ±0,30 | ±1,50 | 1,50 |

La dernière colonne du tableau indique la résolution relative. Celle-ci est une propriété inhérente à la machine et ne peut en général pas être modifiée.

En effet, la résolution est soit l'intervalle de force le plus petit qui peut être lu sur l'échelle de la machine, soit l'intervalle d'instabilité le plus grand de l'indication, c'est-à-dire l'intervalle à l'intérieur duquel les fluctuations de la lecture (affichage numérique) sont comprises. Elle est donc exprimée en unité de force.

Dans le cas d'une échelle analogique (cadran gradué), l'épaisseur des traits doit être uniforme et la largeur de l'aiguille indicatrice ne peut pas être plus grande que celle-ci.

Pour obtenir la résolution r, l'intervalle entre deux traits successifs de la graduation de l'échelle peut être divisée par estimation de la façon suivante : un demi-intervalle si la distance entre les deux traits est inférieure ou égale à deux millimètres et un cinquième, si elle est supérieure à deux millimètres.

Dans le cas d'une échelle numérique (digitale), la résolution r est soit égale à un incrément de l'échelle, soit égale à l'étendue de la fluctuation.

La résolution relative de l'appareil indicateur de charge "a" est définie par la relation :

$$a = \frac{r}{F_L} \times 100$$

dans laquelle $F_L$ est la charge réelle correspondant à 20% de la portée maximale de l'échelle considérée.

### 4.1 Mode opératoire de l'étalonnage

Un temps suffisant doit être alloué pour que le dispositif d'étalonnage soit à une température stable.

L'effort doit être appliqué axialement sur le dynamomètre du système d'étalonnage et, à trois reprises, il est mis en charge par la machine à vérifier jusqu'à la charge maximale correspondant à l'échelle considérée. Après chaque mise en charge, l'indicateur de charge de la machine et celui de la chaîne d'étalonnage sont remis à zéro.

Cette opération a pour but d'apprécier le fonctionnement correct tant de la machine que du système d'étalonnage.

Pour effectuer l'étalonnage, nous disposons de deux méthodes : soit on applique une charge donnée $F_i$ lue sur l'indicateur de charge de la machine et l'on note la charge réelle F indiquée par le système d'étalonnage; soit on applique une charge dont la valeur réelle F est indiquée par le dispositif d'étalonnage et l'on note la valeur $F_i$ correspondante indiquée par l'indicateur de charge de la machine.

La norme prescrit d'effectuer trois séries de mesures sous charges croissantes selon une des deux méthodes envisagées ci-dessus. Cependant, la préférence est donnée à la première. Dans ce qui suit, nous nous limiterons donc à l'utilisation de celle-ci.

En principe, chacune des échelles de la machine est vérifiée sur toute sa portée utile, c'est-à-dire entre 20% et 100% de son étendue nominale $F_N$. La vérification doit porter sur au moins cinq niveaux de charge régulièrement répartis sur toute l'étendue utile de l'échelle.

Après chaque série, le dynamomètre d'étalonnage est si possible tourné d'au moins 90° ou, sinon, sorti et replacé dans la machine pour la série suivante. Cette nouvelle série de mesures doit comporter les mêmes niveaux que la précédente. Si l'on souhaite utiliser la machine à des charges inférieures à 20% de l'étendue $F_N$, il est nécessaire de vérifier la machine jusqu'à la limite inférieure souhaitée.

A chaque série de mesures, le zéro de l'échelle est, si nécessaire, réglé. Après chaque déchargement, on note l'indication résiduelle $F_{i0}$ et on calcule l'erreur relative de zéro "$f_0$" en % à l'aide de l'expression :

$$f_0 = \frac{F_{i0}}{F_N} \times 100$$

Pour la plus petite des échelles, la norme prescrit d'effectuer une mise en charge avec les accessoires branchés afin de constater d'éventuels frottements qui influenceraient le fonctionnement du lecteur de charges de la machine.

L'erreur relative de justesse due à la présence des accessoires calculée à l'aide de l'expression :

$$\frac{F_i - F_D}{F_D} \times 100$$

où $F_D$ est la charge lue sur le dispositif d'étalonnage lorsque les accessoires sont branchés, doit être inférieure aux valeurs indiquées au tableau 2, et l'erreur globale due à tous les accessoires et calculée selon l'expression :

$$\frac{F_D - \langle F \rangle}{\langle F \rangle} \times 100$$

où $\langle F \rangle$ est la moyenne des trois lectures obtenues lors des trois mises en charge consécutives, doit, elle, être inférieure aux valeurs indiquées à la colonne "erreur relative due aux accessoires".

S' il s'agit d'une machine hydraulique, où la mesure de la pression dans le vérin de chargement sert de mesure de la force appliquée, il convient de vérifier, pour la plus petite échelle de la machine, l'influence de la position du piston. Pour ce faire, on fera une vérification si possible, pour une position correspondant à 25% et une position correspondant à 75% de sa course totale.

La détermination de l'erreur relative de réversibilité n'est imposée que si la machine est appelée à travailler en charge décroissante.

Cette erreur est déterminée lors de la troisième mise en charge pour la plus petite et la plus grande des échelles de la machine. Après cette mise en charge, le dynamomètre d'étalonnage est progressivement déchargé et une lecture F ' est faite aux mêmes valeurs de la force $F_i$, lue sur le lecteur de charge de la machine, que lors de la montée en charge.

Pour chacune des cinq valeurs $F_i$ régulièrement réparties sur la partie utile de l'échelle, on notera donc la force F dans le sens croissant et la force F ' dans le sens décroissant et l'on calculera l'erreur relative de réversibilité "u" par l'expression :

$$u = \frac{F - F'}{F} \times 100$$

Toutes les mises en charge étant effectuées, on calculera l'erreur relative de justesse "q" selon l'expression :

$$q = \frac{F_i - \langle F \rangle}{\langle F \rangle} \times 100$$

et l'erreur relative de fidélité "b" selon l'expression :

$$b = \frac{F_{max} - F_{min}}{\langle F \rangle} \times 100$$

où $F_{max}$ et $F_{min}$ sont respectivement pour chaque valeur de $F_i$, la plus grande et la plus petite des lectures de F.

Il ne reste dès lors plus qu'à rédiger le rapport d'étalonnage dont le contenu est également prescrit par la norme.

## 5 Conclusions

Nous constatons donc que le tarage ou étalonnage des machines d'essai en industrie offre parfois quelques difficultés particulières dues souvent à l'obligation d'obtenir un rendement immédiat. Or, en général, une machine convenablement conservée et entretenue ne voit pas ses caractéristiques initiales se dégrader au cours du temps.

Pour remédier à cet état de choses et afin d'éviter les discussions relatives à la valeur réelle d'un résultat d'essai, il s'est avéré nécessaire de normaliser la procédure de vérification des machines d'essai. L'étalonnage proprement dit n'est qu'une partie, quoique importante, de cette procédure.

# 13 CALIBRATION OF CONRETE LABORATORIES. CASE STUDY FROM LIBYA

S.Y. BARONY
University of Raya-Khadra, Tripoli, Libya
Z.A. HATOUSH
University of Nasser, Tripoli, Libya

Abstract
Concrete represents one of the dominant building materials in Libya for which the construction industry is relatively new, and no well recognized and defined national standards and technical specifications are available. The produced concrete compressive strength represents the dominant quality control aspect. The study concentrates on the calibration of concrete compressive strength testing machines, where an appraisal of the testing procedures and result verifications are also presented. Concluding remarks are forwarded at the end.

## 1 Introduction

During the last two decades, and particularly with the discovery of oil and its subsequent large financial returns, the authorities in the Libyan Arab jamahiriya have embarked on very large development schemes with very large sums of money being spent on construction projects for building the infrastructure of the country. These projects have covered the areas of agriculture, housing, education, health, power supply, transportation, communication, water supply, drainage and sewerage, and diverse industrial projects.
The construction industry is relatively new in Libya, where no well recognized technological procedures and well defined national standards and specification are available. As any developing country driving towards fast development and working very hard to bridge the under development gap, multi million projects have been designed and constructed by different international companies in accordance with diverse specifications. The huge investments coupled with the vast area of the country, low population density, and very low number of national technically trained staff for the construction industry led to hundreds of thousands of foreign personel to work in the construction industry as consultants, site engineers, technicians and general labour force.

Due to the fact that concrete represents one of the most dominant building materials in the country, this paper represents one part of an extensive research programme dealing with an assessment of the produced concrete strength in Libya.

The paper highlights the existing conditions of concrete testing laboratories in which more than eighteen laboratories have been visited and inspected, these laboratories constitute national reference, construction site, and general concrete products laboratories. The study concentrates on the calibration of concrete compression machines, where an appraisal of testing procedures and verification of results are also examined.

## 2 Laboratory Equipment

From the laboratories visited and investigated we were able to obtain enough information on a limited number of laboratories, these laboratories are divided into 3 categories as follows :

a) four laboratories are considered as national reference laboratories, where only one of them is responsible for calibration.
b) twelve laboratories are considered as site construction laboratories, where most of them they are construction companies owned and serving more than one construction site.
c) two laboratories are not directly related to company construction but ready mixed concrete and precast concrete production.

Though the laboratories under investigation contain testing equipment for basic constituent materials of concrete, the study was concentrated on concrete compression testing machines which test the sampled cube or cylinder specimen. The machines are quite diverse in the capacity, origin and loading system, where some of these machines are manual and the general documentation procedures are lacking.

## 3 Technical Staff

Most of the technical staff in the construction laboratories are non-national and they are quite different in their academic and technical qualification. Most of them are aware of routine procedure of testing and how to handle and operate the machine. They are not well versed with the proper quality control procedures or the general background of the technical specification and standards. This is thought to be an outcome of the lack of proper supervision from the first party (the owner) and the generally low technological level and awarness of the importance of applying quality control procedures in the country.

## 4 Testing Procedures

Based on the authors experience and the testing results accumulated from different construction project sites and testing laboratories, it can be clearly seen that there exist a diversity of testing report sheets. Few of them are short in information while others have very limited information filled up.

It was understood that all the laboratories visited, except one, investigate the spontaneous and individual results of concrete compressive strength with no concept of verification. Laboratory files and testing results do not indicate any assessment of the produced concrete on interval or statistical bases to show the general quality of the concrete produced.

## 5 Calibration

As a reflection and outcome of what was discussed in the previous part of the paper, calibration of testing equipments should be raised and discussed as a measure and very important issue. This is due to the fact that on the levels of management, technical staff, and machine operators, the concept of calibration in general and testing machine calibration in particular is not well understood, and discussions reflect that they think that calibration which is carried out on installations is good enough for the machine to operate with no serious malfunction. For this matter there exists one national laboratory which has the capability and calibration equipment and carries out the calibration where and when requested to do so.

### 5.1 Type and Range of Calibration

The study covered a period of ten years for which the calibration records of the machines in visited laboratories and others are collected. These records indicate that one calibration is carried out for all the machines except two of them have been calibrated twice, while the records for the original calibration after installation cannot be traced for all the laboratories visited exept for one machine.

The particular example of the selection results of two calibrations of one machine are given in table (1) to present and document the variations which may occur on a machine reliability, bearing in mined that different factors combined or separately may contribute to the produced errors such as operating and maintenance conditions as well as machine age and its operating periods. It was also noticed that steps of load increments and the range of calibration are not constant and do not follow a standard pattern.

Table 1. Two different calibration results of testing machine

| Machine | Capacity (2000) KN | | Capacity (800) KN | |
|---|---|---|---|---|
| Load (KN) | 82 | 85 | 82 | 85 |
| 200 | -0.9 % | -8.207 % | -0.34% | -6.773% |
| 400 | -0.18 | -4.351 | -2.22 | -5.691 |
| 600 | -0.62 | -3.721 | -0.80 | -3.833 |
| 800 | - | -3.231 | -0.74 | -2.952 |
| 1000 | -0.41 | -2.531 | | |
| 2000 | - | -2.032 | | |

Table 2. Calibration errors of two machines in percentage

| Load(KN) | 100 | 200 | 400 | 600 | 800 | 1000 |
|---|---|---|---|---|---|---|
| machine 1 | -38.66 | -18.12 | -10.14 | -7.95 | -5.98 | -5.73 |
| machine 2 | 11.74 | 8.04 | 5.13 | 4.51 | 4.41 | 2.94 |

Table (2)is presented to indicate the extreme calibration readings (errors) and in turn how far the testing results obtained from such a machine may go wrong, and how they may affect the decision making (the engineer's decision on the produced concrete quality in a positive or negative way. The calibration results of these two machines have been excluded from the comparisons due to their extreme error readings.

## 5.2 Calibration Errors

The Calibration data obtained for different machines were studied to give an overview concerning the reliability of the test results obtained using these machines. Fig (1) shows different machine errors

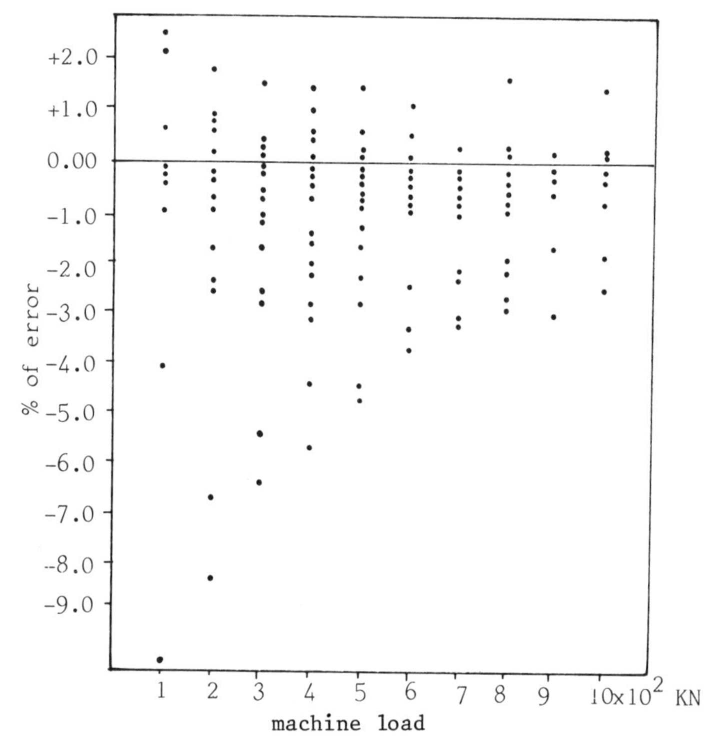


Fig.1. Different machine calibration errors

obtained from Calibration being presented in a limited range of 1000 KN, where it indicates that the majority of machines have more reliable readings at higher level of forces and tend to be in the negative side more than in the positive one. Taking into consideration that the generally produced concrete strength in the country ranges between 20 to 40 N/mm$^2$, which requires a compressive force of 450 to 900 KN to be exerted on a standard cube, then the average indicative error is less than 2%. While if the machines are used to test specimens which require a force of 200 KN or less, then the error becomes relatively high and may cause an adverse effect of rejecting the samples or reducing the factor of safety on the real structure.

Fig(2) represents a statistical form of the values given in Fig(1), which indicate how many readings are available below and above the zero error line, it shows that the errors obtained are more in the negative side than in the positive one which may be looked upon as a favorable condition to the owner.

Fig(3) represents graphically the absolute maximum error of different machines which indicate that about half of the machines have an error of less than 1.5% and only two mashines have an error of more than 6% which is unacceptable, and proper maintenance, repair and adjustment of such machines is a must.

Fig(4) shows the distribution of the error levels of all the obtained machine calibration, which indicates the cluster of the majority of readings arround the small percentage error and the Scatter as the error increases in the negative or positive sides.

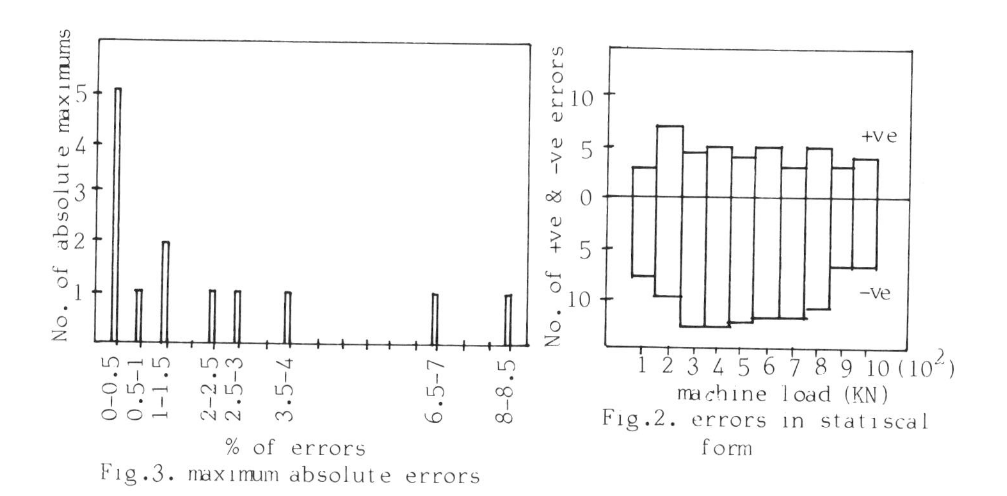

Fig.3. maximum absolute errors

Fig.2. errors in statiscal form

## 6 Discussion and Conclusion

In this paper an overview of the state of construction laboratories in Libya is highlighted. Due to the fact that concrete is considered one of the major construction materials in the country and quality control

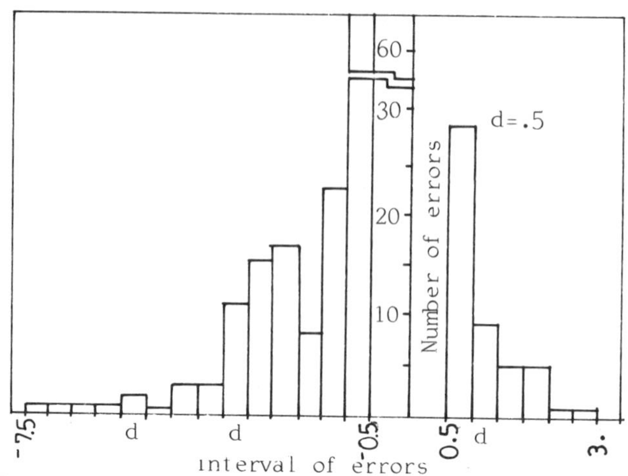

Fig.4. Statistical form of calibration readings

aspect, attention has been focussed on the calibration of concrete compressive strength testing mashines. In this matter a general questionnaire has been forwarded to different laboratories, in addition to the actual visits to the laboratories around Tripoli, which were generally constituting the major international and national consulting and construction companies, as well as the national reference laboratories at universities and research centres, from which some based records for machine calibrations were reviewed and used in this study. In conclusion :

- Technical staff and machine operators in the construction industry are under qualified and the general awareness for regular machine calibration is limited.
- To date, there is no national body or organization which supervises, directs, and controls the testing laboratories, and the recently set up National Center for Specifications and Standards did not assume its responsibilities in this side as yet.
- The limited number of calibration data was not carried out on a regular and organized basis but as an outcome of sporadic requests by the supervising engineers.
- The calibration errors for high exerted forces are generally less than 2% which can be considered reasonably small, but one should not forget the accumalative errors from the other stages of concrete production.
- It was noticed that a few calibrated machines produced relatively high errors as well as the indicated difference in the error for the machines which were calibrated more than once, this necessitates a national scheme for regular maintenance with defined operating conditions and standard calibration schedule of laboratory testing machines and equipments.
- The continuous expansion in the construction industry in general and concrete production in particular justifies and requires the creation of technical institutes and training centers for operation and maintenance of laboratory testing equipments and machines.

# 14 FIABILITÉ DES RÉSULTATS D'ESSAIS DE RÉSISTANCES SUR BÉTON (Reliability of test results for concrete strength)

G. COQUILLAT
Centre Expérimental de Recherches et d'Études du Bâtiment et des Travaux Publics, Saint-Rémy-Lés-Chevreuse, France

**Résumé**

L'organisation par le C.E.B.T.P. pour le compte du R.N.E. de campagnes d'essais d'intercomparaison sur béton durci montre globalement qu'un tiers, environ, des laboratoires, obtient des résultats statistiquement suspects, en compression. Cette proportion ne varie guère selon les années. En revanche, la proportion de suspects atteint la moitié en fendage, ce qui rend, semble-t-il, aléatoire la validité de ce type d'essai.

L'interprétation de résultats d'essais destructifs doit donc toujours être faite avec circonspection.

## GENERALITES

Les essais de résistance au fendage (NF.P-18.408) et surtout en compression (NF.P-18.406) par rupture d'éprouvettes cylindriques ($\Phi$ = 16, h = 32 cm) sont parmi les essais les plus fréquemment réalisés tant par les laboratoires spécialisés que par les bureaux de contrôle , voire par les entreprises elles-mêmes. Ils attestent, en effet, de deux des principales caractéristiques des bétons et sont, de ce fait, contractuels dans de nombreux règlements et cahiers des charges.

Chaque jour, il est donc effectué, en France, plusieurs milliers d'essais de cette nature et il est donc important de connaître le degré de validité des résultats obtenus, car les conséquences de valeurs erronées peuvent être techniquement et économiquement très importantes .

Dans le cadre de son programme d'accréditation de laboratoires de Génie Civil, le Réseau National d'Essais (R.N.E.) prévoit des opérations régulières d'évaluation des capacités techniques des organismes désireux de voir celles-ci reconnues par le R.N.E. Parmi ces opérations, figure la participation à des Campagnes d'Intercomparaisons qui, pour les essais sur bétons durcis, a été confiée au C.E.B.T.P (Unité Bétons et Matériaux Nouveaux) de Saint-Rémy-Lès-Chevreuse.

La participation à ces campagnes d'essais est obligatoire pour les laboratoires accrédités R.N.E., mais elles sont ouvertes à tout autre organisme désireux de s'y associer et de disposer d'éléments objectifs et précis pour la vérification globale des paramètres intervenant dans les résultats de ses essais (procédures, matériels..).

La quatrième campagne a été organisée en 1988-89 et a abordé les deux types d'essais (compression et fendage) alors que les trois premières campagnes n'avaient été relatives qu'à l'essai de compression.

Plus de quatre vingt laboratoires ont participé à cette 4ème Campagne (environ 60 sur chaque essai), soit un nombre équivalent aux campagnes précédentes.

La cinquième campagne aura lieu en 1990-1991.

## MODALITES PRATIQUES SOMMAIRES

Compte tenu d'évidentes contraintes budgétaires, chaque laboratoire ne dispose que de six cylindres par type d'essai. Ceci ne permet de faire les essais d'intercomparaison que sur un seul béton. Il ne s'agit donc pas d'essais interlaboratoires au sens que donne à ceux-ci aussi bien la Norme Française (NF.X-06.041) que la Norme Internationale (ISO S 275) qu'il s'agisse des objectifs ou de la méthodologie. En effet, un seul béton est confectionné (donc un seul niveau de performance est testé, une seule série, un seul âge,...etc).

Toutes les éprouvettes (867) ont été confectionnées lors de deux gâchées (une par type d'essais) dans des moules **métalliques**. Elles sont conservées après démoulage en salle humide ($\theta$ = 20 ± 1°C - H.R. > 95 %). Au 56ème jour, **tous** les cylindres sont auscultés en essais non destructifs (masse, module d'élasticité dynamique et vitesse de propagation du son).

Au 80ème, (± 5) jours, les laboratoires devaient retirer leurs éprouvettes et les entreposer dans leurs locaux jusqu'à la date prévue pour les essais (90ème jour). Les conservations, comme les méthodes d'essais, devaient être celles prévues par les normes correspondantes.

Au 90ème jour, chaque laboratoire procède à la rupture de ses éprouvettes selon ses modalités propres (théoriquement celles normalisées).

## DISPERSIONS DES CARACTERISTIQUES DES BETONS

Les dispersions des caractéristiques non destructives sont appréhendées par les essais non destructifs réalisés au 56ème jour. Elles ont toujours été sensiblement égales quels que soient la gâchée, le béton (niveau de résistance 30 à 50 MPa), et les campagnes d'essais.

| Caractéristiques | Ecart-type | Coef.de variation (%) |
|---|---|---|
| Masse | 60 à 80 g. | 4 à 5,5 |
| Module dynamique | 400 à 600 MPa | 10 à 15 |
| Vitesse du son | 25 à 40 m/s | 5 à 10 |

Les histogrammes ci-après présentent quelques exemples de caractéristiques non destructives à 56 jours.

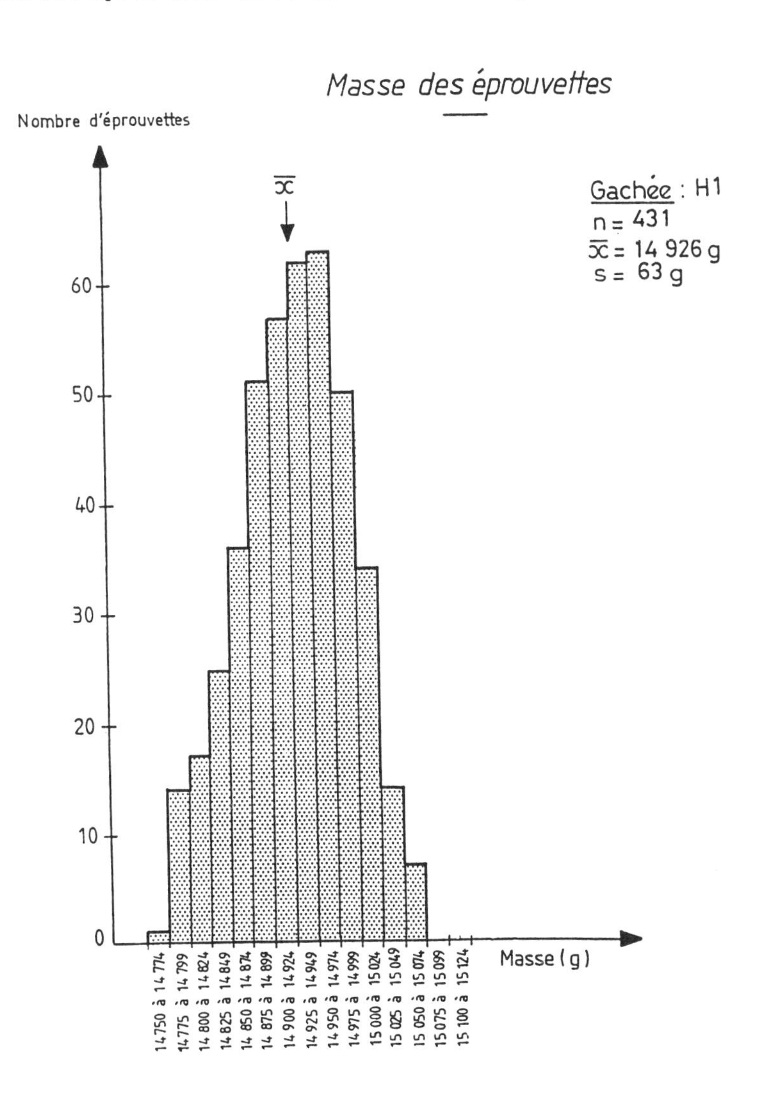

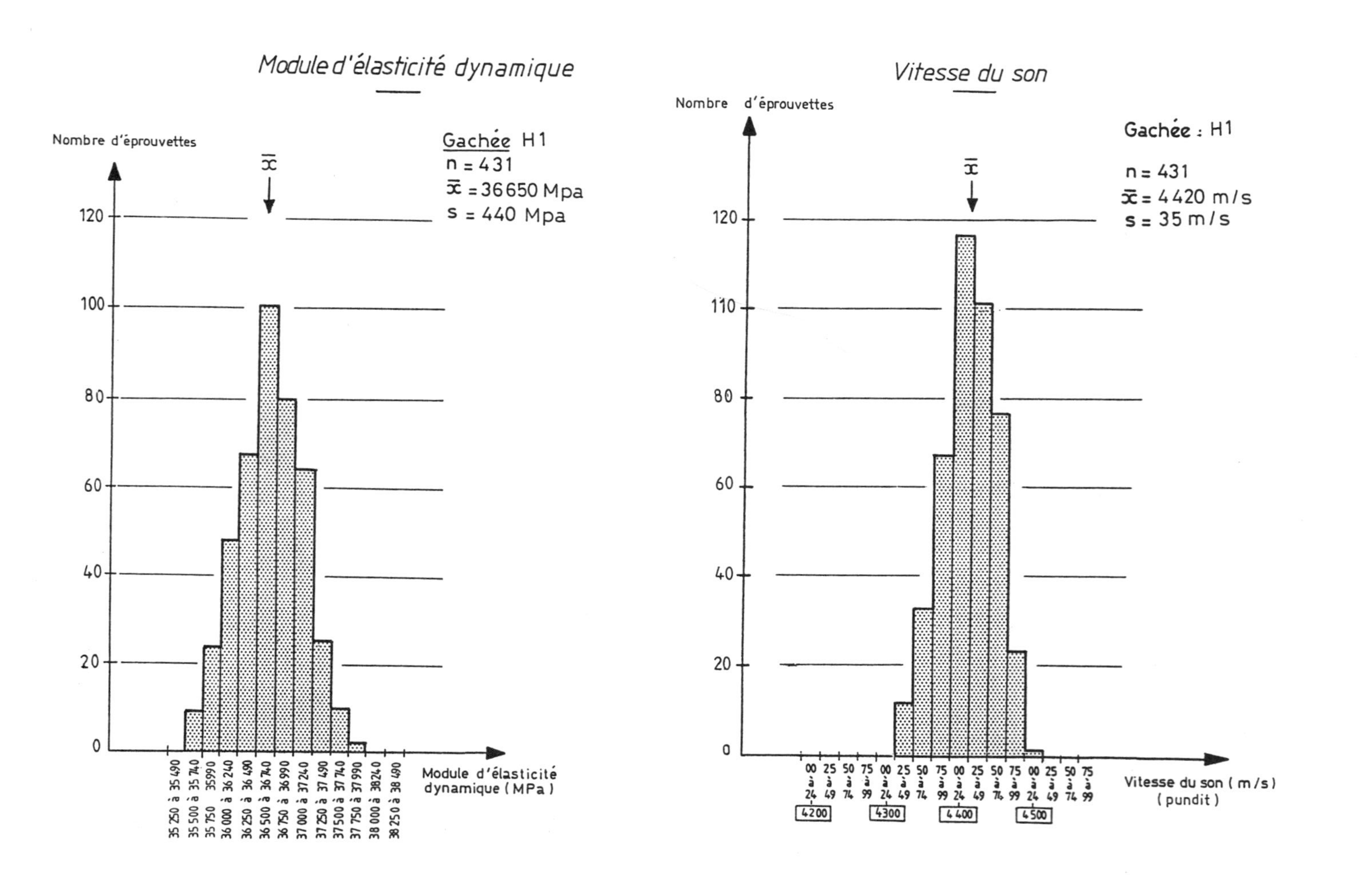

Module d'élasticité dynamique
Nombre d'éprouvettes
Gachée H1
n = 431
x̄ = 36650 Mpa
s = 440 Mpa
x̄
120
100
80
60
40
20
0
35 250 à 35 490
35 500 à 35 740
35 750 35990
36 000 à 36 240
36 250 à 36 490
36 500 à 36 740
36 750 à 36 990
37 000 à 37 240
37 250 à 37 490
37 500 à 37 740
37 750 à 37 990
38 000 à 38 240
38 250 à 38 490
Module d'élasticité dynamique (MPa)
Vitesse du son
Nombre d'éprouvettes
Gachée : H1
n = 431
x̄ = 4420 m/s
s = 35 m/s
x̄
120
110
80
60
40
20
0
00 à 24
25 à 49
50 à 74
75 à 99
4200
4300
4400
4500
Vitesse du son (m/s) (pundit)

Les éprouvettes dont les caractéristiques non destructives sont situées en dehors de la plage de confiance (seuil 97,5 % bilatéral, soit x ± 2,24 s) sont extraites et conservées par le laboratoire organisateur. Il en est de même des éprouvettes dont l'aspect et la forme les rendent suspectes.

**METHODES D'ANALYSES STATISTIQUES** (résistances mécaniques)

De nombreuses modalités d'examens statistiques des résultats peuvent être retenues (celles prévues pour les normes de la série NFX. étant comme on l'a rappelé, non applicables en toute rigueur).

Avant toute exploitation statistique, il est vérifié la normalité des populations puis procédé à l'extraction des valeurs **individuelles** de résistances aberrantes (test de DIXON et test de GRUBBS).

Il est ensuite testé les dispersions (variance) et les décalages (moyennes) par deux méthodes basées sur la comparaison (tests du $\chi^2$ et de STUDENT) des résultats de chacun des participants :

- soit à ceux obtenus par le laboratoire de Saint-Rémy qui a toujours rompu de 30 à 50 éprouvettes selon la campagne, le béton ou le type d'essai (1ère méthode).
- soit à l'ensemble des résultats obtenus par tous les laboratoires définis comme "non suspects" par la méthode précédente. Les valeurs de références résultent alors de la rupture de 150 cylindres environ (2ème méthode).

**Remarque :**
Les campagnes précédentes ont montré qu'une exploitation globale retenant l'ensemble des résultats ne conduit pas à des conclusions satisfaisantes car, par exemple, elle prend en compte, au même titre que les autres, des résultats manifestement non corrects bien que non aberrants au sens statistique.

Le tableau ci-après présente les plages des résultats "non suspects" (moyenne et variance) de la 4ème campagne dans les seuls cas suivants :

- laboratoires n'ayant pas obtenu de valeurs aberrantes ( n = 6),
- calculs effectués selon la seconde méthode (ensemble des résultats des laboratoires "non suspects"),
- niveau de probabilité de 90 %.

| | | Moyenne | Variance |
|---|---|---|---|
| Compression (MPa) | | | |
| | - mini | 36,26 | 0,21 |
| Plage de tolérance | - maxi | 39,04 | 4,69 |
| | - étendue | 2,78 | 4,48 |
| Fendage (MPa) | | | |
| | - mini | 3,27 | 0,013 |
| Plage de tolérance | - maxi | 3,96 | 0,29 |
| | - étendue | 0,69 | 0,28 |

Les histogrammes suivants présentent quelques exemples caractéristiques de résultats obtenus :

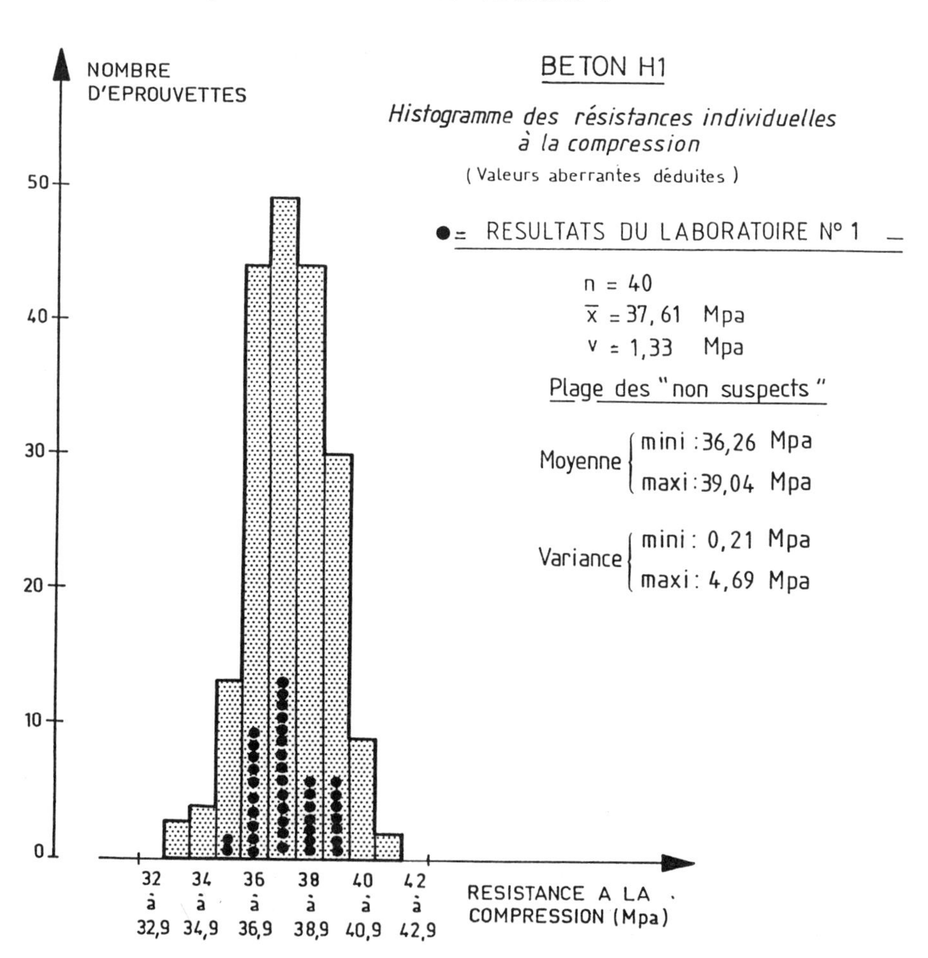

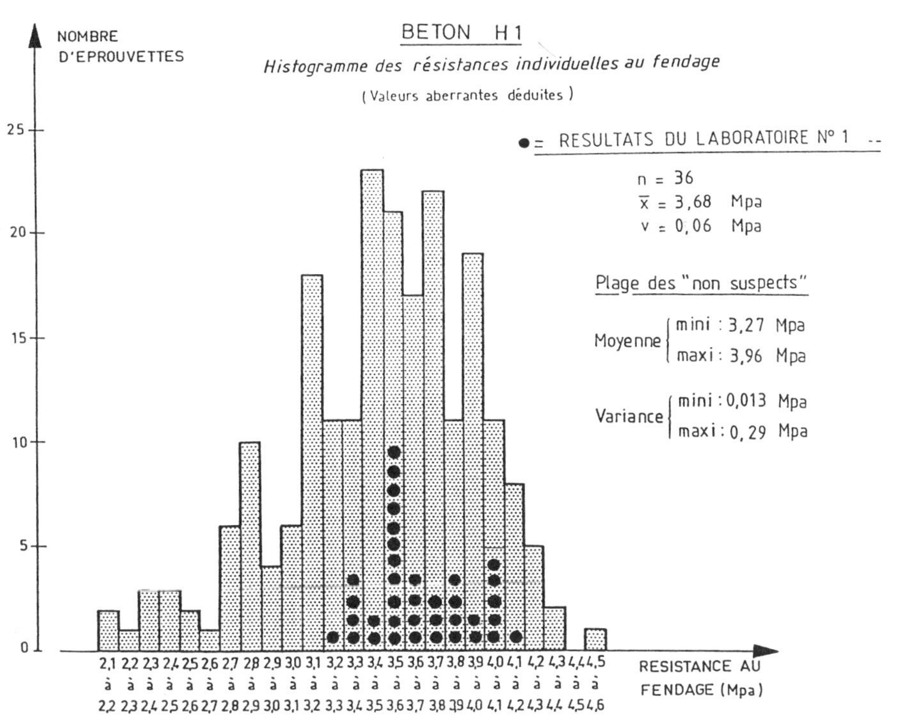
NOMBRE
D'EPROUVETTES
BETON H1
Histogramme des résistances individuelles au fendage
(Valeurs aberrantes déduites)
● = RESULTATS DU LABORATOIRE N° 1
n = 36
x̄ = 3,68 Mpa
v = 0,06 Mpa
Plage des "non suspects"
Moyenne
mini : 3,27 Mpa
maxi : 3,96 Mpa
Variance
mini : 0,013 Mpa
maxi : 0,29 Mpa
25
20
15
10
5
0
RESISTANCE AU FENDAGE (Mpa)

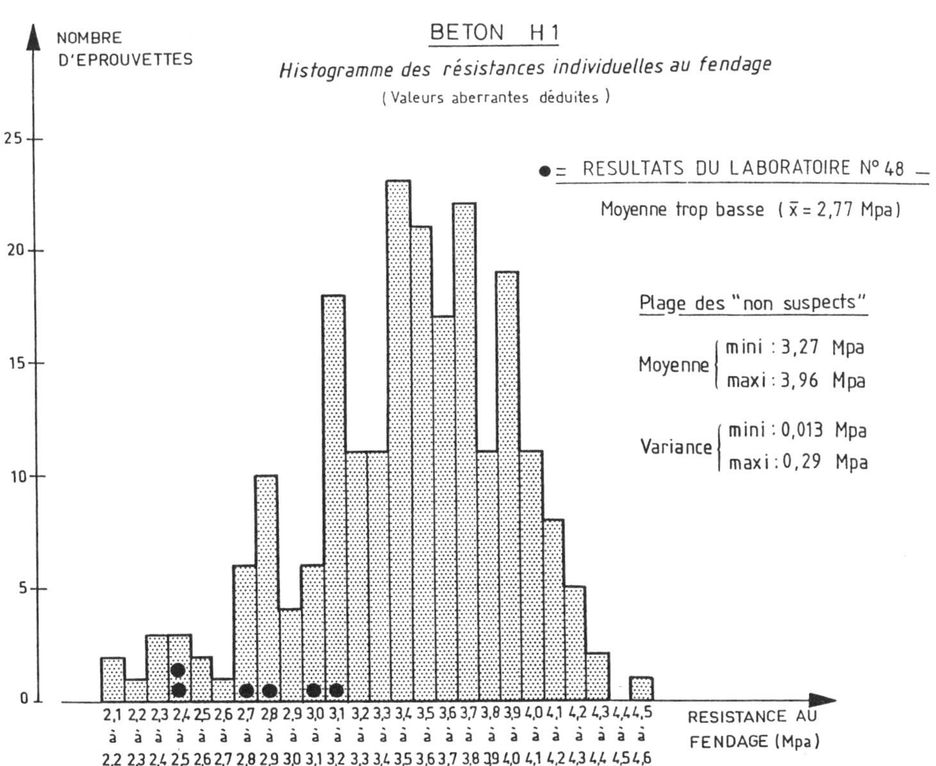
NOMBRE
D'EPROUVETTES
BETON H1
Histogramme des résistances individuelles au fendage
(Valeurs aberrantes déduites)
● = RESULTATS DU LABORATOIRE N° 48
Moyenne trop basse (x̄ = 2,77 Mpa)
Plage des "non suspects"
Moyenne
mini : 3,27 Mpa
maxi : 3,96 Mpa
Variance
mini : 0,013 Mpa
maxi : 0,29 Mpa
25
20
15
10
5
0
RESISTANCE AU FENDAGE (Mpa)

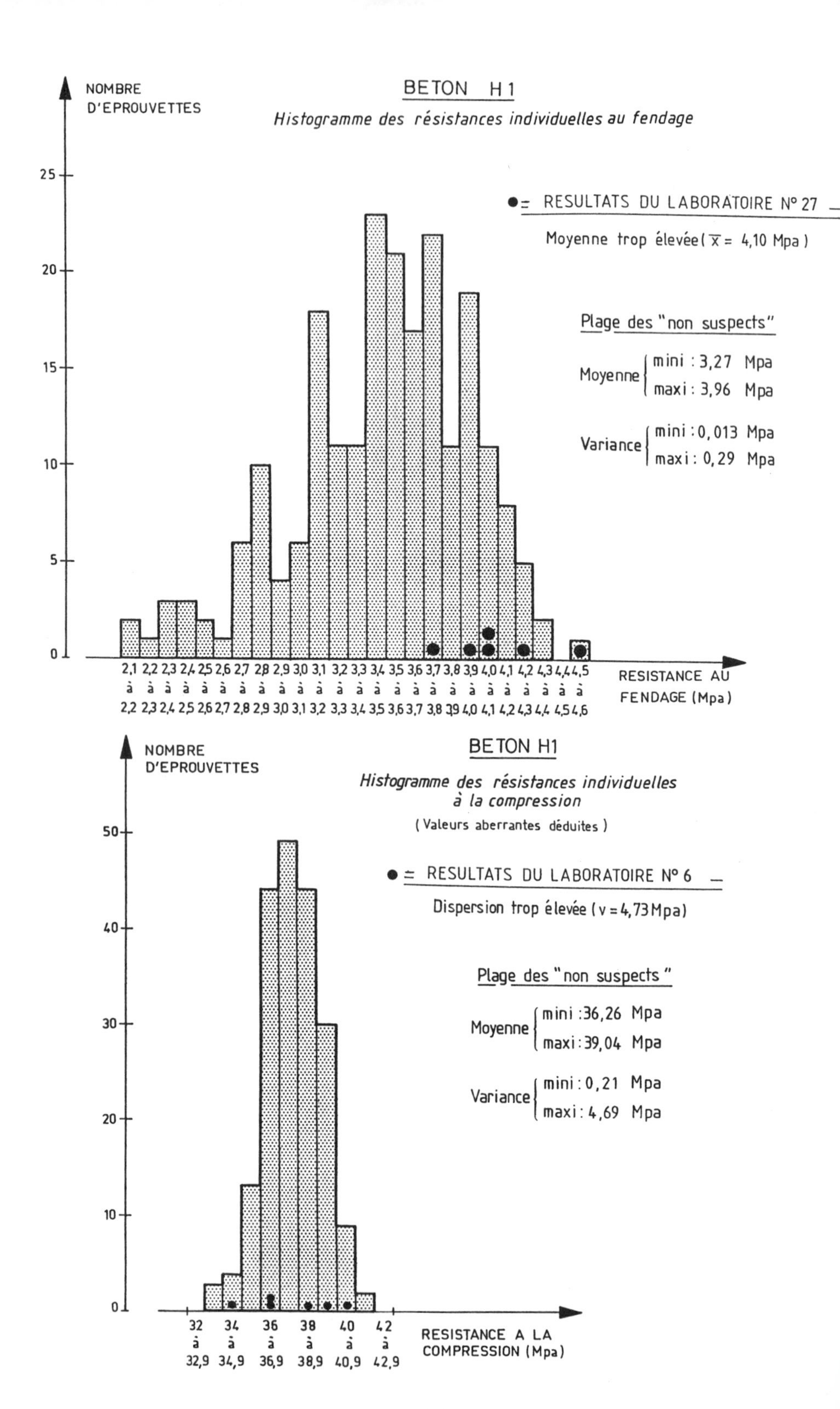
NOMBRE
D'EPROUVETTES
BETON H1
Histogramme des résistances individuelles au fendage
● = RESULTATS DU LABORATOIRE N° 27 —
Moyenne trop élevée ( x̄ = 4,10 Mpa )
Plage des "non suspects"
Moyenne mini : 3,27 Mpa
maxi : 3,96 Mpa
Variance mini : 0,013 Mpa
maxi : 0,29 Mpa
RESISTANCE AU
FENDAGE (Mpa)
NOMBRE
D'EPROUVETTES
BETON H1
Histogramme des résistances individuelles
à la compression
( Valeurs aberrantes déduites )
● = RESULTATS DU LABORATOIRE N° 6 —
Dispersion trop élevée ( v = 4,73 Mpa )
Plage des "non suspects"
Moyenne mini : 36,26 Mpa
maxi : 39,04 Mpa
Variance mini : 0,21 Mpa
maxi : 4,69 Mpa
RESISTANCE A LA
COMPRESSION (Mpa)

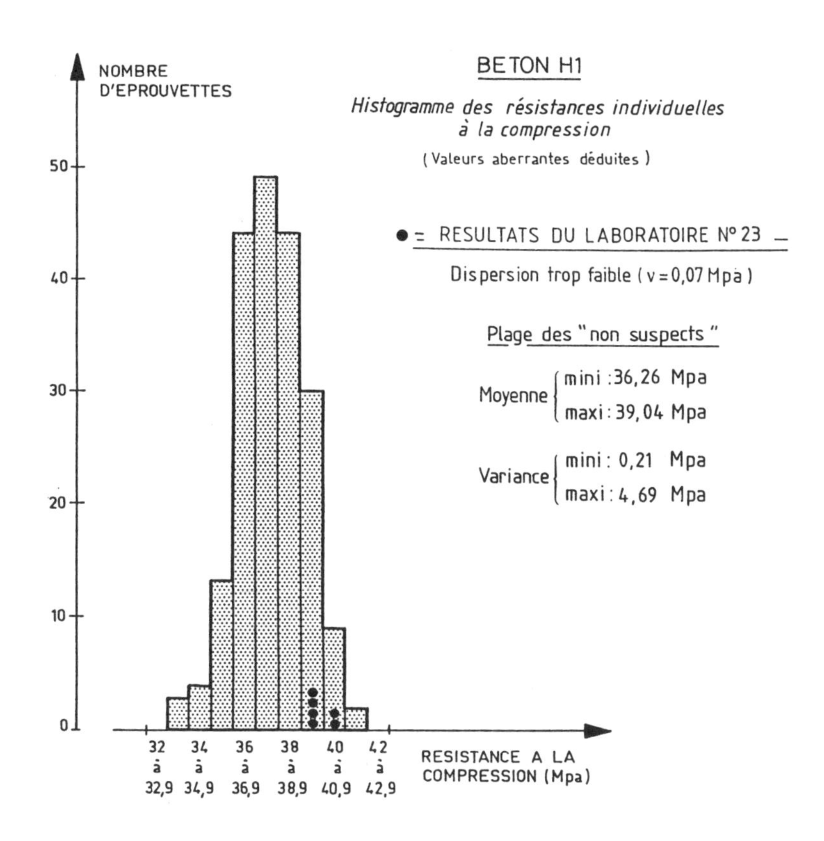


## RESULTATS DE LA QUATRIEME CAMPAGNE

Les conclusions de l'interprétation statistique des résultats **moyens** (par rapport à la valeur de référence calculée par la seconde méthode) sont résumées dans le tableau suivant :

| Essai | Compression | | Fendage | |
|---|---|---|---|---|
| Critère | Moy. | Var. | Moy. | Var. |
| Ensemble des participants | | | | |
| Nombre total | 61 | 61 | 62 | 62 |
| Nombre de "suspects" | 16 | 9 | 21 | 6 |
| Pourcentage de "suspects" | 26 | 15 | 34 | 10 |
| Laboratoires R.N.E. | | | | |
| Nombre total | 21 | 21 | 21 | 21 |
| Nombre de "suspects" | 2 | 5 | 7 | 4 |
| Pourcentage de "suspects" | 10 | 24 | 33 | 19 |

Globalement, s'il n'est pas fait de différenciations selon le type d'anomalie (sur la moyenne ou sur la dispersion), les conclusions deviennent, pour l'ensemble des participants à la 4ème campagne:

| | Compression | Fendage |
|---|---|---|
| Nombre de laboratoires "suspects" | 20 | 28 |
| Pourcentage de labo. "suspects" | 33 | 45 |

La proportion (33 %) de laboratoires ayant obtenu des résultats "suspects" en compression est quasiment identique à celle obtenue lors de la 3ème Campagne (35 %). Il convient de rappeler que cette dernière avait fait apparaître une nette réduction de "suspects" par rapport aux deux premières campagnes (> 40 %).

Pour les laboratoires R.N.E., la proportion détenant des résultats "suspects" est légèrement inférieure (29 %).

Les conclusions, sur les résultats de l'essai de fendage (première organisation), sont beaucoup moins satisfaisantes. En effet, près de la moitié (46 %) des participants ont obtenu des résultats "suspects". Comme les résultats des laboratoires R.N.E. sont tout aussi médiocres, il paraît raisonnable d'estimer que le mode opératoire décrit dans la norme n'est pas assez précis ou fiable.

**CONCLUSION :**

Les anomalies ne peuvent être reliées à des problèmes d'étalonnages de presse, en effet, ceux-ci sont toujours effectués régulièrement par les laboratoires participants et sont satisfaisants (presses de classes A, B ou C selon norme NFP-18.411).

De même, il a été mis en évidence lors de la première campagne que les éventuels défauts dans les ambiances de conservation (trop sèches), après 85 jours n'ont pas de répercussions significatives sur les résistances à 90 jours.

Les principales causes d'obtention de résultats "non corrects ", si on fait abstraction d'erreur de lecture, de transcription ou de non respect flagrant des processus normalisés (vitesse de chargement,... etc.) sont à rechercher dans le fonctionnement dynamique des presses et leurs éventuels défauts (rotules bloquées, plateaux non plans, axes de poussée et de réaction non parallèles ou décalés) ou dans la préparation des cylindres (surfaçage non correct pour la compression ; positionnement en fendage,...).

Par ailleurs, il convient de rappeler que toute interprétation statistique est liée aux seuils choisis

(P = 90, 95, 99 %...). De plus, le nombre nécessairement limité d'essais (6 éprouvettes seulement et un seul béton testé) peut conduire à des conclusions trop ponctuelles.

Les laboratoires concernés ne doivent percevoir l'obtention de résultats "suspects" que comme une alerte conduisant à effectuer les enquêtes, investigations et réglages, éventuellement nécessaires avant début de la 5ème Campagne d'essais (1990-1991).

**REFERENCE**

Statistique appliquée à l'exploitation des mesures - MASSON 1986.

# 15 REPEATABILITY AND REPRODUCIBILITY OF TEST RESULTS WITH BIG NORDIC VERTICAL FURNACES

P.J. LOIKKANEN
Technical Research Centre of Finland, Fire Technology Laboratory, Espoo, Finland

Abstract
Fire safety tests usually require flames to which structures or specimens are exposed. This, as well as differences in environment and equipment causes a special situation for the repeatability and reproducibility of these tests. The features associated with fire tests and circumstances in laboratories of various countries are examined. These are illustrated introducing a case of the Nordic round robin test series project where big test furnaces of different construction and fuel were applied in four laboratories for a specially designed specimen.
Keywords: Fire tests, Repeatability, Reproducibility, Test equipment, Furnaces.

## 1 Introduction

The Nordic countries, i.e. Denmark, Finland, Iceland, Norway and Sweden have been in close collaboration in the field of fire technology for decades. This collaboration was intensified in 1973 by the founding of Nordtest, a joint Nordic organization to harmonize and promote testing in various fields. As a result of the cooperation between the official fire laboratories and authorities in question 40 joint Nordtest Fire methods now exist and the background philosophy as well as the required level of safety and product rules concerning fire behaviour of materials and building components are practically identical in these countries.

The results of tests performed according to Nordtest methods in the official fire laboratory of one country are accepted in the other Nordic countries. This principle requires almost continuous round robin tests in the laboratories to maintain equipment and to check the test procedures followed in each laboratory on such a level that reasonable reproducibility between test results is achieved. At the end of the 1970's Nordtest financed two investigations concerning the test results of method NT Fire 005 by using big vertical furnaces in the official fire laboratories of Denmark, Finland, Norway and Sweden. One investigation concerned non-combustible specimens and the other combustible specimens.

## 2 Fire tests in general

The bearing capacity of building components for various loading cases can regularly be determined by calculations. In this respect the fire resistance which is required to ensure a prescribed level of safety in the case of a fire forms an exception. Only some criteria of the fire resistance of just some types of structures can, according to the present level of knowledge, be shown with acceptable accuracy by calculations. In addition the permissibility of the calculation procedure in the structural codes varies from country to country.

This means that in a number of cases we have to resort to experimental tests to determine the fire resistance. The fire resistance tests require special facilities, the most remarkable part of which is a fire test furnace where the specimens are exposed to fire circumstances. In these furnace tests the reproducibility of test results between different laboratories is currently a great concern.

The differences in results between tests in one laboratory and between two laboratories can be considered to be due to deviations or differencies in

- phenomena which are measured or on which the method is based
- properties of materials
- structure of specimens
- environment
- equipment
- calibration of equipment
- materials other than those under test (i.e. fuels)
- human performance.

Environmental conditions in the test laboratory which may have a considerable effect on the test results are exemplified by air flow in the vicinity of the specimen or flames or around the gauges for the air flow is, as a rule, difficult to control.

## 3 Nordic research with combustible specimens

### 3.1 Test furnaces

The primary properties of the vertical furnaces in the fire laboratories of Dantest, Denmark, Technical Research Centre of Finland (VTT), SINTEF, Norway, and National Testing Institute (SP), Sweden, are given in Table 1.

Table 1. Primary properties of vertical furnaces

| Property | Laboratory | | | |
|---|---|---|---|---|
| | Dant. | VTT | SINTEF | SP |
| Inside dimensions (mm) | | | | |
| width | 2450 | 4000 | 2500 | 3000 |
| height | 3250 | 3000 | 3000 | 3000 |
| depth | 1520 | 2000 | 1530 | 1800 |
| Volume ($m^3$) | 12.1 | 24.0 | 11.5 | 16.2 |
| Area of walls ($m^2$) | 33.3 | 52.0 | 31.8 | 39.6 |
| Inside covering stone | | | | |
| trade name | 1) | 1) | 2) | 1) |
| weight by volume ($kg/m^3$) | 850 | 850 | 2100 | 850 |
| thermal properties at 400 K | | | | |
| thermal conductivity (W/mK) | 0.28 | 0.28 | 0.80 | 0.28 |
| specific heat, J(kgK) | 978 | 978 | 900 | 978 |
| thermal inertia ($Ws^{\frac{1}{2}}/Km^2$) | 482 | 482 | 1230 | 482 |
| Fuel | | | | |
| quality | 3) | 4) | 4) | 4) |
| thermal value | 14.7 $MJ/Mm^3$ | 41.5 MJ/kg | 41.5 MJ/kg | 41.5 MJ/kg |

1) Hiporos 850 2) Höganäs Krona 2100 3) "Manufactured gas" oil based 4) Diesel oil

In Finland and Sweden the furnaces have a volume about 100 % and about 35 % larger than in Denmark and Norway respectively. The inside coating of the furnace was in Denmark, Sweden and Finland of the same Hiporos brick, but in Norway of heavy stone with a 2...3 -fold thermal conductivity over the others. The fuel used was gas in Denmark, but diesel oil in the other laboratories.

**3.2 Test material and specimens**

The test material used in the specimens consisted of specially made wooden particle boards, which had all been made with the same machine and from the same raw material batch. The thickness of the boards was 30 mm ± 0.3 mm. The weights of the 51 particle boards used ranged from 20.45 to 21.16 $kg/m^2$ with standard deviation 0.93 %.

The specimens comprised a wall construction consisting of a frame of locking bars with the 30 mm thick particle board on the side which was exposed to fire.

The construction of the specimens is shown in Fig. 1. In the middle of the specimen a so-called test board (weight 20.75...20.86 $kg/m^2$, width 1225 mm) was placed, and on both its sides equally wide-sided boards of the test material. Measurements and observations concerning the specimen itself were concentrated on the test board at the middle of the specimen.

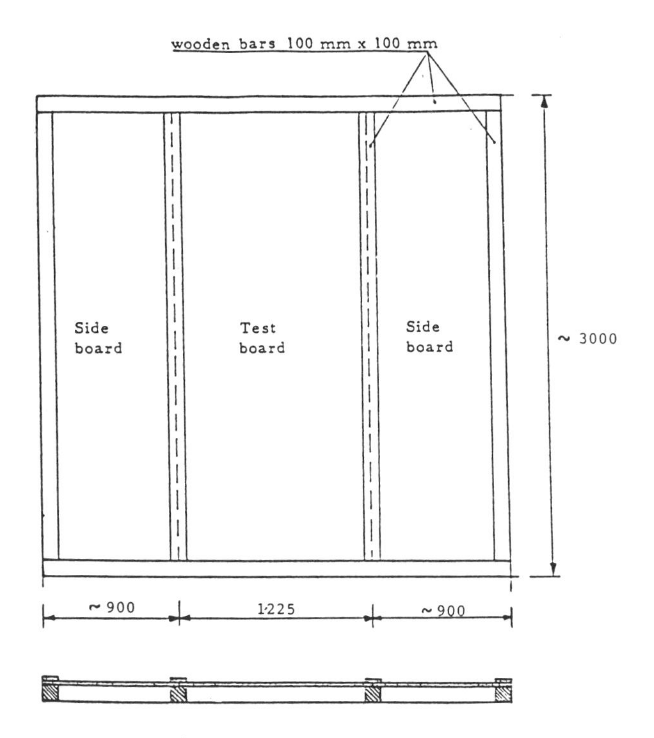


Fig.1. Construction of specimen

The specimens were conditioned to the state of equilibrium moisture ranging from 8.0 to 9.7 % of the dry constant weight at 105 °C.

### 3.3 Measurements and observations

Altogether fifteen tests were conducted, three in Dantest, SINTEF and SP, and the remaining six in VTT. The tests were performed according to the normal routine procedure used in each laboratory. During each test furnace temperature, surface temperature on unexposed side, furnace pressure conditions, oxygen and carbon dioxide content in the furnace, time to ignition on exposed side, colour changes on unexposed side, glow and holes in specimen, flames on the unexposed side and weather were measured or recorded.

### 3.4 Results of surface temperature measurements

Only the surface temperatures, which define the fire resistance of the tested wall construction, are treated here. According to NT Fire 005, which is identical to the international standard ISO 834 the fire resistance of a non-loadbearing specimen is defined by the average temperature rise 140 °C or the maximum temperature rise 180 °C on the unexposed surface or the loss of integrity of the specimen. The following time intervals from the commencement of each test are handled:

time $t_1$ when the exposed side ignites
time $t_2$ when the average rise of temperature on the unexposed surface is 140 °C
time $t_3$ when the maximum temperature on the unexposed surface is 180 °C
time $t_4$ when the unexposed side ignites.

The ignition of the unexposed side means the burn-through of the specimen. The burn-through time can be illustrated as in Fig. 2.

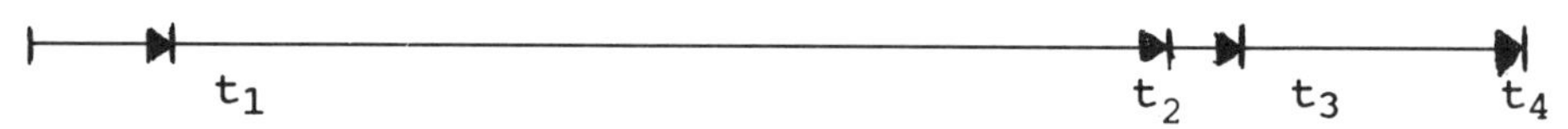


Fig.2. Parts of burn-through time

Parts of the burn-through times are analyzed in Table 2. The fire resistance time of the specimen is defined in practice by $t_2$ or $t_3$ or burn-through time $t_4$ is also possible. As seen in Table 2 the mean values in each laboratory follow that sequence although in some individual tests $t_3$ has been reached before $t_2$.

The repeatability of the test series in each laboratory has been characterized by the standard deviation of mean calculated from Student's t-distribution. The area $\mu$ which contains the results with probability of 95 % has been calculated in Table 2 from formula

$$\bar{x} - t \cdot s/n^{\frac{1}{2}} \; \bar{x} \quad \leq \mu \leq \bar{x} + t \cdot s/n^{\frac{1}{2}} \tag{1}$$

where $\bar{x}$ is the mean value of results, s is the standard deviation and t is a coefficient which depends on the number of tests n. As seen the area ranges between 1.6 % and 6.4 % of the mean for $t_2$, between 1.8 and 6.3 % for $t_3$ and between 0.8 and 9.4 % for $t_4$. The calculation of the repeatability and the reproducibility of all tests according to ISO 5725 gives values which are shown on the bottom lines of Table 2.

**4 Discussion**

In the New Approach document produced by the European integration process fire safety is one of the essential safety requirements. That means that the fire test methods and classifications which are in use in the European countries must be harmonized.

Table 2. Parts of burn-through times (min. sec) in the tests with notations of fig. 2.

| Lab | Test | $t_1$ | $t_2$ | $t_3$ | $t_4$ |
|---|---|---|---|---|---|
| Dantest | 1 | 5.10 | 47.40 | 47.00 | 49.45 |
| | 2 | 5.30 | 47.50 | 48.00 | 52.30 |
| | 3 | 5.15 | 47.30 | 48.30 | 52.30 |
| | m.v. | 5.18 | 47.40 | 47.50 | 51.35 |
| | s.d. | 5.97 | 0.64 | 2.94 | 5.67 |
| VTT | 1 | 4.20 | 45.25 | 47.10 | 53.10 |
| | 2 | 4.30 | 46.40 | 47.40 | 53.25 |
| | 3 | 4.40 | 46.10 | 45.25 | 52.55 |
| | 4 | 4.40 | 44.40 | 46.30 | 52.55 |
| | 5 | 4.35 | 46.00 | 46.30 | 52.20 |
| | 6 | 4.45 | 46.30 | 47.10 | 53.30 |
| | m.v. | 4.35 | 45.54 | 46.42 | 53.03 |
| | s.d. | 4.95 | 1.62 | 1.88 | 0.80 |
| SINTEF | 1 | 2.30 | 42.30 | 42.30 | 48.30 |
| | 2 | 2.55 | 43.10 | 43.35 | 49.40 |
| | 3 | 3.40 | 42.25 | 42.35 | 49.50 |
| | m.v. | 3.02 | 42.42 | 42.53 | 49.20 |
| | s.d. | 36.0 | 1.78 | 2.58 | 2.70 |
| SP | 1 | 3.30 | 44.30 | 46.30 | 52.30 |
| | 2 | 3.15 | 43.30 | 43.30 | 48.00 |
| | 3 | 3.30 | 45.30 | 44.30 | 52.30 |
| | m.v. | 3.25 | 44.30 | 44.50 | 51.00 |
| | s.d. | 7.75 | 4.13 | 6.27 | 9.37 |
| Total m.v. | | | 45.20 | 45.50 | 51.35 |
| Repeatability | | | 1.46 | 2.35 | 5.17 |
| Reproducibility | | | 6.01 | 6.20 | 6.27 |

m.v. = mean value $\bar{x}$ (min.sec.)
s.d. = standard deviation $t \cdot s/(n^{\frac{1}{2}} \cdot \bar{x})$ from equation (1) (%)

Among these methods the standard to determine the fire resistance of building components is of prime importance. The international standard ISO 834 is perhaps the most thoroughly investigated and carefully prepared standard method among the fire safety tests. It is also fairly widely applied in various countries.

The biggest problem associated with ISO 834 is in particular the repeatability and the reproducibility of the test results. The main reason for this is the big test furnaces which are used. The furnace constructions vary between laboratories. The main differences are:

dimensions of furnaces and their proportions
refractory lining of furnace interiors
number and location of burners
kind of fuel used by burners
location of flue
regulation of overpressure
measuring instrumentation.

In addition, very important factors are also the supporting of the specimens corresponding to their boundary conditions in practice, conditioning of specimens and application of loading.

The repeatability and reproducibility of results in the Nordic research described in paragraph 3 above were actually good. We have to remember, however, firstly that the furnaces of the participating laboratories quite closely ressembled each other. Secondly the specimens were very simple without any ambiguities which are more common in real building elements. Thirdly there was no loading thus excluding a number of divergencies.

The big test furnaces are a big investment and it is most unlikely that laboratories can afford larger rebuilding of their furnaces. Hence the only possibility to get mutually comparable results is to perform the test to a standard as accurately prescribed as possible. The most important questions which are still open in the revision of ISO 834 are the measurement of radiation and the calibration of furnaces. The most promising ways to solve these questions come at the moment from research projects financed also by Nordtest. The need for harmonized test procedures to determine the fire resistance of structures is greater now than ever.

## 5 References

Holm, C. and Loikkanen, P. (1981) Joint investigation of vertical furnaces in Nordic countries. Technical Research Centre of Finland. Research notes 56/1981, 52 + app. 6 p., Espoo, Finland

ISO 834. Fire-resistance tests - Elements of building construction (1975). International Organization for Standardization, 16 p.

ISO 5725. Precision of test methods - Determination of repeatability and reproducibility by interlaboratory tests (1981), International Organization for Standardization, 42 p.

Nordtest Method NT FIRE 005. Elements of building construction: fire resistance (1976), Nordtest, 1 + 16 p.

# 16 THE ORGANISATION AND RESULTS OF AN INTERLABORATORY PROGRAMME OF CONCRETE COMPRESSION TESTING, TRINIDAD AND TOBAGO, WEST INDIES

R.W.A. OSBORNE
Department of Civil Engineering, The University of the West Indies, St Augstine, Trinidad and Tobago, West Indies

Abstract
The paper reports on the organisation and results of a programme executed using centrally prepared concrete compression specimens. It reviews the specimen-preparation procedures adopted to reduce variability, examines the differences between nominally identical batches (necessitated by limitations of mixer capacity), and also reports on the interlaboratory comparisons.

The within-mix, inter-batch variability points to the need for improvement in specimen repeatability; the interlaboratory comparisons indicate the effects of choice of testing,machine and type, size and compressive strength of specimen.
Keywords: Interlaboratory Concrete Tests, Compression Testing Machines, Cubes, Cylinders.

## 1 Introduction

In many developing countries, resources and systems for quality assurance in compressive strength testing of concrete, such as the Frame Stability Tester (or Strain Cylinder), BS1881: Part 115 (1986), DIN 51302: Part 2 (1983), are often not available. This does not remove the need for some means of checking the reliability of testing machines used for the compressive testing of concrete specimens.

It is well known that accuracy of force-application alone is a necessary but insufficient criterion for the reliability of compression testers intended for cube-testing. Symmetry of load application, particularly with cube specimens which possess materials-properties inhomogeneity, remains a critical requirement. Born and Struchtrup (1985) record recent serious performance deficiencies at well-known German institutes.

As part of a larger programme of research by Osborne (1989) it became necessary to have a performance reference for the machine used for all the compressive testing. A Frame Stability Tester was not available; an interlaboratory test programme was therefore prepared and carried out.

## 2 Experimental Procedure

Moulded concrete compression specimens were prepared in the Structures and Concrete Laboratory, Department of Civil Engineering, UWI, St. Augustine, Trinidad (West Indies). Three mixes were made with fine and coarse Trinidad Guanapo aggregate (a siliceous subrounded gravel containing approximately 99% quartz) and locally-manufactured Ordinary Portland Cement, as follows:

| Mix Code | W/C Ratio | A/C Ratio | Fine:Coarse Aggregate Ratio |
|---|---|---|---|
| R8 | 0.55 | 5.0 | 40:60 |
| R11 | 0.45 | 4.0 | 40:60 |
| R22 | 0.75 | 6.0 | 40:60 |

The coarse aggregate grading was as follows: cumulative percentages passing the 25, 20, 10, 5, 2.36, and 1.18-mm sieves were respectively 100, 99, 52, 5, 1, 0. The fine aggregate satisfied BS882: 1973 Zone 2 fine aggregate. The aggregate's absorption is approximately 1.09%; all aggregate was batched air-dry.

Specimens of different sizes, types, and strengths test different aspects of compression-machine performance (particularly as regards stiffness and symmetry, and ball-seating behaviour); two sizes each of cubes and cylinders of three different strength-ranges were tested at each laboratory, as follows:

| Cubes | Cylinders (2:1 height/dia.) |
|---|---|
| 100 mm | 101.6 mm dia. |
| 152.4 mm | 152.4 mm dia. |

This is similar in general concept to Clause 4 of RILEM Recommendation CPC.17 (1986). Cylinders were despatched uncapped; the resulting variations in strength-values therefore reflect the effects of end-capping and final curing, as well as those of specimen-making and testing.

Because of small mixer capacity, each mix had to be batched more than once to make the necessary number of specimens. For each batch, moulds were filled and vibrated on a 200-cm x 76-cm vibrating table, and coded.

The procedure was repeated until the required number of specimens had been prepared. All filled moulds were stored in air at 20±2°C and >95% r.h. for the first 24h±1 hrs., then demoulded, their marks transferred to the specimens, and these were stored in water indoors at 26.5±1°C until tested, or until despatched (not before 10 days' age, via road, all journeys < 1 hr.) to the other participating laboratories. (Water-in-shade temperatures indoors for the laboratories ranged from 25 to 28°C). Specimens were enclosed in water-soaked furniture foam wrapped in plastic sheeting to reduce damage and maintain curing. The distribution of samples for a typical mix is shown schematically in Fig. 1.

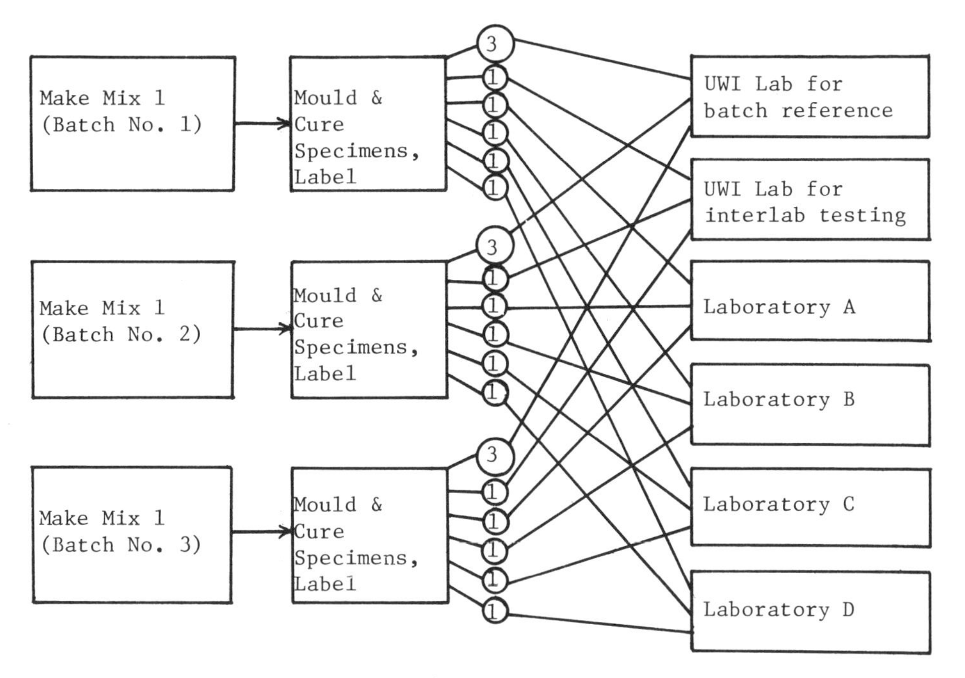


Fig. 1. Distribution of Samples in the Interlaboratory Test Series

Note: Encircled numbers (viz. (3), (1) refer to the quantity of specimens despatched.

## 3 The Laboratories and Their Equipment

The laboratories were chosen because locally they were those with the largest volumes of concrete specimen-testing over the preceding 15 years, and therefore those which most affected the values of concrete compressive strength in local use.

A detailed questionnaire requesting information on (a) equipment design and specifications (b) calibration, service, and maintenance history (c) qualifications and experience of laboratory personnel, was completed by each laboratory. Only the briefest summaries of these data are provided in Table 1.

## 4 Procedural Safeguards

These may be summarised as follows:

(i) Careful briefing of laboratory personnel.

(ii) Specimens tested at above 42 days' age, to minimize strength-gain-rate effects

(iii) All specimens measured before testing, to produce corrected strength values.

(iv) All specimens tested damp within the same six-hour period on the same day.

Table 1. Laboratories Involved in the Test Series

| Name | Person in Charge | Machine | Force Capacity | Approx. Date of Mnfr. | Load Pacer |
|---|---|---|---|---|---|
| UWI | Author; Civil Engr | ELE MK XII four-column (UK) | 3000 kN | 1980/81* | Yes |
| Trintoplan | Civil Engr | Forney QC-200 DR two-column (USA) | 400,000 lbf (1800 kN) | Not Known; before 1977 | No |
| CARIRI | Civil Engr | ELE EL31-092 four-column (UK) | 1560 kN | 1978 | No |
| Readymix | Tech Mngr | CONTEST GD-10A (UK) | 2000 kN | 1979/80* | Yes |
| Trinidad Cement | Works Chemist | Avery-Dennison Model 7112 CCG (UK) | 250 ton (2500 kN) | 1967 | Yes |

(v) Each laboratory received one specimen of each type and size from each batch.

(vi) All specimens from each batch vibrated together on vibrating table.

## 5 Results

### 5.1 Interlaboratory Comparisons

The results are summarised in Table 2 (in which results from individual laboratories are shown divided by the overall interlaboratory mean strength for the particular specimen and mix); Table 3 shows the general patterns of the effect of the choice of specimen on indicated strength.

### 5.2 Inter-batch Strength Variability

Due to limited numbers of other moulds, only 100-mm cubes were used to measure inter-batch variation. The results are summarised in Table 4.

## 6 Conclusions

(i) Except CARIRI, laboratory (mean) results are typically within ±5% of the overall interlaboratory mean. Born and Struchtrup (1985) reported a 12.6% relative variation of mean values between premier German institutes in 1980.

(ii) Internal dispersion about respective laboratory means (last line, Table 2) is relatively constant, 3.2% to 4.6%. The CARIRI

*Likely to have satisfied the Frame Stability criterion as originally manufactured.

Table 2. Summarised Results of Interlaboratory Testing

| MIX | (1) UWI | (2) TRINTO-PLAN | (3) CARIRI | (4) READYMIX | (5) TRINIDAD CEMENT | INTERLABORATORY MEAN INDICATED STRENGTH, (MPa) |
|---|---|---|---|---|---|---|
| | | 100-mm cubes (based on means of 3) | | | | |
| R8 | 0.97 | 0.95 | 1.01 | 1.04 | 1.01 | 47.8 |
| R11 | 0.97 | 0.98 | 0.96 | 1.04 | 1.03 | 65.7 |
| R22 | 0.97 | 0.97 | 1.00 | 1.06 | 1.00 | 29.6 |
| | | 152.4-mm cubes (based on means of 3) | | | | |
| R8 | 1.01 | 1.02 | 0.93 | 1.08 | 0.96 | 44.8 |
| R11 | 1.01 | 1.02 | 0.87* | 1.08 | 1.02 | 58.5 |
| R22 | 1.04 | 0.99 | 0.96 | 1.06 | 0.96 | 24.7 |
| | | 101.6-mm cylinders (based on means of 2) | | | | |
| R8 | 0.95 | 1.02 | 0.585? | 0.99 | 1.03 | 39.8** |
| R11 | 0.94 | 1.03 | 0.50 ? | 0.97 | 1.06 | 50.6** |
| R22 | 0.91 | 0.94 | 0.72 ? | 1.07 | 1.08 | 17.3** |
| | | 152.4-mm cylinders (based on means of 2) | | | | |
| R8 | 1.00 | 1.03 | 0.96 | 1.01 | 1.01 | 40.0 |
| R11 | 0.85*** | 1.01 | 1.04 | 1.04 | 1.02 | 51.4 |
| R22 | 1.07 | 0.97 | 0.90 | 1.04 | 1.02 | 16.4 |
| Mean = | $0.985_{11}$ | $0.99_{12}$ | $0.925_{10}$ | $1.04_{12}$ | $1.017_{12}$ | |
| $\sigma_{n-1}$ | 0.046 | 0.032 | 0.083 | 0.035 | 0.034 | |

* Asymmetrical failure of 2 of 3 cubes
** Mean of 4 labs only; CARIRI excluded
*** Unusual failure evident in 1 of 2 cylinders. Excluded from mean-value calculation

Table 3. Interlaboratory Mean Strengths-Effect of Specimen Choice

| MIX | R8 | R11 | R22 |
|---|---|---|---|
| Large cyl / Large cube | 0.89 | 0.88 | 0.66 |
| Small cyl / Small cube | 0.83 | 0.77 | 0.58 |
| Small cube / Large cube | 1.07 | 1.12 | 1.20 |
| Small cyl / Large cyl | 0.995 | 0.98 | 1.05 |

Table 4. Inter-Batch Variations

| Batch | Mean $f_c$, $\sigma_{n-1}$ (MPa) | Batch | Mean $f_c$, $\sigma_{n-1}$ (MPa) |
|---|---|---|---|
| R11 No. 1 | 63.4, 2.32 | R22 No. 1 | 28.5, 0.59 |
| 2 | 63.1, 3.14 | 2 | 26.0, 0.47 |
| 3 | 62.7, 1.69 | 3 | 25.0, 0.25 |
| 4* | 63.8, 0.80 | R8 No. 1 | 52.3, 3.44 |
| 5* | 64.8, 2.50 | 2 | 48.4, 1.56 |

Note: Stressing rate: 150 kN/min = 15 MPa/min, BS1881: Part 116: 1983
*Produced on a different date from the other R11 batches.

value is 8.3%.

(iii) The CARIRI data clearly demonstrate the divergences of indicated strength which can occur when using machines of lower rigidity and modest ball-seating design.

(iv) The divergence of the CARIRI mean value is always negative from the interlaboratory mean, and is always worst for R11 (the strongest mix); 152.4-mm cubes were down by 13%.

(v) The Trintoplan cube failures all exhibited marked asymmetry, but the strengths agree well with UWI, Readymix, and TCL. The reasons for this are unclear.

(vi) 100-mm cubes were less sensitive to machine choice than 152.4-mm cubes, and show closer agreement between laboratories.

(vii) The effect of concrete strength. Regardless of specimen type or size, the interlaboratory dispersion of strength is largest for the strongest mix, R11.

(viii) Effect of specimen size. The effect of a change in specimen size is less for cylinders than for cubes (last line, Table 3). This tends to confirm the fact that for the same concrete, cylinders are less demanding on testing-machine performance, i.e. are less sensitive to end-conditions and are suitable for testing on a wider variety of compression testing machines; Sigvaldason (1966).

## 7 Closing Comments

In construction communities with limited facilities, such exercises are desirable to define interlaboratory precision being attained under normal conditions.

## 8 References


Born, D. and Struchtrup, H.H. (1985), RILEM International Conference on Destructive Testing Equipment, 18/19 April 1985, Dubendorf/Zurich.

Osborne, R.W.A. (1989), Factors Affecting Selected Methods of Testing Concrete for Compressive Strength, Ph.D. Thesis, Department of Civil Engineering, University of the West Indies, St. Augustine, Trinidad.

Sigvaldason, O.T. (1966), Mag. Conc. Res., Vol. 19, No. 57, Dec. 1966, 197-205.

# 17 TEST QUALITY OF THE OFFICIALLY APPROVED TESTING LABORATORIES FOR CONCRETE IN FINLAND

K. VAITTINEN
Technical Research Centre of Finland, Building Materials Laboratory, Espoo, Finland

ABSTRACT

Currently there are 11 officially approved testing laboratories for concrete in Finland. The testing quality of these laboratories, which arises from their cooperation, was examined by means of comparative tests between them during 1977 - 89.

## DEFINITIONS

Repeatability is the value below which the absolute difference between two single test results obtained with the same method on identical test materials and under the same conditions (same operator, same apparatus, same laboratory, and a short interval of time), may be expected to lie with a specified probability.

Reproducibility is the value below which the absolute difference between two single test results obtained with the same method on identical test material, under different conditions (different apparatus, different laboratories and/or different time), may be expected to lie with a specified probability.

## INTRODUCTION

A testing laboratory for concrete is officially approved by the Ministry of the Environment. The Technical Research Centre of Finland (VTT) is by law the approved laboratory and is responsible for the coordination and technical inspection of all approved laboratories. VTT also acts as technical advisor to the Ministry. The main items for approval certification are well educated and trained staff, the calibration of testing machines, unified testing methods, comparative tests and inspection visits.
Using it's own load cells, VTT calibrates every machine in the approved laboratories once a year. These load cells are calibrated at the Finnish National Measurement Centre of Force situated in VTT's Metals Laboratory. The calibration chain continues through VTT to international calibration in Germany.
VTT makes an inspection visit to the approved laboratories once a year to ensure that all items for approval certification are fulfilled. The outcome of the visits is reported to the Ministry.
Also once a year, VTT arranges comparative cube tests between all approved laboratories as required in their approval certificates. The number of approved laboratories has risen from five to 11 from 1977 to 1989. Most of the laboratories participating in comparative tests have two machines, a primary one and so-called vice-machine.

## CONCRETE MIXES AND TEST SPECIMENS

Concrete mixes are selected from a number of VTT's standard mixes. The strength of the concrete varies from year to year from 20 to 70 MN/m2, as can be seen in Table 1. Comparative cube tests are performed using 150 mm cubes prepared from the same carefully mixed batch. After demoulding the specimens are cured in VTT's curing room at 20±2 °C and >95% relative humidity.

## TRANSPORTATION AND TESTING

Before dispatch to the participating laboratories, the specimens are randomly selected into groups of six. After carefully packing one group is sent to each testing machine with an accompanying letter describing the testing procedure. All packages are opened at the same time and the specimens placed in curing rooms. All tests are performed according to the Finnish standard on the same day and at the same time in each laboratory. The compressive strength and density are calculated for every specimen.

## TEST RESULTS

The test results are sent to VTT's research scientist, who compiles them for statistical calculation. The final results are imparted to each participating laboratory. Normally the mean values and standard deviations of the results from each machine are calculated and compared with earlier results from previous years. In this report the quality of testing is analysed statistically by other procedures and comparisons of the test results. Repeatabilities and reproducibilities are calculated according to International Standard ISO 5725.
Table 1 shows the results of comparative cube tests performed during years 1977 - 1989. The yearly results are given separately for the primary machines and vice-machines. Primary machines are used in every day testing in each laboratory, and their technical quality is roughly the same. Vice-machines are additional equipment and are not nesessarily used on a daily basis. Several of these are fairly old and their technical quality and load ranges are more variable than with primary machines. Table 1 (columns 1) and 2)) shows the yearly number of participating laboratories and vice-machines used. Also shown are the mean values and standard deviations yearly, and the total numbers of test specimens used in comparative tests. Between 1977 and 1989, a total of 95 primary and 82 vice-machines were used in comparative cube tests on a total of 1062 used test specimens.

Table 1. Results of comparative cube tests during 1977 - 1989.

| PRIMARY MACHINES | | | | VICE-MACHINES | | | |
|---|---|---|---|---|---|---|---|
| year | 1) | mean | S.D. | 2) | mean | S.D. | total No. of specimens |
| 1977 | 5 | 44.1 | 1.6 | 5 | 43.7 | 1.6 | 60 |
| 1978 | 5 | 43.2 | 1.2 | 5 | 42.9 | 1.2 | 60 |
| 1979 | 5 | 48.0 | 1.7 | 5 | 47.7 | 1.5 | 60 |
| 1980 | 6 | 32.2 | 1.5 | 5 | 31.8 | 1.1 | 66 |
| 1981 | 6 | 27.7 | 0.8 | 5 | 27.3 | 0.7 | 66 |
| 1982 | 6 | 60.2 | 2.6 | 5 | 61.8 | 2.9 | 66 |
| 1983 | 7 | 43.0 | 2.1 | 7 | 42.0 | 2.0 | 84 |
| 1984 | 8 | 43.2 | 1.0 | 7 | 43.0 | 1.2 | 90 |
| 1985 | 9 | 23.2 | 1.2 | 7 | 23.3 | 1.2 | 96 |
| 1986 | 9 | 34.5 | 1.0 | 8 | 34.7 | 1.6 | 102 |
| 1987 | 9 | 63.5 | 2.3 | 7 | 62.0 | 4.4 | 96 |
| 1988 | 9 | 46.5 | 2.2 | 8 | 45.7 | 2.2 | 102 |
| 1989 | 11 | 26.7 | 1.4 | 8 | 26.6 | 0.9 | 114 |
| Total | 95 | | | 82 | | | 1062 |

1) Number of participating laboratories
2) Number vice-machines used.

Figure 1 shows the correlation between the variation coefficients of primary machines and the strength of concrete. Figure 2 shows the same for the vice-machines. The regression line slopes slightly downwards with the primary machines and slightly upwards with the vice-machines. The variation coefficient in both groups is roughly the same (4%) when the strength of concrete is below 50 MN/m2, but above 60 MN/m2 the results from the vice-machines are more variable and the variation coefficients also higher than with the primary machines.

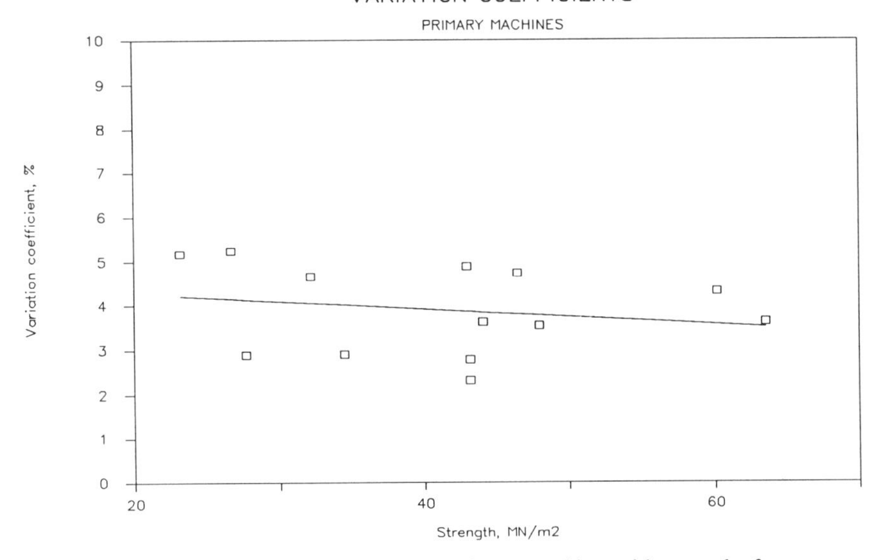


Figure 1. Correlation of variation coefficient of primary machines with strength of concrete.

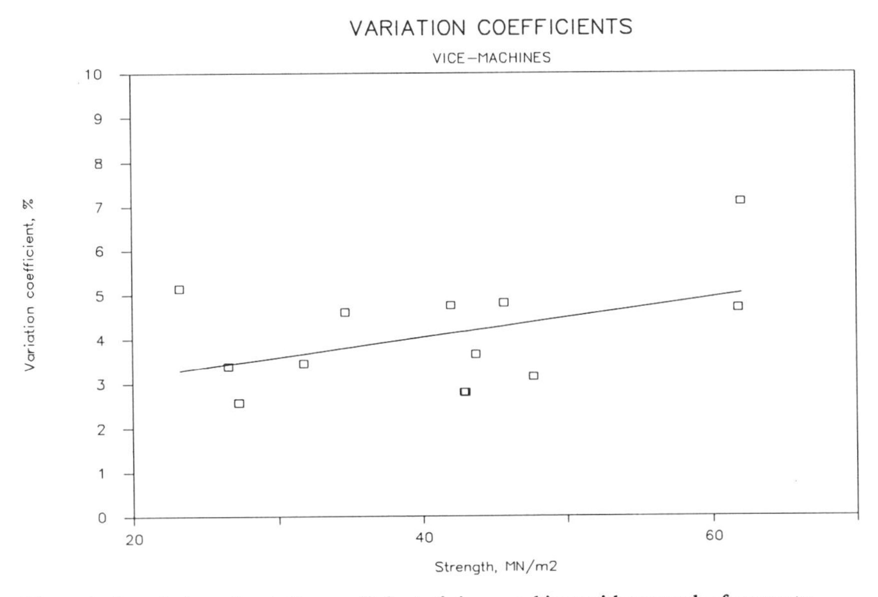


Figure 2. Correlation of variation coefficient of vice- machines with strength of concrete.

Figure 3 shows the correlation of calculated repeatabilities and Figure 5 the calculated reproducibilities of the primary machines with the strength of concrete. Figures 4 and 6 show the same values for the vice-machines. Figures 3 and 5 show that the slope of the regression line for reproducibility in primary machines is greater than that for repeatability. Thus the reproducibility of even technically high standard compressive machines is more strongly dependent on the strength of concrete than is repeatability. The same tendency can be observed in Figures 4 and 6 concerning vice-machines. Here the difference between the slopes is more significant. The slopes of the regression lines are greater for the vice-machines in both cases, indicating that repeatability and reproducibility are more dependent on the stength of concrete if the technical quality of the testing machines is lower.

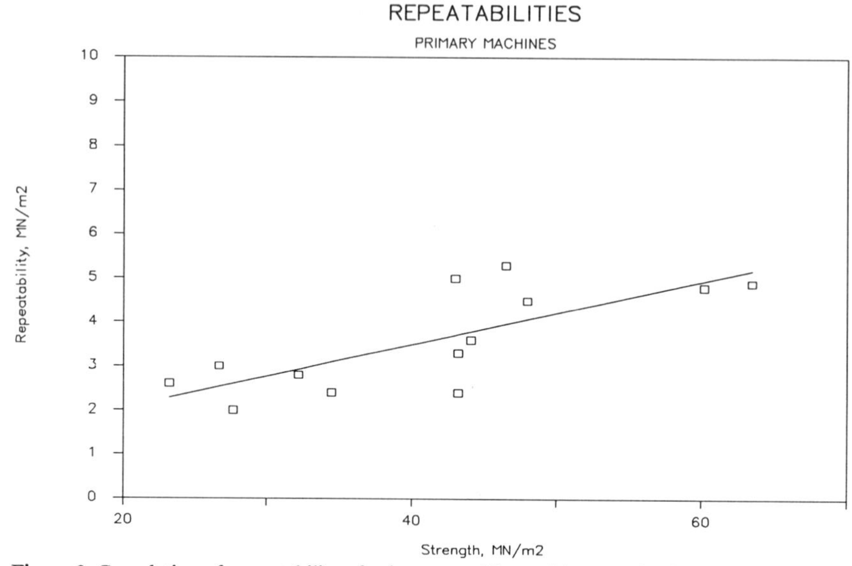


Figure 3. Correlation of repeatability of primary machines with strength of concrete.

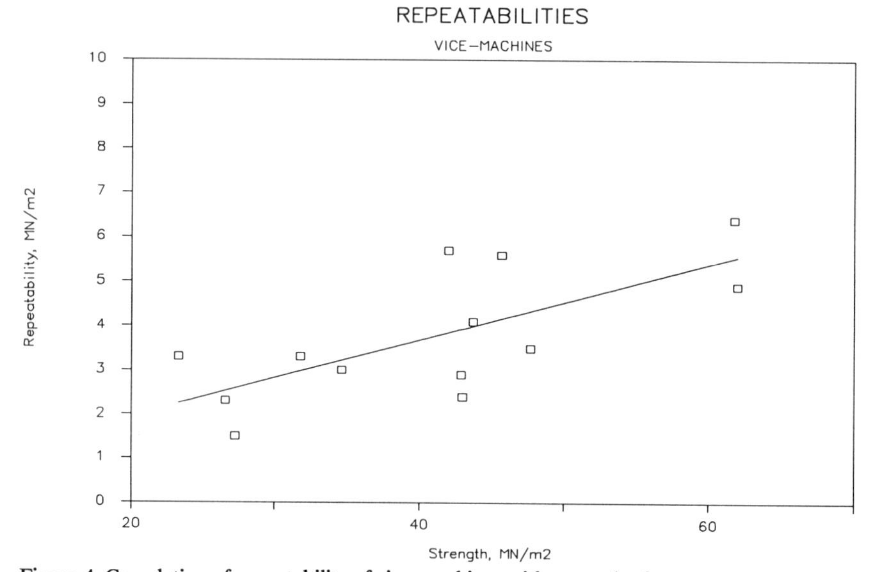


Figure 4. Correlation of repeatability of vice-machines with strength of concrete.

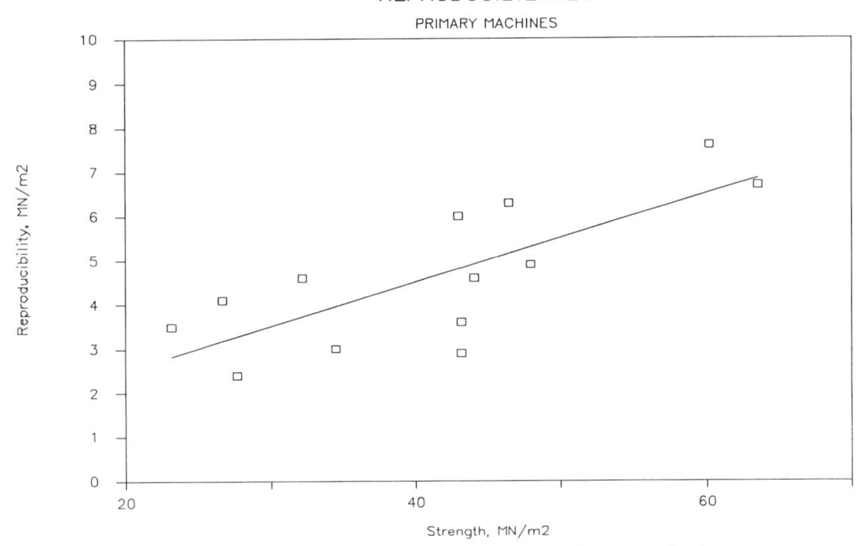


Figure 5. Correlation of reproducibility of primary machines with strength of concrete.

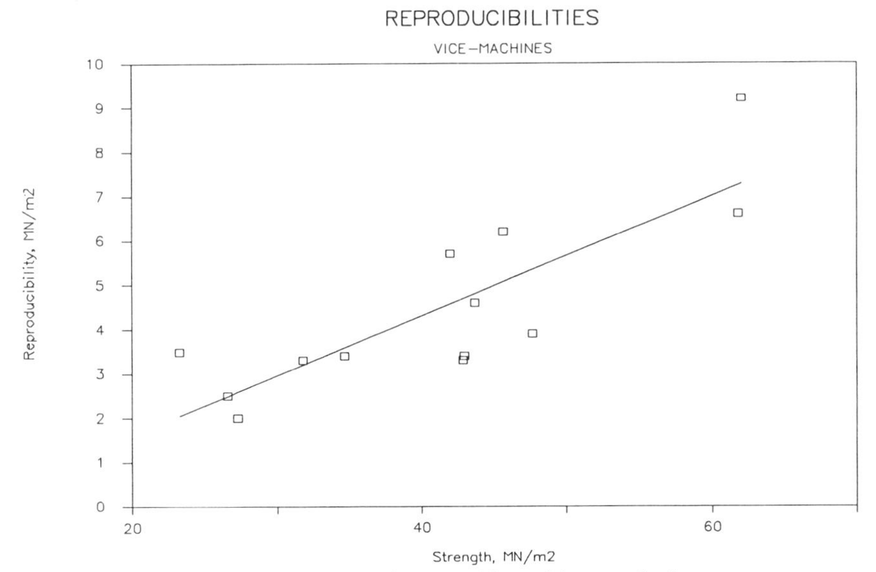


Figure 6. Correlation of reproducibility of vice-machines with strength of concrete.

Figure 7 shows the variations in primary testing machines of four laboratories over a period of 5 years. The results are shown as devations from the mean of 100%. This is the mean value of all the results of primary testing machines in each laboratory. The laboratories have the following testing frequencies: A: 15000 specimens per year, B: 9000, C: 3000 and D: 1000 specimens per year. The figure shows all deviations to be in the same range. Thus the quality of testing is not dependent on the testing frequency if the frequency is greater than 1000 specimens per year.

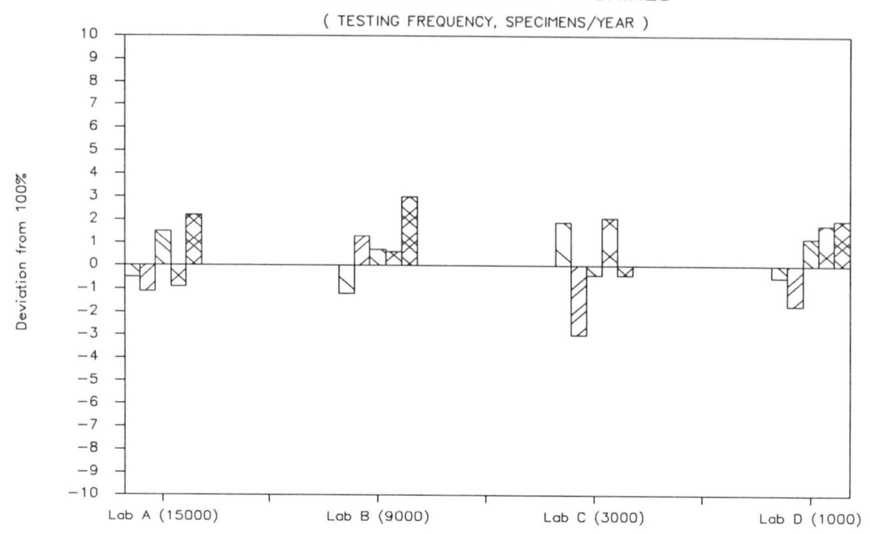


Figure 7. Variations in primary testing machines and test frequencies of four laboratories.

## CONCLUSIONS

The variation coefficients of technically lower standard vice-machines are higher, especially with a high strength of concrete.
The reproducibility is more strongly dependent on the strength of concrete than is repeatability.
Both repeatability and reproducibility are more dependent on the strength of concrete if the technical quality of the testing machines is lower.
The quality of testing is not dependent on the testing frequency if the frequency is greater than 1000 specimens per year.
Correct calibration values (scale error less than 1%) do not guarantee that the results of comparative tests are acceptable if the technical quality of the machines is lower.

# 18 PROGRAMME D'ESSAIS INTERLABORATOIRES BASÉ SUR LA CONSTANCE DU COEFFICIENT DE VARIATION
## (Interlaboratory test programme based on constant coefficient of variation)

J.C. LOPEZ-AGUI
Instituto Español del Cemento y sus Aplicaciones, Madrid, Spain

Résumé
Basé sur l'hypothèse de la constance des coefficients de variation de l'essai des résistances mécaniques du ciment ce travail développe la base théorique de l'analyse statistique d'un programme d'essais interlaboratoires qui permet d'évaluer non seulement les paramètres de la répétibilité et de la reproductibilité de la méthode d'essai mais aussi les composantes de la dispersion globale de chaque laboratoire (erreur analytique et erreur biaisée) pour effectuer leur classement.
Mots clés: Répétibilité, Reproductibilité, Constance des Coefficients de Variation.

Le présent travail aspire à offrir une base théorique sur laquelle on puisse structurer un programme d'essais interlaboratoires autant qu'il soit possible d'accepter l'hypothèse sur la constance des coefficients de variation de la répétibilité et de la reproductibilité.

Le "Instituto Español del Cemento y sus Aplicaciones" (Institut Espagnol du Ciment et ses Applications) coordonne un programme d'essais interlaboratoires avec une périodicité semestrielle auquel participent plus de 50 laboratoires qui réalisent des essais de ciments. Notre expérience des dernières années nous permet d'assurer que l'hypot des coefficients de variation constants est assez approchée dans le cas des essais de résistances mécaniques des ciments, et de ce fait cet institut applique depuis longtemps le modèle décrit ci-dessous pour l'analyse statistique des données.

Tout d'abord on définira les caractéristiques du programme de référence:

Un nombre p de laboratoires.
Un nombre q de niveaux d'essai (q > 1).
Un nombre constant n d'essais réalisés par chaque laboratoire à chaque niveau.

Le résultat d'un essai quelconque sera exprimé par le symbole $Z_{ijk}$, oú l'indice "i" représente le laboratoire générateur du résultat, l'indice "j" le niveau d'essai et l'indice "k" la k-ième réplique de l'essai dans la cellule "ij". D'une façon plus simple, l'indice "j" est un type concret de ciment essayé (il y a q types de ciments différents) et l'indice "k" un des n essais qu'un laboratoire donné réalise avec un type de ciment donné.

Le modèle mathématique sur lequel on va travailler, en supposant qu'on peut négliger l'effet intéraction entre les laboratoires et les niveaux d'essai, est le suivant:

$$Z_{ijk} = \mu_j + \beta_{ij} + e_{ijk} \qquad i = 1....p \quad j = 1....q \quad k = 1....n \tag{1}$$

où $\mu_j$ représente la véritable valeur (inconnue) de la propriété essayée (par exemple la résistance) du ciment du type "j", $\beta_{ij}$ est une variable aléatoire représentant l'effet différentiel ou biais incorporé par le laboratoire "i" aux n répliques réalisées du ciment du type "j" et, finalement, $e_{ijk}$ représente l'erreur expérimentale associée au résultat obtenu par le laboratoire "i" de la réplique "k" du type de ciment "j".

Le procédé au moyen duquel nous incorporons au modèle ci-dessus l'hypothèse de constance dans les coefficients de variation inter et intralaboratoires, en plus de celle de normalité dans le comportement des erreurs, aussi bien biaisées qu'expérimentales, réside dans le postulat que les variables aléatoires $\beta_{ij}$ et $e_{ijk}$ se distribuent normalement suivant:

$$\beta_{ij} \sim N\,(0\,;\,\mu_j^2\,\delta_L^2) \qquad e_{ijk} \sim N\,(0\,;\,\mu_j^2\,\delta_r^2) \tag{2}$$

où $\delta_L$ est le coefficient de variation interlaboratoires (supposé constant mais inconnu) et $\delta_r$ le coefficient de variation intralaboratoires ou de la répétibilité (supposé aussi constant et inconnu).

A part les hypothèses du point (2) on supposera aussi, comme d'habitude, l'absence de corrélation des différents $e_{ijk}$ entre eux-mêmes et de ceux-ci entre les $\beta_{ij}$.

Si maintenant nous incorporons les nouvelles variables aléatoires

$$\beta'_i = \beta_{ij}/\mu_j \qquad e'_{ijk} = e_{ijk}/\mu_j \tag{3}$$

nous connaissons par la théorie statistique que

$$\beta'_i \sim N(0, \delta_L^2)$$
$$e'_{ijk} \sim N(0, \delta_r^2) \qquad (4)$$

le modèle (1) pourra donc s'écrire :

$$Z_{ijk} = \mu_j (1 + \beta'_i + e'_{ijk}) \qquad (5)$$

En introduisant la variable

$$W_{ijk} = \frac{Z_{ijk} - \mu_j}{\mu_j} \qquad (6)$$

le modèle se représente par

$$W_{ijk} = \beta'_i + e'_{ijk} \qquad (7)$$

avec

$$W_{ijk} \sim N(0, \delta_L^2 + \delta_r^2) \qquad (8)$$

ou

$$W_{ijk} \sim N(0, \delta_R^2) \qquad (9)$$

$\delta_R$ coefficient de variation de la reproductibilité

$$\delta_R^2 = \delta_L^2 + \delta_r^2 \qquad (10)$$

Etant donné que $W_{ijk}$ n'est pas observable, nous ne pouvons pas appliquer directement la technique de l'Analyse de Variance au modèle (7). D'abord nous devons définir l'observable

$$w'_{ijk} = \frac{z_{ijk} - \hat{\mu}_j}{\hat{\mu}_j} \qquad (11)$$

où $\hat{\mu}_j$ est l'estimateur de $\mu$ donné par la notation

$$\hat{\mu}_j = \frac{\sum_i \sum_k z_{ijk}}{np} \qquad (12)$$

Maintenant, en utilisant l'observable $w'_{ijk}$ (qui se distribue comme $w_{ijk}$) dans le modèle (7) nous pouvons appliquer la technique de l'Analyse de Variance pour estimer les paramètres $\delta_L^2$, $\delta_r^2$ et $\delta_R^2$.

Le tableau ANOVA qui en résulte pour le modèle (7) corrigé est le suivant:

Tableau 1. Tableau ANOVA

| Origine de variation | Somme des carrés | Degré de liberté | E.M.C. |
|---|---|---|---|
| TOTAL | $S'_{TOT} = \sum_i\sum_j\sum_k w'^2_{ijk}$ | | |
| LABORATOIRE | $S'_L = \frac{1}{nq} \sum_i[\sum_j\sum_k w'_{ijk}]^2$ | p-1 | $\delta_r^2 + nq\,\delta_L^2$ |
| ERREUR | $S'_e = S'_{TOT} - S'_L$ | npq-p-q+1 | $\delta_r^2$ |

et à partir duquel on peut déduire:

$$\hat{\delta}_r^2 = \frac{S'_e}{npq-p-q+1}$$

$$\hat{\delta}_L^2 = \frac{1}{nq}\left(\frac{S'_L}{p-1} - \hat{\delta}_r^2\right) \tag{13}$$

et finalement

$$\hat{\delta}_R^2 = \hat{\delta}_r^2 + \hat{\delta}_L^2$$

Avec ces estimateurs définis on peut donc déduire les expressions de la répétibilité et de la reproductibilité.

$$r_j \approx 2.8\, \hat{\mu}_j\, \hat{\delta}_r$$
$$R_j \approx 2.8\, \hat{\mu}_j\, \hat{\delta}_R \tag{14}$$

lesquelles logiquement sont proportionnelles à la moyenne de chaque niveau, ce qui correspond avec l'hypothèse d'origine sur laquelle on s'est basé, c'est-à-dire supposer constants les coefficients de variation de la répétibilité et de la reproductibilité.

Un cas particulier très utile du modèle développé est celui où l'on essai deux types de ciment ($q = 2$) et où il n'existe qu'un essai par cellule ($n = 1$).

Dans ce cas il est utile d'employer la notation suivante :

$$z_{i11} = x_i$$
$$z_{i21} = y_i \tag{15}$$
$$\mu_1 = \mu_x$$
$$\mu_2 = \mu_y$$

Le modèle (5) peut s'écrire:

$$x_i = \mu_x (1 + \beta'_i + e'_{i11}) \qquad y_i = \mu_y (1 + \beta'_i + e'_{i21}) \tag{16}$$

et on peut démontrer facilement les expressions suivantes :

$$\hat{\mu}_x = \bar{x} \qquad \hat{\mu}_y = \bar{y}$$

$$\hat{\delta}_L^2 = \frac{s_{xy}}{\bar{x}\bar{y}}$$

$$\hat{\delta}_r^2 = \tfrac{1}{2}\left(\frac{s_x^2}{\bar{x}^2} + \frac{s_y^2}{\bar{y}^2}\right) - \frac{s_{xy}}{\bar{x}\bar{y}} \tag{17}$$

$$\hat{\delta}_R^2 = \tfrac{1}{2}\left(\frac{s_x^2}{\bar{x}^2} + \frac{s_y^2}{\bar{y}^2}\right)$$

où $s_x^2$, $s_y^2$ et $s_{xy}$ sont les variances et covariances de l'échantillonnage.

En plus, une fois représentés les deux résultats obtenus par chaque laboratoire sur un diagramme de dispersion, ceux-ci auront une tendance à se regrouper sous une forme élliptique, étant donné que les courbes d'équiprobabilité de la distribution conjointe de la variable aléatoire (X,Y) sont des ellipses de centre M $(\mu_x, \mu_y)$ dont l'équation peut s'écrire:

$$\frac{1}{1-\rho^2}\left[\frac{(x-\mu_x)^2}{\mu_x^2(\delta_L^2+\delta_r^2)} - 2\rho\frac{(x-\mu_x)(y-\mu_y)}{\mu_x\mu_y(\delta_L^2+\delta_r^2)} + \frac{(y-\mu_y)^2}{\mu_y^2(\delta_L^2+\delta_r^2)}\right] = \chi^2_{2,\alpha} \tag{18}$$

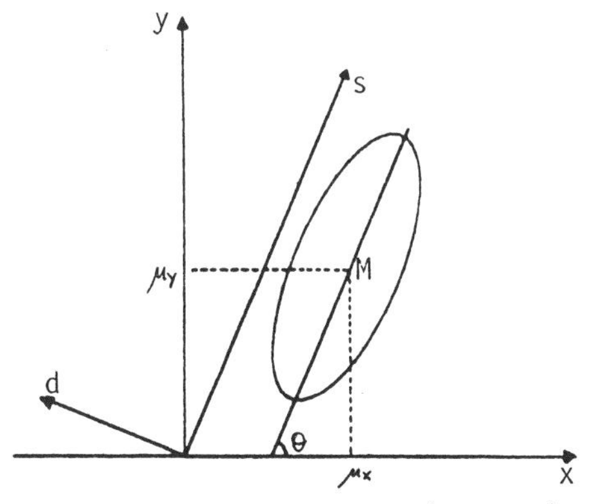


Fig. 1 Diagramme de dispersion

avec $$\rho = \frac{\delta_L^2}{\delta_L^2 + \delta_r^2} \tag{19}$$

ayant avec l'axe principal une inclinaison $\theta$ telle que :

$$\text{tg } 2\theta = \frac{2\mu_x\mu_y \ \delta_L^2}{(\mu_x^2 - \mu_y^2)(\delta_L^2 + \delta_r^2)} \tag{20}$$

Il est possible, à partir de ces idées, de classer les différents laboratoires par la définition de deux mesures de l'erreur présenté par chacun d'eux. A cet effet nous définissons

$$S = x \cos\theta + Y \sin\theta$$
$$D = -x \sin\theta + Y \cos\theta \tag{21}$$

comme des nouvelles variables aléatoires (il est possible de démontrer qu' elles sont non correlées par le fait d'être associées aux axes principaux des ellipses d'équiprobabilité).

Si nous définissons maintenant les mesures des dispersions (pour chaque laboratoire)

$$u_i = \frac{(s_i - \hat{\mu}_S)^2}{\hat{\sigma}_S^2}$$
$$v_i = \frac{(d_i - \hat{\mu}_D)^2}{\hat{\sigma}_D^2} \tag{22}$$

on peut comprendre que $u_i$ est en étroite relation avec la dispersion biaisée présentée par le laboratoire "i", et de même pour $v_i$ avec le biais de l'erreur expérimentale. Une mesure de la dispersion globale sera donc $u_i + v_i$ et on pourra classer les laboratoires sur la base de cette valeur. Si on préfère classer les laboratoires par rapport à chacun des deux types de dispersion séparément , on utilisera individuellement $u_i$ et/ou $v_i$.

# 19 AN INTERLABORATORY TEST PROGRAMME ON THE DBT RAPID TEST TO DETERMINE THE ALKALI REACTIVITY OF AGGREGATES

R.E. OBERHOLSTER and G. DAVIES
Division of Building Technology, CSIR, Pretoria, South Africa

Abstract
The Division of Building Technology's rapid test method for the alkali reactivity of siliceous aggregates takes only 12 days to deliver results in comparison with the ASTM C 227 mortar prism method which takes between 6 and 12 months to give conclusive results with similar types of aggregates. The precision of the method is good and the results obtained in an inter-laboratory test programme indicate that the repeatiblity is good if the prescribed procedures are strictly adhered to.
Keywords: Alkali-aggregate reaction, rapid test method.

## 1 Introduction

Several methods are being employed for the screening of siliceous aggregates for potential alkali reactivity. Most of these tests have disadvantages of one sort or another. The quick chemical test, although requiring only one day to complete, can give unreliable results with quartz-bearing rocks such as greywacke, argillite, granite and quartzite. Petrographic examination and the gel pat test (Jones and Tarleton, 1958) are both reasonably quick but are really only useful as screening tests. The ASTM C227 mortar prism test and the concrete prism test both require between six and twelve months to obtain conclusive results in the case of quartz-bearing aggregates. The rock cylinder test (ASTM C 586) takes at least eight weeks to give conclusive results and significant expansions are only obtained if the cylinders are cored perpendicular to the veins or the rock layers (Brandt et al. 1981). A rapid mortar prism test, which involves storage in saturated NaCl solution at 50 °C, was proposed by Chatterji (1978). Unpublished data of the Division of Building Technology (DBT), however, indicate that this method did not significantly reduce the time required to obtain results with quartz-bearing aggregate.

Van Aardt and Visser (1982) proposed a method which involved the storage of mortar prisms prepared in accordance with ASTM C 227 in a 1M NaOH solution at 80 °C. They obtained expansions of more than 0.10 % within 10 days with deleteriously reactive quartz-bearing aggregates. This method was further developed and criteria established by Oberholster (1983). The test has now generally been accepted by the international community, (Hooton and Rogers 1989; and Hudec and Larbi 1989), is being investigated by the CSA Standards Committee for possible adoption and has been approved by the ASTM on a trial basis for a period of two years (R D Hooton, personal communication). It is also one of the rapid methods for determining the alkali reactivity of aggregates that is currently being evaluated by RILEM Technical Committee TC-106.

The Portland Cement Institute of South Africa (PCI) agreed to collaborate with the DBT on setting up an inter-laboratory test programme for the rapid test method with the aim of standardising the preparation, storage and measurement of specimens in the two laboratories, and of solving any problems that might arise (Davies and Oberholster 1987).

## 2 Description of method

The mortar prisms are prepared in accordance with ASTM C 227-81. The prisms are demoulded after 24 hours and their length measured accurately to 2 μm using a vertical comparator (dial-type strain gauge). They are then immersed in water (at room temperature) in a closed container which is placed in an oven and kept at a constant temperature of 80 °C. After 24 hours in the oven, the prisms are removed to a room at a temperature of 23 °C and, before significant cooling has taken place, their length is measured. This reading is used as the zero reading. The prisms are then immersed in a 1M NaOH solution at 80 °C and stored in an oven kept at 80 °C. Tightly covered polymer containers, large enough for the prisms to be totally immersed, are used because the caustic solution corrodes glass and metal. The prisms are then measured each working day for 14 days in a room at a temperature of 23 °C, quickly enough to avoid significant cooling, and the average expansion of the three prisms is then calculated for each day; if none of the values differs from the mean by more than 15 % the repeatability is considered satisfactory; if not, the test is repeated on a new batch of prisms. The average expansion on the 12th day is taken as the reference value for assessing the potential alkali reactivity of the aggregate.

Details of the method, the influence of the concentration of NaOH, the alkali content of the cement and storage temperature and criteria for the test have been given by Davies and Oberholster (1987).

## 3 Precision of the method

Two experiments were undertaken to establish the precision of the test within the DBT laboratory. In the first, a series of rapid tests was performed on six aggregates, by three different operators. The first operator undertook two independent tests while the other two completed a single test each. Although only four results are available for each aggregate and a rigorous statistical approach is therefore impossible, an indication of the precision of the rapid test for aggregates of different alkali reactivity was obtained.

In the second experiment all the results of a reactive Malmesbury metasediment (A28) used as a reference alkali-reactive sample were pooled. In all, ten rapid tests (including those mentioned above) were undertaken on this sample by the same operator over a period of three years. The larger number of independent results allows a more accurate determination of the precision than in the first experiment.

The two experiments are discussed separately below:

### 3.1 Precision of the test when using different aggregates

Table 1 shows the mean, minimum and maximum expansion, and the coefficient of variation (CV) on the 12th day for the different aggregate samples. The CV was very erratic and large for all the samples until about the 7th day after which it stabilised to a value of between 10 and 21 %. There was a general decline in this value with increasing expansion, which was to be expected. From these results it seems that reactive aggregates (those which expand more than 0,10 % by the 12th day) can be measured with a CV of better than 18 % assuming the aggregate is reasonably homogeneous.

Table 1. Results on the 12th day for six aggregate samples subjected to the rapid test four times by three different DBT operators

| Aggregate Sample | Expansion (%) | | | Coefficient of variation (%) |
|---|---|---|---|---|
| | Mean | Minimum | Maximum | |
| N | 0,010 | 0,008 | 0,012 | 21,1 |
| WS | 0,031 | 0,024 | 0,034 | 15,5 |
| A49 | 0,117 | 0,094 | 0,144 | 17,8 |
| A33 | 0,239 | 0,203 | 0,281 | 14,9 |
| A28 | 0,340 | 0,311 | 0,390 | 10,4 |
| A34 | 0,344 | 0,313 | 0,381 | 10,0 |

### 3.2 Precision of the test when using a single aggregate

The mean, minimum and maximum expansions and CV obtained from ten independent tests completed on a single sample (A28) are presented in Table 2. The CV starts at 72 % on day one and stabilises after the 9th day at about 8 %.

In Figure 1 it is clear that the results obtained by operators 2 and 3 fall within the minimum and maximum values obtained by operator 1 from as early as the second day. Since the test depends on values obtained on the 12th day, inter-operator variation at the DBT laboratory is satisfactorily low.

## 4 Inter-laboratory testing

The first series of tests was undertaken with the aim of

Table 2. Statistical data for the rapid test undertaken on aggregate A28 (10 determinations) by DBT Operator 1

| Day | Expansion, % | | | Coefficient of variation % |
|---|---|---|---|---|
| | Mean | Minimum | Maximum | |
| 1 | 0,006 | 0,000 | 0,012 | 71,9 |
| 2 | 0,027 | 0,016 | 0,039 | 34,5 |
| 3 | 0,056 | 0,028 | 0,089 | 44,7 |
| 4 | 0,095 | 0,035 | 0,139 | 46,5 |
| 5 | 0,157 | 0,082 | 0,211 | 27,0 |
| 6 | 0,196 | 0,127 | 0,243 | 18,6 |
| 7 | 0,233 | 0,181 | 0,276 | 13,9 |
| 8 | 0,267 | 0,220 | 0,302 | 11,2 |
| 9 | 9,294 | 0,248 | 0,325 | 8,9 |
| 10 | 0,313 | 0,288 | 0,345 | 6,6 |
| 11 | 0,326 | 0,304 | 0,361 | 6,5 |
| 12 | 0,346 | 0,295 | 0,390 | 8,0 |
| 13 | 0,364 | 0,314 | 0,397 | 6,9 |
| 14 | 0,374 | 0,328 | 0,409 | 7,0 |

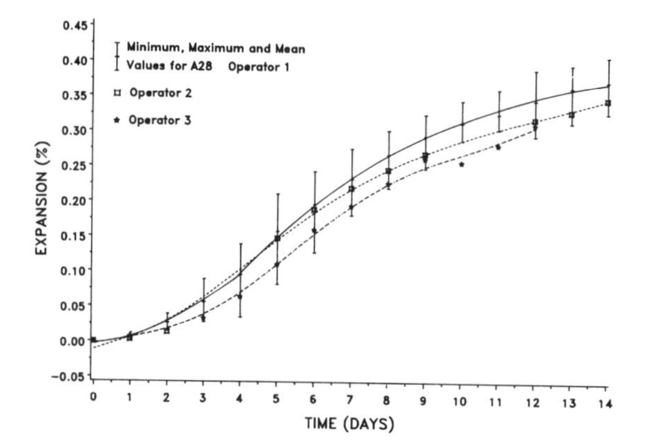


Fig. 1. Expansion versus time curves for accelerated tests undertaken by three DBT operators on aggregate sample A28. The curve for operator 1 was obtained from 10 independent tests. Fewer readings were taken by operator 1 on the 3rd, 4th, 10th and 11th days since they usually fell over a weekend

establishing whether there were any large discrepancies between the expansions measured in the two laboratories. Only a single operator was used by the PCI laboratory while the DBT contribution to this test was carried out by operator 1. Instructions regarding the test procedure, as well as samples of graded aggregate, and cement were provided by DBT. The results are presented in Table 3. The expansions recorded by PCI were consistently lower than those obtained by DBT in all but one case (an innocuous norite aggregate, N). With this difference in mind, a number of additional tests were undertaken with aggregate A28 .

Firstly, two sets of prisms were made simultaneously at each laboratory by the operator at that laboratory. After 24 hours' curing, one set from each laboratory was interchanged. Each laboratory thus had one set of prisms made by the DBT operator and one set made by the PCI operator. This was undertaken to determine whether the measuring and exposure conditions in the two laboratories were the same.

A second experiment was then carried out where the two operators went to each laboratory in turn, and together made two sets of four prisms each. The test was then completed by the operator of that particular laboratory. This was to determine whether the procedure for making the prisms differed in the two laboratories.

From Table 4 it is clear that differences remained, with consistently higher expansions being recorded with the prisms made by the DBT operator. Because of the scatter of results it is difficult to evaluate the test data. The differences between the procedures in the two laboratories can best be judged by comparing the results for the DBT-made prisms, with the same aggregate-cement mix and tested at DBT. The mean expansion measured for the DBT-made prisms in this series of tests and the coefficient of variation compare well with four independent tests done at the DBT laboratory by three different operators.

The prisms made by the PCI operator showed a greater expansion than in the early experiments and although the expansions were consistently lower than with the prisms made at the same time by the DBT operator, the last four results in this particular series of tests were within the range of the DBT measurements.

The fact that prisms made by the DBT operator and cured and tested at PCI gave DBT-like results indicated that the curing,

Table 3. Expansion (measured on the 12th day) for the first suite of aggregates tested at the PCI and DBT laboratories

| Aggregate | Expansion, % | | | |
|---|---|---|---|---|
| | PCI | DBT | | |
| | | Mean | Minimum | Maximum |
| N | 0,010 | 0,010 | 0,008 | 0,012 |
| WS | 0,011 | 0,031 | 0,024 | 0,034 |
| A28 | 0,241 | 0,346 | 0,295 | 0,390 |
| A33 | 0,153 | 0,239 | 0,203 | 0,281 |
| A34 | 0,214 | 0,344 | 0,313 | 0,381 |
| A49 | 0,068 | 0,117 | 0,094 | 0,144 |

Table 4. Comparative expansions (on the 12th day) of prisms made at PCI and DBT

| | Expansion (%) | | Difference in expansion | Difference as a % of DBT results |
|---|---|---|---|---|
| | PCI prisms | DBT prisms | | |
| Test run at PCI* | 0,247 | 0,295 | 0,048 | 16,2 |
| Test run at DBT* | 0,218 | 0,324 | 0,106 | 32,7 |
| Test run at PCI** | 0,293 | 0,319 | 0,026 | 8,1 |
| | 0,286 | 0,352 | 0,066 | 18,7 |
| Test run at DBT** | 0,328 | 0,359 | 0,031 | 8,6 |
| | 0,333 | 0,371 | 0,038 | 10,2 |
| Mean | 0,284 | 0,337 (0,340)# | 0,053 | 15,7 |
| Minimum | 0,218 | 0,295 (0,311) | 0,026 | 8,1 |
| Maximum | 0,333 | 0,371 (0,390) | 0,106 | 32,7 |
| Coefficient of variation | 15,9% | 8,5% (10,4%) | | |

* Experiment in which two sets of prisms were made by the operator at each laboratory and one set was interchanged

** Experiment in which two sets of prisms were made by the operators at the same laboratory

\# Results in brackets are from 4 independent tests undertaken on A28 by three different operators in the DBT laboratory

measuring and exposure conditions in the two laboratories were not responsible for the differences in results. The discrepancy was therefore attributed to the procedures for making the prisms.

The method employed by the PCI operator was, on close scrutiny, found to differ only in the compaction of the wet mortar in the prism moulds. Instead of punning the mortar with a tamping bar, a trowel was used to chop the sample. To test if this discrepancy caused the consistently lower expansion recorded by the PCI operator, a set of prisms was made according to the PCI method by the DBT operator, and vice versa and the expansions measured in the PCI laboratory. The results are presented in Table 5. The PCI-made prisms were now within the range obtained by the DBT while for the first time the DBT-made prisms gave an expansion well below the range normally obtained in the DBT laboratory.

It was deduced from this that the manner in which the prisms were made influenced the permeability or porosity of the mortar.

## 5 Follow-up experiments

It is known that the permeability and porosity of mortar is influenced by the water/cement ratio of the mix. In order to determine the influence of the water/cement ratio on expansion, an

Table 5. Results of an experiment in which the DBT operator made prisms using the PCI operator's method, and vice versa

| Operator | Expansion on 12th day, % | Results of all the previous tests in the series, % | | |
|---|---|---|---|---|
| | | Mean | Minimum | Maximum |
| DBT | 0,232 | 0,337 | 0,295 | 0,371 |
| PCI | 0,339 | 0,284 | 0,218 | 0,333 |

experiment was undertaken using aggregate A28 with a water/cement ratio varying from 0,47 to 0,55.

The results of this experiment are shown graphically in Figure 2. There is a significant increase in expansion with an increase in the water/cement ratio. The trend is almost linear although there is a flattening off of the curve at higher values. The plot of expansion versus flow (Figure 3) shows a similar trend, but in this case the curve appears sigmoidal. A plot of expansion versus flow for all samples for which data are available is presented in Figure 4. The data show greater scatter than in Figure 3, but the trend is still clearly the same.

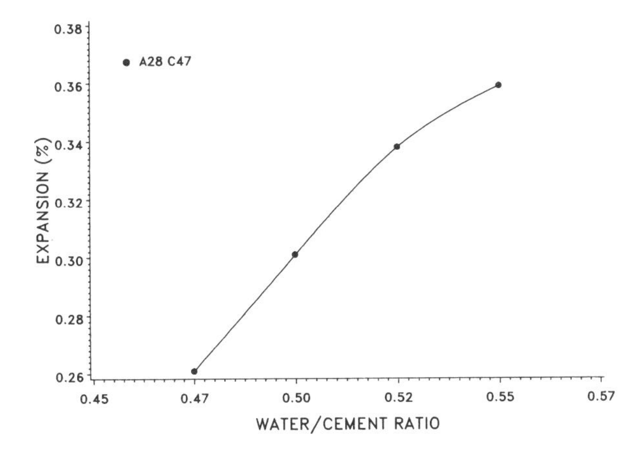


Fig 2. Expansion versus water/cement ratio for prisms made with aggregate A28

Finally, an experiment was undertaken to determine the effect of compaction on expansion. The degree of compaction is difficult to quantify, so a number of prisms made with the same cement and aggregate A28, were weighed by the DBT and PCI operators and the mass of the prisms was plotted against expansion on the 12th day. Since standardised moulds were used, the mass of the prisms should be approximately proportional to their bulk density, which should in turn be related to the degree of compaction. Figure 5 is a plot of expansion versus mass for 12 prisms made with the same mix by the DBT and PCI operators. It was clear that expansion decreased with increasing compaction. On average, the DBT prisms were lighter and showed greater expansion than those produced by the PCI operator.

The results presented above suggest that the permeability of the mortar prisms is an important parameter with regard to expansion. For aggregate A28 at least, expansion increases with an increasing water/cement ratio and decreasing bulk density. The reason appears fairly obvious, in that the higher water/cement ratio and lower bulk density increase the permeability and porosity of the mortar, making it easier for the solution containing the alkali hydroxides to reach the reactive aggregate fragments within the mortar prisms. The lower strength of the samples with the higher porosity could also be important in terms of their higher expansions.

## 6 Summary and conclusions

Factors influencing the repeatability of the test are summarised below:

1. the temperature at which the prisms are stored during the test
2. the concentration of the alkali solution
3. the water/cement ratio of the mortar mix
4. the degree of compaction of the mortar during the making of the prisms.

The first two factors can be controlled by using an oven in which the temperature can be regulated, and by careful laboratory practice. The water/cement ratio is more problematic since different aggregates and cements require different amounts of water to produce the same workability. Since workability influences the compaction of a mixture (which is also an important factor), the

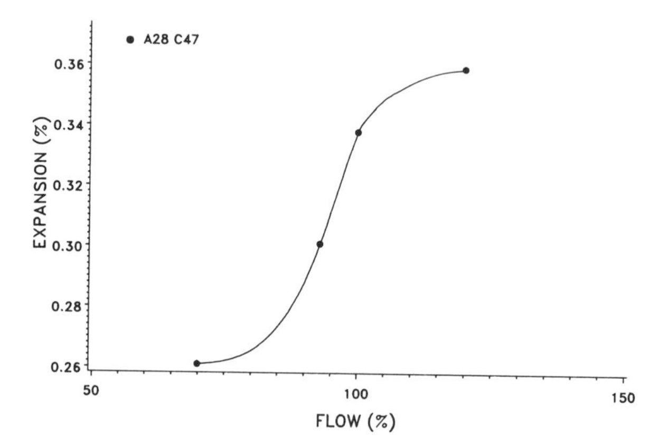


Fig. 3. Expansion versus flow of the mortar mix used in prisms made with aggregate A28. Note the sigmoidal shape of the curve

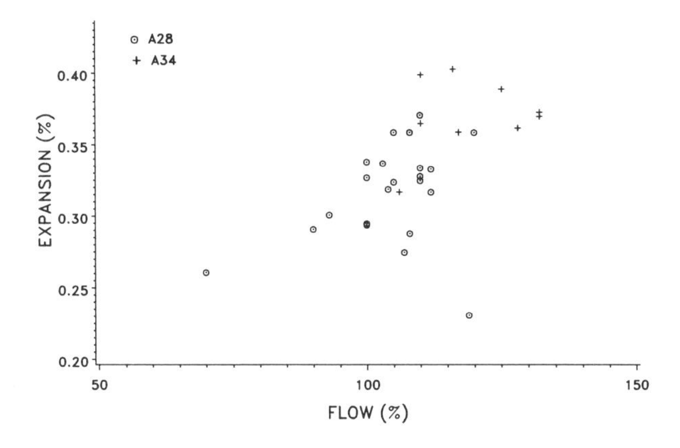


Fig. 4. Expansion of a number of different prisms as a function of the flow of the mortar

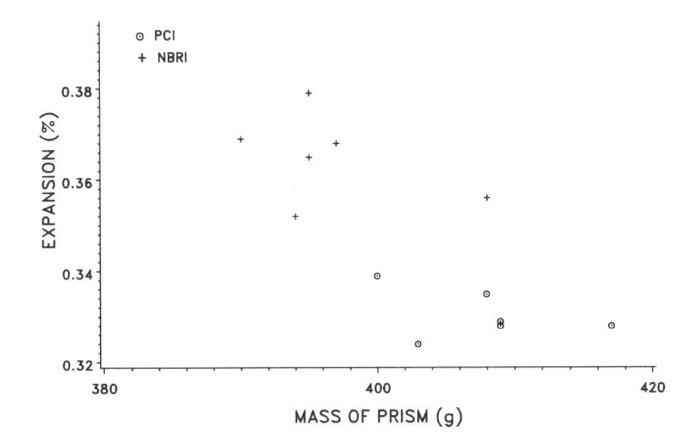


Fig. 5. Expansion of twelve prisms made by the DBT and PCI operators with aggregate A28, as a function of the mass of the prism

water/cement ratio cannot be kept constant. The present system of defining a workability range (100 to 120 % flow on a flow table) appears to be the most practical. The data presented earlier for sample A28 indicate that sensitivity of expansion to flow is greatest in the range of 90 to 105 %. This means that small variations in flow result in high variations in expansion. The range of flow recommended in the ASTM C227-81 specification is between 100 and 120 %, which partially overlaps the sensitive portion of the expansion-versus-flow curve. It appears from the experimental data that greater precision can be obtained if the flow range is re-specified, to fall between 105 and 120 %. However, it should also be borne in mind that data are available for a single aggregate (A28) only and that it is possible that other aggregates would behave differently.

Of the above factors the most difficult to control is the making of the prisms, which must exactly follow the procedure specified in ASTM C227. Even apparently small deviations can have a significant effect on the expansion. The results of the inter-laboratory test programme demonstrate that the DBT's rapid test can be used by another laboratory without a significant loss of accuracy or precision.

## 7 References


Brandt, M.P. Oberholster, R.E. and Westra, W.B. (1981) A contribution to the determination of the potential alkali reactivity of Tygerberg Formation aggregates. **Proc Fifth Int Conf on Alkali-Aggregate Reaction in Concrete.** S252/11, Cape Town.

Chatterji, S. (1978) An accelerated method for the detection of alkali-aggregate reactivities of aggregates. **Cement Concr Res,** 8, 647-650.

Davies, G and Oberholster, R.E. (1987) An interlaboratory test programme on the NBRI accelerated test to determine the alkali reactivity of aggregates. **NBRI Special Report BOU 92,** CSIR, Pretoria.

Hooton. R.D. and Rogers, C.A. (1989) Evaluation of rapid test methods for detecting alkali-reactive aggregates. **Proc 8th Intern Conf on Alkali-Aggregate Reaction in Concrete,** 439-444

Hudec, P.P and Larbi, J.A. (1989) Rapid test methods of predicting alkali reactivity. **Ibid,** 313-320

Jones, F.E. and Tarleton, R.D. (1958) Reactions between aggregates and cement. **Nat Build Studies Res Paper No 25,** BRE, London.

Oberholster, R.E. (1983) Alkali reactivity of siliceous rock: Diagnosis of the reaction, testing of cement and aggregate and prescription of preventive measures. **Proc 6th Intern Conf on Alkali-Aggregate Reaction in Concrete,** 419-433, Copenhagen.

Van Aardt, J.H.P. and Visser, S. (1982) Reactions between rocks and the hydroxides of calcium, sodium and potassium. **Progress Report Part 2. CSIR Research Report BRR 577,** Pretoria.

# 20 FIRE MATERIALS TESTING COORDINATION IN JAPAN

I. NAKAYA, M. YOSHIDA, T. GOTOH, Y. HIRANO,
S. KOSE, Y. KITAGAWA, Y. MIMURA
and S. KOIZUMI
Building Research Institute, Ministry of Construction,
Tsukuba, Japan
K. SUZUKI and K. INOUE
Housing Bureau, Ministry of Construction, Tokyo, Japan

Abstract
Fire preventive material's certification is done by the Minister of Construction according to the Building Standard Law. The Building Research Institute is responsible for the control of the testing done by other specified testing institutes in cooperation with the Building Disaster Prevention Section of the Ministry of Construction. The Building Research Institute participates in the ISO round robin tests and other international activities for the development of the new testing and evaluation system. In the Testing and Evaluation Department of BRI, the researches to establish a new evaluation system are also conducted. These researches will help to improve and simplify the current system.
Keywords: Appraisal, Building Materials, Evaluation, Fire Prevention, Testing.

## 1 Introduction

In some Japanese buildings or houses, the use of fire preventive materials in the structure and/or surface finish of walls/ceilings is required for the prevention of rapid fire spread. Most of the fire preventive materials are classified into the three categories: non-combustible, semi-non-combustible, fire-retarded. The classifications of fire preventive materials are done according to the test results except the typical materials such as concrete, brick, steel, glass etc., which are explicitly designated as the "non-combustible materials" in the Building Standard Law.

The certification of the fire preventive materials is done by the Minister of the Construction. Its management is done by the Disaster Prevention Section of the Ministry of Construction (MOC) with the assistance of the Building Research Institute (BRI) and the Building Center of Japan (BCJ).

Five test methods for the classification of the fire preventive materials are defined by the Ministry of Construction. They are Non-combustibility Test, Surface Combustibility Test, Heating Test of Perforated Material, Toxicity Test, and Box Test. The combination of the tests necessary for the certification depends on the class.

## 2 The testing and classification

### 2.1 Tests

The Non-combustibility Test is done to testify the contribution of the material itself to the temperature rise of the gas in the hot furnace. The amount of the temperature rise is the criteria of the test.

The Surface Combustibility Test is done to testify the performance of the material in a small size under the similar heating condition as that in the real fire. Assessment criteria are the deformation, the crack size, the amount of heat release, the smoke density, etc.

The Heating Test of Perforated Material is done to testify the combustibility of the material (heat release and smoke emission phenomena) under the same heating condition as the Surface Combustibility Test. The test piece has three perforations to evaluate the burning behavior of whole material.

The Toxicity Test is done to testify the combustion gas toxicity under the nearly same heating condition as the Surface Test but the ventilation is controlled. The toxicity of the combustion gas of red lauan is used for the critical one.

The Box Test is done to testify the heat release behavior (the amount and the rate of the heat release) of the material in a medium size under the similar fire condition as the pre-flashover fire.

The details of the test methods are designated in the Notifications #1828 (1970) and #1231 (1976) of MOC.

### 2.2 Classifications

The classification is done based on the combination of the above two or three test results. The combinations of the tests required are as shown in Table 1. The rank of the class may reflect the heat and smoke emmissivity.

| Class | Tests | | | | |
|---|---|---|---|---|---|
| | N.C. | S.C. | H.P. | Tox. | Box |
| Non-Comb. | R | R | NR | NR | NR |
| Semi-non-Comb. | NR | R | R | R | R |
| Fire Retarded | NR | R | NR | R | NR |

Table 1 Combination of the tests
(R:Required, N.R.:Not Required)

N.C.:Non-combustibility Test
S.C.:Surface Combustibity Test
H.P.:Heating Test of Perforated Material
Tox :Toxicity Test
Box :Box Test

## 3 Certification system in Japan

### 3.1 Group and individual certifications

The certification of the fire preventive materials is done by the Minister of Construction according to the Building Standard Law with the support of the Building Research Institute (BRI) and the Building Center of Japan (BCJ). There are two types of certification: the Group Certification and the Individual Certification.

The Group Certification is a system to certify the group of the same materials except for the difference in thickness or shape etc., whose difference may not affect so much their fire performance. The associations of the manufacturers can apply for this type of certification but the manufacturers themselves cannot do so. In this case, the associations are required to be responsible for the quality of the materials.

The Individual certification is a system to certify the specific product, which will be on sale. In this case, manufacturers apply for the certification individually and have to make the tests for the quality control individually. The test results are required to be reported to the Minister of Construction every year.

## 3.2 Application of the Group Certification

To acquire the Group Certification, the association has to submit the request form for the test with the application form of the certification to BRI. The tests are done in the Testing and Evaluation Department of BRI. If the material passed the tests, the obtained test results are sent to the Housing Bureau of MOC with the former application form. Then the certification may be given to the association.

The associations which got the certification can make request to add or to change products names and/or manufacturers names. they cannot submit application form directly to MOC but have to do it through BRI.

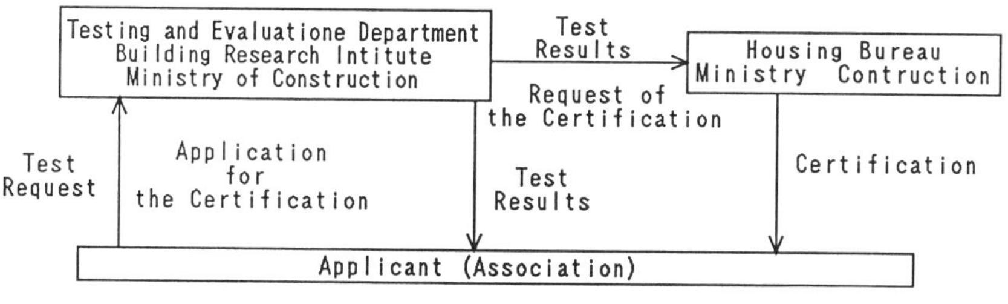


Figure 1 Procedure for the group certification

## 3.3 Application of the Individual Certification

To acquire the Individual Certification, the manufacturers have to apply for the technical appraisal to the Building Center of Japan (BCJ) with the test results issued by one of the public testing institutions which are specified by the Building Guidance Division of MOC. The testing institutions currently specified for fire preventive materials tests are the General Building Research Corporation, the Japan Testing Center for Construction Materials, and BRI.

BCJ has an appraisal committee of fire preventive materials. The committee consists of professional people of universities, testing institutions and private companies. The application submitted with test results is examined in this committee.

Concerning the certification of the materials produced outside Japan, an official guideline was set up in 1987. According to this guideline, the test results issued by the foreign public testing institutions can be used to get the appraisal if these tests are considered as equivalent to those specified by MOC. After the manufacturers get appraisal from BCJ, they can apply for the certification. After they got the certifications, the change of the name of the product or the factory etc. will be accepted through BCJ.

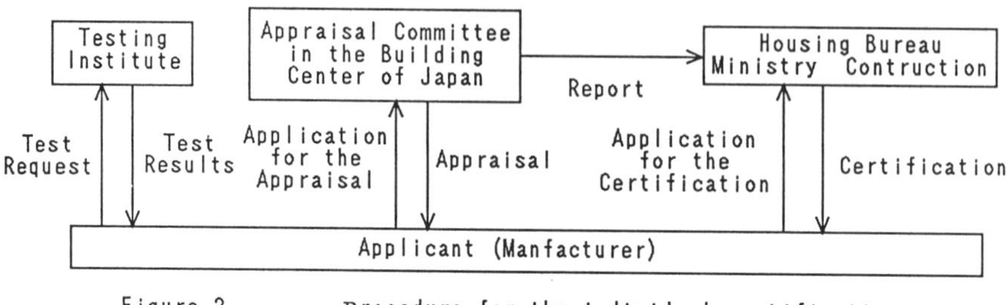


Figure 2 Procedure for the individual certification

## 4 BRI's role for the testing coordination

### 4.1 Testing coordination among the testing institutions

BRI convenes a committee for the information exchange and the testing coordination among BRI and the other two testing institutions specified by MOC. The members of this committee are the staffs of BRI and the institutions. The officers of MOC and BCJ also attend the committee as observers.

Since the main purpose of this committee is to avoid the contradiction among the test results, the test schedule and the test results are reported by the two institutions. The discussion on the test procedure for the non-typical specimen is also the main topic. The supplemental rules on the test methods are sometimes designated by the committee.

### 4.2 Seminar for the testing procedures

For the quality control of the materials, most of the manufacturer associations and some of the manufacturers possess their own testing apparatuses. Although the coordination of the test results obtained by these apparatuses is not our duty, BRI commenced the education and training service for the technicians engaged in the testing.

BRI has been giving them one or two-day seminar every fiscal year since 1988. More than 200 persons attended the last seminar. As there is still a strong demand of the associations for this kind of seminars, the seminar will be held annually.

### 4.3 Calibration of the test apparatuses

More than 1000 sets of the Non-combustibility Test apparatuses and the Surface Combustibility Test apparatuses are used in the public testing centers and the testing sections of the manufacturers associations in addition to BRI and the other officially specified testing institutions. These apparatuses except the ones for the official testing are used for quality control and pre-screening of the newly developed materials.

It is now unsatisfactory to neglect calibration.
Although there are chances to check the test results each other among the test apparatuses of the official testing institutions, there has been no check system concerning the other test apparatuses. Recently, BRI proposed to build up a calibration system of the testing apparatuses and the preparation for the enforcement is proceeding.

## 5 BRI's contribution to ISO

BRI has four ISO type fire test apparatuses: the Flame Spread Test, the Ignitability Test, the Cone Calorimeter and the Room Corner Test. Among these, BRI is participating to the round robin tests of the Flame Spread and the Room Corner. BRI's fire researchers also have a close contact with the ISO TC92/SC3 (Combustion Gas Toxicity).

As we understand the unity of the fire tests is very important, we are always ready to participate in the round robin tests and to review the new test methods proposed by international organizations.

## 6 BRI's research on testing

The purpose of the testings is to evaluate the fire safety of the materials. The best way to do so is to conduct the real fire tests in the real use and occupancy. As this is almost impossible to do, we are gathering the information for the fire safety evaluation through the following five types of tests.

| | |
|---|---|
| Real fire tests | Using the real house, the full scale fire experiments have been done every two or three years. |
| Large scale tests | Using the full size model room, the fire experiments have been done. |
| Medium scale tests | Box tests and some medium size model fire experiments have been done. |
| Small scale tests | The tests using the small piece of the material like ISO ignitability and flame spread tests in addition to the official tests have been done. |
| Computer simulation | Computer simulation model for the fire spread, smoke spread and the charring process of wood have been studied. |

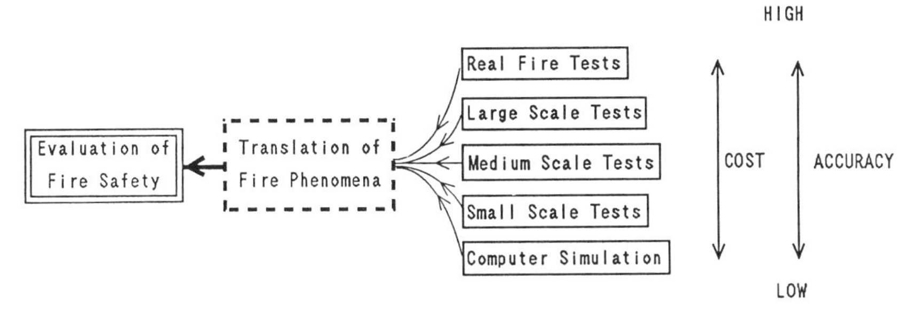


Figure 3 Testing and evaluation

To evaluate the fire safety, the test results should be translated into the real fire phenomena at first. Then the phenomena can be translated into the size of the fire damage. Finally, the fire safety can be evaluated. In BRI, the evaluation system itself is also an important research subject.

## 7 Summary

In Japan, the regulations for the materials used in the buildings are done according to the Building Standard Law. The certification of the fire preventive materials is done by the Minister of Construction. The materials are classified according to the results of the official tests done by BRI or the official testing institutions.

BRI contributes to the coordination of the test results among the testing institutions, centers etc. ISO's work for the unity of the testing is BRI's great concern and BRI sometimes sent researchers to the meetings. BRI is also participating to the round robin tests.

The improvement of the testing and evaluation system is the great concern, too. The researches related to these are being done in the Testing and Evaluation Department.

## 8 References

Building Guidance Division and Urban Building Division, Housing Bureau, the Ministry of Construction (1987) (Eds.) **Introduction to the Building Standard Law**, Building Center of Japan.

Ditto., **Outline of the Approval & Certification System Under the Building Standard Law**, ditto.

Ditto., **Introduction to the Technical Appraisal**, dito.

Ditto., **National Building Regulations of Japan**, ditto.

# 21 CEMENT QUALITY CONTROL AND ASSURANCE – YUGOSLAV EXPERIENCE AND SPECIFIC CHARACTERISTICS

K. POPOVIĆ, N. KAMENIĆ, M. MAKJANIĆ,
D. DIMIC, S. MILETIĆ, LJ. SEŠUM
and B. VARLAKOVA
Civil Engineering Institute, Zagreb, Split; Institutes for Testing Materials, Ljubljana, Belgrade, Banja Luka, Skopje, Yugoslavia

Abstract
Control and assurance of cement quality in Yugoslavia is based upon results of testing from producers' control laboratories and at the same time upon the results obtained in supervising institutes.
Statistical methods are used for evaluating acceptability of strength testing results obtained from cement factories and the strength conformity with standard requirements.
The paper gives a short presentation of ten years experience with such control system and also some peculiarities of Yugoslav cement assortment consisting mostly of blended cements.
Keywords: Cement, Quality, Control, Assurance, Testing, Compliance, Statistic, Certificate.

## 1 Introduction

The assortment of portland cement in Yugoslavia is somewhat uncommon. Only about 5 per cent of roughly 10 million tons of cement produced yearly, does not contain any mineral admixture. This fact is the consequence of a long term tendency to reduce the production costs i.e. energy consumption for clinker burning and to palliate the cement shortage. Blast furnace slag, natural pozzolanas and fly ashes are commonly used; the variety of these materials and of their qualities have caused a rather complicated classification and definitions in our cement standards. Cements containing up to 30 per cent of mineral admixtures are designated as portland cements and marked PC. Those containing more than 30 per cent of blast furnace slag are marked M (metallurgic), and those with more than 30 per cent of natural pozzolana or fly ash are marked P (pozzolanic). The quantity and the type of admixture also has to be designated after the mark PC. For instance, the designation PC 15 z 45 S means that the cement contains up to 15 per cent of slag, or if the cement contains e.g. more than 15 but less than 30 per cent of pozzolana it is identified by PC 30 p 35 B. When different admixtures are added to cement (composite portland cement), the addition is marked with "d", and the prevailing admixture must be noted. So, the designation PC 15 dz 45 B, means that up to 15 per cent of slag and pozzolana is present together, and that slag (z) prevails.

The next mark indicates the class of cement, i.e. nominal compressive strength of standard cement mortar (in $N/mm^2$) determined after 28 days. Classes 25, 35, 45 and 55 are covered by Yugoslav standard. According to the early strengths classes 35 and 45 are divided to the subclass marked "B" at the end of designation (higher early strength), and the other marked "S" (lower early strength).

Since the addition of some pozzolanas causes increased water demand, there is a special requirement in Yugoslav standard which does not exist in other international standards i.e. water for standard consistency of cement paste is limited to 30.0or 32.0 per cent for cements containing up to 30, or more than 30 per cent of pozzolanic admixture respectively. Also there is no limit for insoluble residue if the cement contains pozzolana.

Other physical and chemical requirements are similar to the requirements of most international standards, as well as the methods of testing. Mechanical properties are e.g. determined using standard silica sand sized 0.09 to 2.00 mm, W/C ratio 0.50 and the equipment identical to ISO R 679.

## 2 Control system

In 1981, a new conformity procedure for quality control of cement using statistical methods was introduced by special Yugoslav government "Order". This document defines quality control and quality assurance of cement production. Only six institutions in the country are authorized to supervise the cement quality in accordance with that "Order". Every cement manufacturer is obliged to make a long term arrangement (contract) with one of these authorized institutions.

The certificate is issued on the basis of testing the results obtained in the manufacturer's control laboratory (autocontrole) and at the same time in the supervising institution's laboratory as follows:

Water requirement for standard consistency, setting time, soundness and mechanical properties have to be determined in the manufacturer's laboratory on daily average samples and results have to be sent to the supervising institution.

The supervising institution collects samples by chance in the cement factory several times a month (spot sample). If the production of the considered cement is under 8000 tons per month, one spot sample is taken, for cements produced in quantity of 8000 to 36000 tons per month two spot samples are taken, and for cements in quantity of more than 36000 tons per month the supervising institution has to take three spot samples per month. Each sample unit is divided into three parts - one for testing in the cement manufacturer's laboratory the second is for testing in the supervising institution, and the third for retest (retained sample).

Chemical, physical and mechanical characteristics of these samples are determined in both laboratories. The manufacturer is obliged to send these results together with strength results of daily samples to the supervising institution. On the basis of all these results over the period of three months the supervising institution issues the certificate valid for the next three months.

In order to satisfy conditions for obtaining the certificate, all chemical characteristics (L.O.I., insoluble residue, $SO_3$, MgO and $Cl^-$) as well as physical characteristics (0.090 mm sieving residue, Blaine value, setting time, water for standard consistency and soundness) have to meet the standard quality requirements. Exceptionally, one result of a particular quality parameter does not have to comply with standard, but such cement sample has to be followed with two samples exhibiting correct value of the particular characteristic.

## 3 Conformity procedure

The most interesting part of this system of quality control and quality assurance is the evaluating of mechanical properties. Mechanical properties of cements are evaluated by means of statistics. Three sets of 28 days strength data are formed. Results obtained daily in the manufacturer's laboratory form set $N_1$, strengths of spot samples taken by the supervising institution twice or three times a month determined in the manufacturer's laboratory form set $N_2$, and those determined in the supervising laboratory form the set $N_3$. The conformity procedure for the above mentioned control period of three months starts with calculating the mean values and standard deviations of populations for sets $N_1$ and $N_3$. Subsequently F test is used for estimating differences between standard deviations of sets $N_1$ and $N_3$. If the difference is not significant, mean values of $N_1$ and $N_3$ are compared by means of T-test, and if that test also shows that the difference in insignificant, set $N_1$ (manufacturer's laboratory results) is used to estimate the compliance of cement strength with conformity criteria $x_{char} \leq \bar{x} - K.s$

In case there is significant difference between mean values or standard deviations of these two sets, the same test procedure is applied for sets $N_2$ and $N_3$, and in the

Table 1. THE FIVE YEARS REVIEW OF STRENGTH SETS WHICH WERE THE BASES FOR CONFORMITY PROCEDURE

| | Cement | Set | n* |
|---|---|---|---|
| INSTITUTION A | | | |
| Manufacturer 1. | | | |
| | PC30z45S | N1 | 15 |
| | | N2,3 | 10 |
| | | N3 | 6 |
| Manufacturer 2. | | | |
| | PC30dz45S | N1 | 10 |
| | | N3 | 2 |
| INSTITUTION B | | | |
| Manufacturer 1. | | | |
| | PC30dz45S | N1 | 26 |
| | | N2,3 | 7 |
| | | N3 | 1 |
| Manufacturer 2. | | | |
| | PC30z45B | N1 | 17 |
| | | N2,3 | 8 |
| | | N3 | 5 |
| Manufacturer 3. | | | |
| | PC30dz45S | N1 | 13 |
| | | N2,3 | 3 |
| | PC30dp35S | N1 | 12 |
| | | N2,3 | 3 |
| Manufacturer 4. | | | |
| | PC30p45S | N1 | 9 |
| | | N2,3 | 5 |
| | | N3 | 1 |
| INSTITUTION C | | | |
| Manufacturer 1. | | | |
| | PC30dz35S | N1 | 2 |
| | | N2,3 | 3 |
| | | N3 | 5 |
| Manufacturer 2. | | | |
| | PC30p45S | N1 | 7 |
| | | N2,3 | 4 |
| | | N3 | 8 |
| Manufacturer 3. | | | |
| | PC30dz45S | N1 | 13 |
| | | N2,3 | 2 |
| | | N3 | 3 |
| Manufacturer 4. | | | |
| | PC45B | N1 | 1 |
| | | N2,3 | 1 |
| | | N3 | 1 |
| | PC30dp45S | N1 | 4 |
| | | N2,3 | 7 |
| | | N3 | 6 |
| INSTITUTION D | | | |
| Manufacturer 1. | | | |
| | PC30z45S | N1 | 3 |
| | | N2,3 | 8 |
| | | N3 | 8 |
| Manufacturer 2. | | | |
| | PC30z45S | N1 | 11 |
| | | N2,3 | 2 |
| | | N3 | 7 |
| Manufacturer 3. | | | |
| | PC30dz45S | N1 | 12 |
| | | N2,3 | 4 |
| | | N3 | 5 |
| INSTITUTION E | | | |
| Manufacturer 1. | | | |
| | PC30dp45S | N2,3 | 3 |
| | | N3 | 12 |
| | PC30z45S | N2,3 | 6 |
| | | N3 | 8 |
| INSTITUTION F | | | |
| Manufacturer 1. | | | |
| | PC30z45S | N1 | 18 |
| | | N2,3 | 2 |
| | | N3 | 3 |

TOTAL:

| | | |
|---|---|---|
| 173 x N1 | 52.1 | per cent |
| 78 x N2,3 | 23.5 | " " |
| 81 x N3 | 24.4 | " " |

n* number of estimations

case there is no significant difference between sets $N_2$. and $N_3$, set $N_{2,3}$ is formed from the results of $N_2$ and $N_3$. Mean value and standard deviation of the new $N_{2,3}$ set serve for evaluating the strength compliance with standard requirements. Finally, if a significant difference between sets $N_2$ and $N_2$ has been established, the estimation is made on the basis of the characteristics of only set $N_3$ using the above expression for conformity criteria. The same expression is also used to estimate the manufacturer's laboratory daily results at the end of every month to detect a possible failure of strength in time.
The acceptability constant K depends on the population of the considered set e.g. for n=6, k=1.92; for n=9, K=1.64; and for n=16, K=1.40.
$x_{char.}$ is the characteristic strength value defined as quality requirement in Yugoslav standard. These minimal compressive strengths of cement mortar after 28 days have to be: 22.0 N/mm² for class 25, 31.0 N/mm² for class 35, 40.0 N/mm² for class 45, and 49.0 N/mm² for class 55 cement.

It is obvious that the cement mortar strengths have to be around the nominal cement class to satisfy the expression $x_{char.} \leq \bar{x} - K.s.$

The intention of this system of strength control was to give the manufacturer an opportunity to base quality control upon his own laboratory results whenever possible. On the other hand, such system makes it impossible to evaluate incorrect results of the cement manufacturer's laboratory.

Obviously for cements produced in small quantities (up to 8000 tons per month), three specimens sampled by the supervising institution over the control period of three months are not sufficient for statistical evaluation of strength. For such cases the same system is applied as for the accepting results of chemical and physical parameters.

Finally if chemical physical and mechanical characteristics of the cement comply with the described procedure, the supervising institution issues the certificate accompanied by the test report for the three months period.

Ten years experience shows a significant variation in the set of strength results which were the basis for estimation. The different procedures are influenced by different supervising institutions, different manufacturers and also by the type of admixture, composition and class of cement as presented in table 1.

The described system has helped to assure better cement quality i.e. average higher strength level as well as the strength uniformity.

# 22 STATUS OF STANDARDIZATION OF CONSTRUCTION MATERIALS IN BANGLADESH

A.U. AHMED
Housing and Building Research Institute, Dhaka, Bangladesh

Abstract
This paper reviews the organizational position of standard testing as a professional discipline of science, assessment of manpower and scientific occupation, analytical research and service facilities and their linkage with other research and development activities. A comparative study on the standardization of building materials as per Bangladesh standard to that of the other International standard has also been discussed. Some aspects of building materials, manufactured in the country are outlined in the paper. The paper also focuses on the interaction between research results-standardization-and in-field application in the construction system. In this context the achievements of the Institutions in Bangladesh have been mentioned.
Keywords: Standardization, Test Quality, Professional Disciplines, Quality Assurance, Analytical Standardization, Physico-Chemical Standardization, Mechanical Standardization.

## 1 Introduction

As a developing country of the Third World, Bangladesh emerges into the great task of infrastructural development. Major activities in this sector include the construction of buildings, roads and bridges.This land is predominantly a great deltaic flood plain with frequent natural calamities like cyclone, flood, sea-surge etc. It also lies in the seismic zone where earthquakes occasionally hits some of its region. Considering the average per-capita income (about US$ 120 only), it could be categorized as extremely poor nation having a population of 110 millions. Thus, infrastructural development poses a challenge to ingenuity of even the most talented engineers, planners and architects because of the severe ceilings that are set on the cost of the construction. Therefore, for the interest of the individuals, there is always a tendency to use sub-standard materials which must be controlled through some regulations - and this has been done by the quality test and proclamation of quality assurance. Generally, for the assurance of the quality of materials for construction, one has to follow some standard method of test procedure along with standard apparatus. In a broad sense the test methods for the control of the construction materials may be designated by:

Physico-Chemical standardization - this conforms with the physical and chemical properties of the materials.
Analytical standardization - this conforms with the chemical composition of the materials.
Mechanical standardization - this conforms with the mechanical properties of the materials.

The reliability of test results is the function of precision i.e. reproducibility and accuracy of standard results. This reliability can be achieved by the following procedures:

Analysing a particular sample for a given parameter with as many methods as possible.
Analysing standard closely similar reference materials with the test sample materials.
Participating in analytical inter-comparison studies for a sample.

Further, any quality assurance programme should be based on a recognized source of quality assurance requirements, such as a nationally established standard. It provides a strong basis for defending the reliability of the results. In the application of science and technology for national development and thereby to improve the quality of life, the state-of-art of testing laboratories in Bangladesh is briefly mentioned here in. These are Bangladesh Standards and Testing Institute (BSTI), Housing & Building Research Institute (HBRI), Bangladesh Council of Scientific and Industrial Research (BCSIR), Bangladesh University of Engineering and Technology (BUET),Bangladesh Atomic Energy Commission (BAEC), the Universities ( 5 ) and Analytical Standardization in Industries.

## 1.1 Bangladesh Standard and Testing Institute (BSTI)

This has been constituted by integration of the Central Testing Laboratory (CTL) and the Bangladesh Standard Institution (BSI). It is an Institution for certifying standards prepared in Bangladesh and testing of materials of various descriptions, primarily administered by the Ministry of Industries and Trade and Commerce. The testing procedures are the routine prescriptions given by BS and IS, wherever possible and those suggested by BSI. So far, it has drafted 318 standard specification for the local materials in accordance with either BS or IS out of which about 15 are of construction materials.

## 1.2 Housing & Building Research Institute (HBRI)

This is the Institution engaged in the research activities in construction materials, housing structures and techniques. Besides the research works it also perform the quality tests of the construction materials through its Analytical, Physical and Engineering Laboratories with modest equipments. Quite a good numbers of qualified scientists, engineers and architects are now working at this Institute.

### 1.3 Bangladesh Council of Scientific and Industrial Research (BCSIR)

The BCSIR is responsible mainly for chemical and technological research related to the development of process industries and products based on indigenous raw materials. It has the largest number of laboratories equiped with modern instruments and trained manpower. Besides the research activities, it also performs the quality test according to the standard specification.

### 1.4 Bangladesh University of Engineering and Technology (BUET)

Facilities for analytical chemistry, technical analysis and testing of materials exist in many departments of BUET. For example, the quality test or the analytical standardization are conducted by the Department of Chemistry or Department of Chemical Technology. On the other hand, technical analyses or mechanical behavior of the engineering materials are carried out both by the Civil Engineering Department or Department of Metallurgy of the University.

### 1.5 Bangladesh Atomic Energy Commission (BAEC)

Analytical testing is a recognized scientific discipline in BAEC. It has both research and service aspects. The R & D programme in analytical science in the BAEC has the added advantage that in addition to chemical principles, they can incorporate nuclear principles which in fact have given a new dimension in the field of analytical chemistry.

### 1.6 University Laboratories

Only analytical standardization of the materials could be carried out in some laboratories of the Universities. The involvement of the Universities is limited mainly to academic research and test activities.

### 1.7 Analytical Standardization in Industries

The largest industrial units in Bangladesh that require the service of analytical quality testing are Cement Manufacture Factories, Steel Mills, Insulation and Sanitary Material Manufacture Industry etc. The nature of chemical analysis and quality control required in these organizations is concerned with raw materials, finished products and trouble free operation of the unit processes.

## 2 Inter-comparison studies of test quality

There is a great need for standardization of analytical methods and calibration of instruments in production of construction materials. For this purpose a wide variety of reference materials are needed. The quality tests competence developed in any laboratory requires validation to provide accurate information within acceptable limits. This is possible by testing a particular sample for a given parameter with as many methods as possible. The best agreement of these measurements would be the expected value. The same validation of a particular methodology is possible by analysing a standard reference material, similar to the sample. In Bangladesh, intercomparison studies or cross-check of the results for the quality assessment is a new exercise. A few laboratories were selected for participation in cross-check studies in some selected areas. The results are given in the Table 1 and 2 which represent the analytical determination(1987) of drinking water and mechanical standardization of mortar respectively.

Table 1. Analytical determination of Ca & Mg in drinking water

| Element | Lab.code | Technique adopted. | Number of measurement | Reported value ± SD (ppm) |
|---|---|---|---|---|
| Ca | BW 10 | TT | 5 | 85.2 ± 1.10 |
| | BW 11 | AAS | 1 | 52.0 ± X |
| | BW 12 | AAS | - | 15.6 ± X |
| | BW 13 | AAS | 4 | 15.7 ± 0.08 |
| Mg | BW 10 | TT | 5 | 72.0 ± 2.20 |
| | BW 11 | AAS | 1 | 18.0 ± X |
| | BW 12 | AAS | - | 22.7 ± 0.03 |
| | BW 13 | AAS | 3 | 10.6 ± 0.06 |

Where, BW 10, BW 11, BW 12 and BW 13 indicates different participants involved in the test work and TT for Titrimetry, AAS for Atomic Absorption Spectrophotometry.

It is observed from the Table 1 that there is a wide variation with the change of person although the same instrument was in use for the test. Therefore, the skill of the operator of the Instrument must be ensured.

Similarly from the Table 2, it is observed that there is a great deviation of the compressive strength of building materials, although the test samples were prepared from the same materials and under similar conditions. This may be due to the lack of proper calibration and

Table 2 . Results of cross-check for 50 mm cube mortar, having the cement sand ratio 1:2.75 carried out in the Laboratories of BSTI, HBRI and BUET at different ages.

| Serial No. | Age of curing | Name of Lab | No.of tests performed | Reported compressive strength ± SD (MPa) |
|---|---|---|---|---|
| 1. | 11 days | BSTI | 6 | 18.8 ± 3.2 |
| 2 | | HBRI | 6 | 16.7 ± 1.4 |
| 3. | 29 days | HBRI | 3 | 19.3 ± 0.8 |
| 4 | | BUET | 6 | 16.1 ± 2.6 |

use of reference materials, differences in test procedures and absence of skill in operation. This sort of ambiguity should be eliminated for the assurance of test quality.

## 3 Necessity of test quality and quality assurance

The test quality and quality assurance of construction materials is to check the quality of test results provided by different laboratories engaged in testing the materials. The overall quality assurance of the materials is vested on quality control and quality assessment. The quality control is an established mechanism to control specification and this is achieved through the test methods while quality assessment is the mechanism used to verify that the test results comply with acceptable limits according to the standard specification provided either by manufacturers or by the designers.

The success of any civil engineering project depends on quality control tests before, during and after construction. The quality of the materials should be ensured according to the specification of the manufacturers and the test procedures must be in accordance with the relevant procedures established through standardization within the laboratory or by some other internationally accepted method.

## 4 How quality assurance is achieved

According to W.L Delvin (1984), analytical chemists have been concerned for a long time about the quality of the measurement data they produce. They use practices in their laboratories that have been developed over the years for controlling activities to assure the reliability of their data. That fact is apparently true for other fields of quality assurance and workers also. There are some elements which have direct influence in laboratory quality assurance and as many as nine elements have been identified.These are:

Organization
Quality assurance programme
Training and qualification
Procedures
Laboratory records
Control of records
Control of equipments and materials
Control of measurement
Deficiencies and corrective actions.

Fulfilment of these elements is essential for developing test quality and quality assurance for construction materials and structures, failure of which may cause adverse effect to the quality.

## 5 Capability of beneficiaries for using standard products

A significant number of beneficiaries of the Third World Nations are ignoring the quality test or standard specification in their construction for the reason of poverty. It is true even where a standard code of practice does exist. The economic condition has forced the beneficiaries to use the sub-standard materials and construction method.

## 6 Conclusion

Standards and specifications are basic regulatory instruments for promotion of acceptable products on the market and in the context of building materials, they ensure economy, durability, safety and health in construction. The adoptating of standards through quality control measures of the construction materials is an effective means of ensuring quality of production and sometimes cost reduction. In the developing countries advances in production of indigenous materials are facing constraint due to the low quality and in some cases, for lack of standard specification and quality assurance. Therefore, it should be the prime objective of the developing countries to formulate their own standard based on reliable data and ensure . test quality and quality assurance. To rationalize specification, if necessary different institutions of multiple discipline would be integrated together with a joint programme.

## 7 References

Khan, A.H. (1987) Report on cross-check measurements of some elements in some selected field samples, presented in the first UNESCO National Workshop on Analytical Standardization, held in Dhaka, 2-4 August, 1987.

Delvin, W.L. (1984) Assuring the quality of data through laboratory Quality Assurance, Journal of Testing and Evaluation, JTEVA, Vol.12, No.5, Sept. 1984, pp. 268 - 272.

# 23 INTERPRETATION AND REPRODUCIBILITY OF CONCRETE STRUCTURE LABORATORY TESTS AND TESTS DURING CONSTRUCTION *IN SITU*

T. JÁVOR
Research Institute of Civil Engineering, Bratislava, Czechoslovakia

**Abstract**
This paper points out some problems of interpretation and reproducibility of testing modulus of elasticity,shrinkage,creep and hydration heat of concrete.Laboratory tests applied to the bridge structures give good results only in analysis of temperature effects.As far as the evaluation of actual stresses in structure is concerned the aplication of laboratory results is made difficult by the number of effects on volume changes of concrete.In future research special attention should be given to the influence of reinforcement to shrinkage or to the influence of creep to modulus of elasticity.
Keywords:Shrinkage,Creep,Modulus of elasticity of concrete, Stress,Calculated modulus of elasticity,Hydration heat, Temperature gradient of concrete.

## 1 Introduction

Procedures for laboratory testing of material constants of concrete structures are generally well set down by international regulations and national standards.The RILEM recommendations often deal with methodology of testing or with process of utilisation of different plant instrumentation.Testing concrete elements or parts of structure should comply with the principle of model similarity.If these are concrete elements there are usually no special problems,although it is not so easy to comply with the requirement to preserve adequate reduction of bending stiffness of cross sections or the sections of prestressed reinforcement or even the stress of the reinforcement by a model size reduction of the elements.Comparing results of laboratory testing with testing in situ the stress value appears as the basic characteristic of a structure for which the corresponding building was designed,i.e.calculated.Starting from the simple Hooke's law,stress is expressed as a product of Young modulus of concrete elasticity and strain, there is a problem of reproducibility of laboratory material characteristics to building conditions not only the characteristics of modulus of elasticity but also of creep, shrinkage, temperature effect etc. which have either direct or indirect effect on the strain.

## 2 Modulus of concrete elasticity

The modulus of elasticity is the basic material characteristic because its value affects directly a calculated stiffness of a structure. The value of the modulus of elasticity is very important for calculations of deformations and it affects internal forces distribution in statically indeterminate structures. Checking the state of the stress or strain of a structure in situ it might be interesting to set the optimum way of interpreting a laboratory determined modulus of elasticity, of analyzing a modulus of concrete elasticity determined by an ultrasound or other nondestructive method in comparison with some standard or calculated modulus of elasticity. By gradual loading of concrete samples a stress-strain diagram is obtained which is generally nonlinear /Fig. 1/. Not to mention a number of effects as concrete age, temperature conditions of sample testing, sample shape etc., history of loading, speed of loading, stress levels for which the modulus of elasticity is being determined as e.g. initial secant or tangent modulus, or secant modulus in special point must not be omitted.

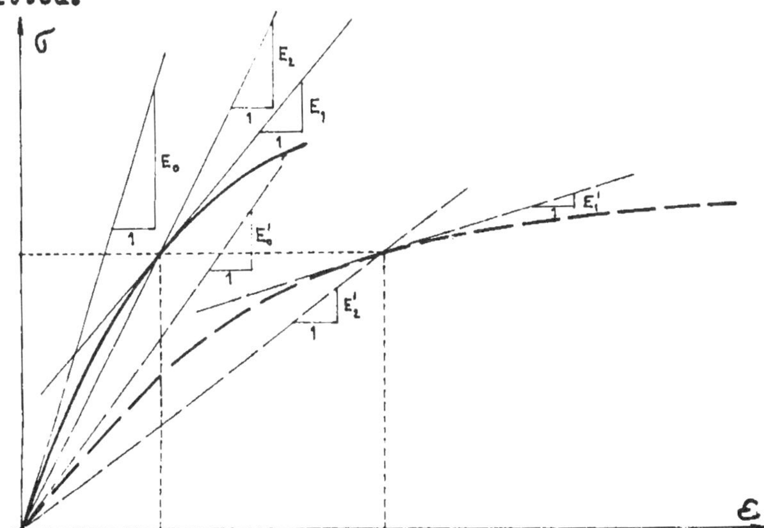


**Fig. 1.** Stress-strain diagram of concrete / $E_0$-initial tangent modulus, $E_1$-tangent modulus in point 1, $E_2$- secant modulus of concrete/.

Though on the condition that stress is lower than 40% of concrete strength a linear dependence of stress and deformation in a physical sense is presumed,a sudden stress loading of a structure or a sample causes a deformation increasing with time,i.e.causes the effect of creeping.The reproducibility of our standard tests of the modulus of elasticity is questionable because of e.g. cantilever concreted bridges being additionally gradually prestressed where the state of strain is very complicated and different in every cross section.Naturally effects of shrinkage, influence of changes of temperature effects,hydratation heat effects for large cross sections and others must be added.In designing usually only modulus of elasticity in regard to creep is considered,for example in the form of

$$E/t_o/ = \frac{1}{J/t_o + \Delta t,\ t_o/}$$

where $J/t,t_o/$ is a function of concrete creep,
$t_o$ is a concrete age,
$t$ is time of action of the observed load.
The function of creep is adequately defined by so called "Triple Power Law" by Bazant,where basic creep,i.e.creep by constant humidity and temperature,limit secant modulus, constants dependent on concrete strength and composition and on loading time are considered.Even in this way defined modulus of elasticity of concrete is difficult to be confronted with laboratory tests and as far as prestressed concrete segmental bridges concerned I considered it impossible.For bridges built over rivers are differently cooled and tempered by the sun and material characteristics are absolutely disfigured by this.There is only one possibility left and it is a confrontation of concrete samples stored in conditioning chambers with samples placed in a hollow of a bridge and checking of both these groups of samples made of the same concrete by a non-destructive method, best of all by ultrasound and by this way a non-destructive evaluation of moduli of elasticity of a built structure. It must be remembered that the structure was designed for so called calculated modulus of concrete elasticity,i.e. given for a stress of cca 30% strenth of tested body and for a concrete age of 28 days.A confrontation of modulus of elasticity calculated from deformation during a loading test is also interesting.Our tests have proved the Triple Power Law to be a good starting point for this analysis,i.e.for a short-time loading during the load-test.

## 3 Change of volume of concrete

Reproducibility of laboratory tests of shrinkage and creep of concrete is confirmed by illustrations of the course of strains in bottom fibres of nucleus segment of an cantilever concreted cable-stayed bridge over the lake Jordan in the town Tabor.Fig.2 shows courses of the total strains measured by embedded vibrating wire gauges during the construction of this bridge.Fig.3 shows a course of concrete shrinkage /curve 3/ and creep /curve 1,2/ at the stress level of 3.6 MPa taken on beams of the size 15/15/70 cm made of the same concrete as the segment observed at Fig.2. The beams-samples being placed in a hollow of this segment. The temperature course in concrete of this segment measured by embedded vibro-wire thermometers is shown at Fig.4 and it is for the same period of construction,i.e.cca during 1200 days.

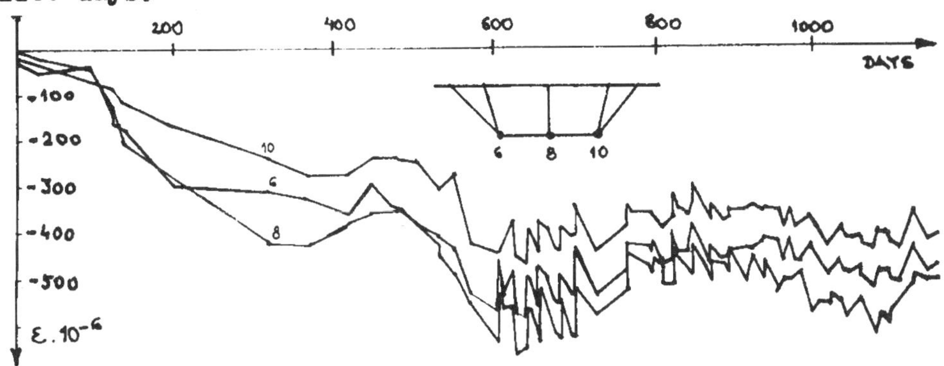


Fig.2. Course of strains of the first casted in place segment near the column of the cable-stayed bridge in the city Tabor

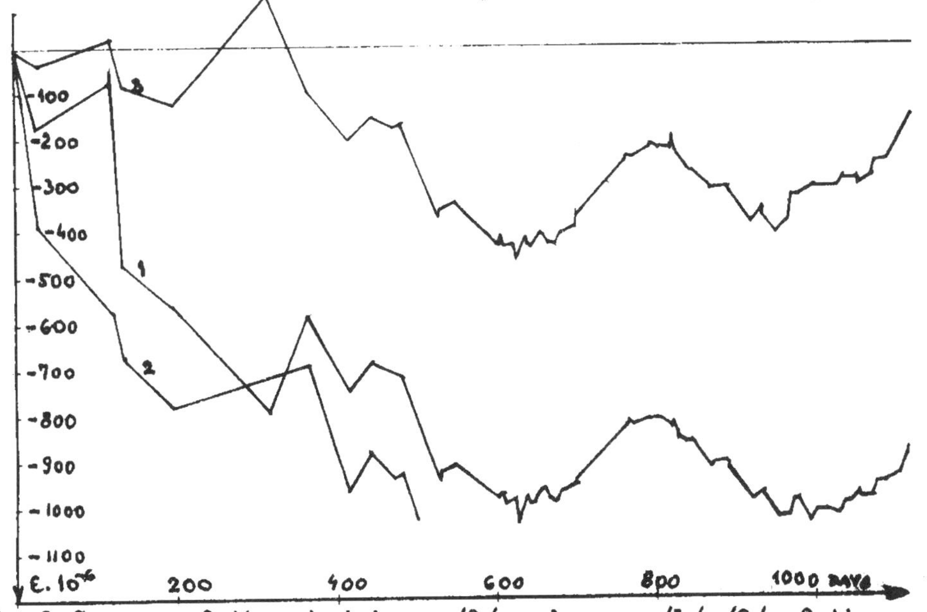


Fig.3.Course of the shrinkage /3/ and creep/1/,/2/ of the cable stayed concrete bridge over the Jordan lake in Tabor

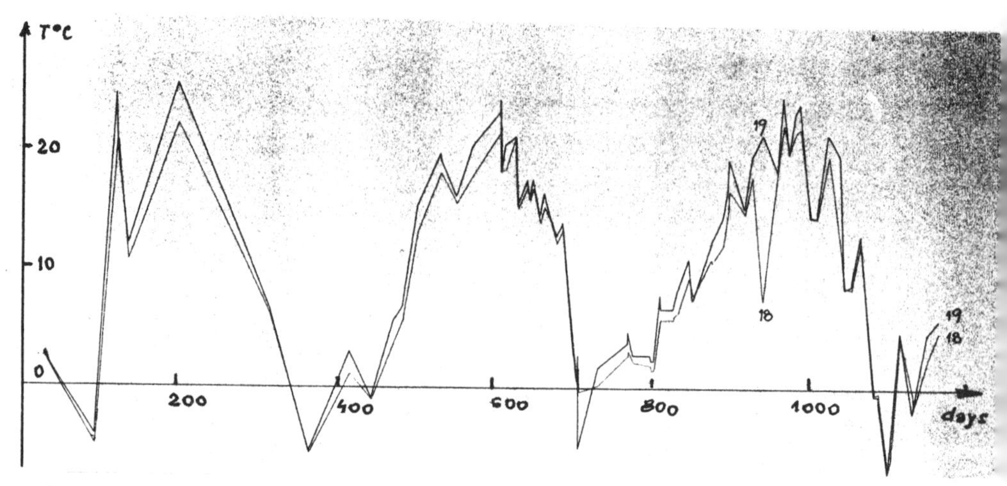


Fig.4.The temperature course in concrete of the first segment of the cable stayed bridge in Tabor measured by the embedded vibrating wire thermometers during 1200 days

It seems to be enough to subtract the values of volume changes of concrete taken on samples-prisms from overall volume changes measured in the structure.Unfortunately only free shrinkage of these prisms reaches strains of the same value as strains given in the segment itself including strains from prestressing or from weight of the cantilever bearing structure itself.Because the theoretical designed values of stresses are during construction between 3.5 - 7.0 MPa,the creep values received from samples are extremely high,but in accordance with the standard assumptions.During this time the shrinkage values received from our samples i.e.$300.10^{-6}$ are in accordance with CEB-FIP Code.If neglected the temperature influences and the dimensions of segments comparing with prisms dimensions,there it can be seen the great influence of the crosssection reinforcing compared with the unreinforced prisms.This circumstance must be taken into account by transformation of the strains in to the stress, which however leads to difficulties by variously reinforced observed segmental box girder crosssections,where the mild steel i.e.non prestressed reinforcing is placed in closed reinforcing cages.Then here acts the active and passive shrinkage as well as the creep,because the reinforcement takes the pressure strain from the volume changes of the concrete.

The influence of the reinforcement on shrinkage and creep was analysed in various institutes.From our results /VUIS,Jerga/ we can see in Fig.5 the characteristic curve of shrinkage of unreinforced prisms of dimensions 15/15/ 70 cm with the shrinkage of simple reinforced prisms and the theoretical values in accordance with the Czechoslovak standards.Already on this simple example was demonstrated the difference of $-120.10^{-6}$ after 100 days,when the shrinkage of the reinforced cross section during the same time was $-250.10^{-6}$ only,what corresponds with the free shrinkage by 20°C temperature and 60 % relative humidity in accordance with the Czechoslovak standard.Similar results were received also during the bending tests of the various reinforced beams.Experimentally received course of shrinkage in depth direction of beams 15x28x340 cm confirmed the great influence of the reinforcing resistance, which in regard to the asymmetrical reinforcement consequented large strain differencies on the upper and lower part of cross section.The concrete stress of the reinforced elements due the shrinkage can receive values equal to the mean tensile strength of the concrete.If we wish to correctly interprete the results from the laboratory by the measuring in constructions,it is necessary to analyse the volume changes,shrinkage and creep of various reinforced prisms.

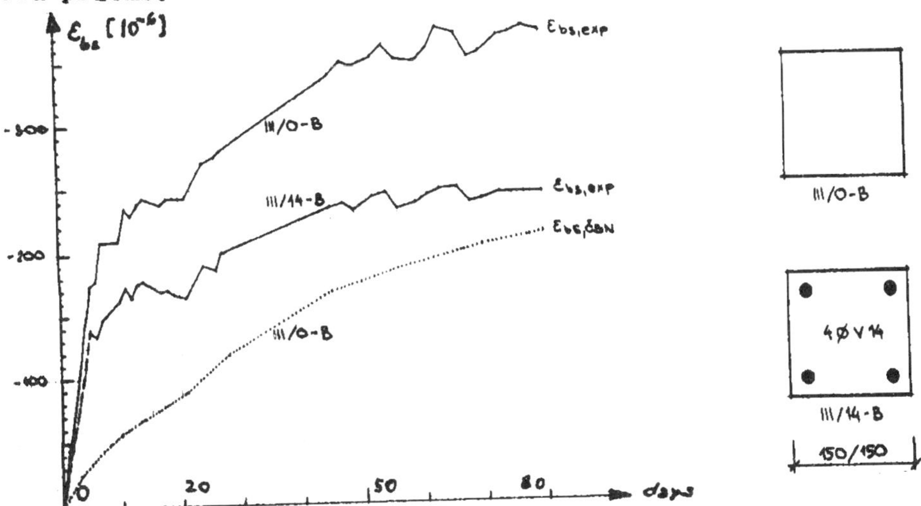


Fig.5.Shrinkage curve of simple reinforced prisms in comparison with the shrinkage of not reinforced prisms and the theoretical values in accordance with the ČSN standards.

## 4 Temperature gradient of concrete

Diagrams at Fig.2,3,4 show an evident influence of temperature effects on concrete strain,but special attention must be given also to an analysis of temperature gradient in cases of bigger thickness of concrete walls,e.g.of box-girder bridges with large span /Fig.6/.For this type of structures laboratory testing of a temperature gradient was done on two enclosed,polystyrene insulated samples 1x1x2 m size/Fig.7/.

Fig.6.Cantilever prestressed concrete bridge over Danube in Bratislava of 170 m span with 2 m thickness of the lower slab of the first box-girder segment

Fig.7.Laboratory testing of a temperature gradient on two enclosed polystyrene insulated samples with embedded vibro-wire gauges and thermometers

The surface of 1x1 m side was without insulation. In these elements 78 thermometers and for purpose of shrinkage measuring 40 vibro-wire strain gauges were fixed in with concrete at three levels and five transversal sections. Due to the same conditions of production and the same insulation the results of both samples were congruent. The highest reached temperatures were of 76°C at the ambient temperature of 23°C, while a theoretical value of 73°C was expected. Gradients are bigger in the centre and smaller at the edges of the element. It was recommended to concrete the actual structure at low temperatures above zero, because surface cracks were expected unless the concrete surface was separately insulated. A laboratory experiment was approved by measuring taken at a building site and the course of temperatures measured by concreted in readers /shown at Fig.8/ told a temeperature gradient of ca max 60°C at temperature of environment of about 5°C. The measuring of the temperature gradient was fully automatised during 140 hours. The laboratory assumptions were confirmed.

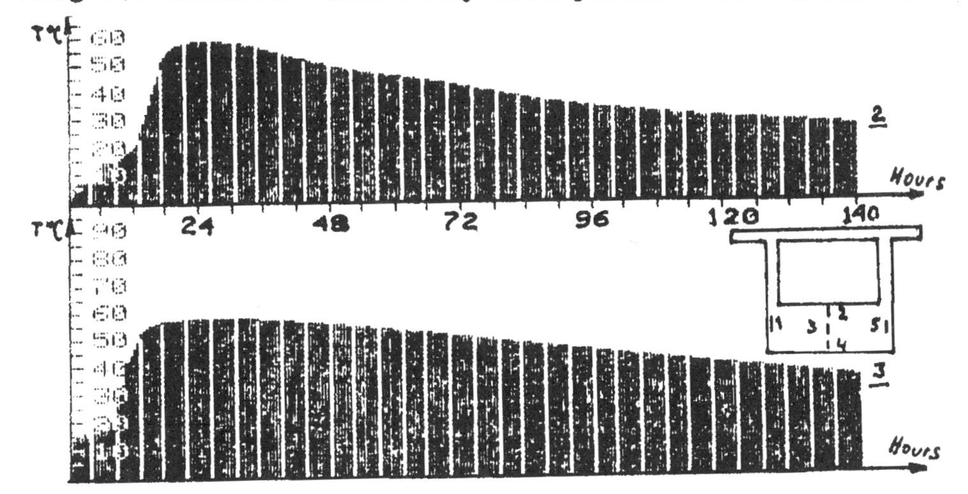


Fig.8. Course of temperatures of the lower slab of the first box-girder segment of the cantilever bridge in Bratislava

## 5 References

Bazant, Z.P. /1985/ Triple Power Law for concrete Creep, Journal of the Engineering Mechanics Division, ASCE, Vol.111, No.1, pp.63-83.

Jerga, J. /1989/ Shrinkage Strains of Reinforced Elements, Research Report of VÚIS, pp.1-35.

Javor, T. /1990/ Experimental Techniques used on Concrete Cable stayed Bridges, Proceedings of the Experimental Mechanics Conference, Lyngby, Denmark, pp.1-8.

PART THREE

# QUALITY ASSURANCE IN TESTING LABORATORIES

# 24 WHY IMPLEMENT QUALITY ASSURANCE IN TESTING LABORATORIES?

D.H. STANGER
Harry Stanger Limited, Elstree, Herts, UK

Abstract
For over a decade governments have been introducing laboratory accreditation schemes in support of their national quality campaigns. Review of national standards and measurement systems has taken place with international competitiveness in mind. The independent testing laboratory owner has been invited to join the voluntary club of accredited laboratories. This paper describes the experiences of one independent laboratory organisation with accredited laboratories in the United Kingdom, Australia and Hong Kong and will also refer to the author's experiences as a member of national and international organisations involved in the development of systems for the assessment of testing and calibration services.
Keywords: Accreditation, Quality Assurance, Total Quality Management, Calibration, Testing, Reports, Specification Standards.

## 1 Introduction

History indicates that testing laboratories emerged around 1860 to support those providing inspection services. A testing laboratory is defined by the International Organization for Standardization (ISO) as "a body disposing of the necessary measuring and test equipment and appropriately qualified persons whose main function is to measure, examine, test or otherwise determine the composition, characteristics or performance of product, materials, components, assemblies or structures."

While a few testing laboratories employ hundreds of qualified staff, over 80% have a staff of less than 25. Until recently, a testing laboratory acquired many of its assignments on the basis of its professional reputation. As product liability and other similar legislation became the commercial way of life, the testing laboratory community absorbed an ever increasing number of assessments by second and third parties. The laboratory owner has reacted to the ever increasing legalistic environment by the implementation of Quality Assurance to a degree that his Quality Control Manual and allied documentation demonstrates a commitment to Total Quality Management. The advent of laboratory accreditation schemes with a national stamp of approval has in recent years significantly reduced the number of assessments and so provided the laboratory owner with national laboratory accreditation schemes that have been written in almost every case by government with international acceptance in mind.

This paper, which formed part of a presentation to a meeting of ASTM Committee E36 in 1988, highlights the experiences of one laboratory group that volunteered to be assessed by national accreditation schemes operated in the United Kingdom, Australia, Hong Kong and Saudi Arabia. The paper examines the demands and attendant costs absorbed by the laboratory owner and his staff at the request of accreditation bodies in exchange for market recognition to provide professional laboratory testing services in support of national quality campaigns at home or overseas.

## 2 National Quality Campaign

The laboratory owner looks to Government and Standard Bodies to provide an enhanced commercial environment within which the laboratory's quality assurance policy will meet the "General Requirements for the Technical Competence of Testing Laboratories" as laid down in ISO/IEC Guide 25. This standard along with the European EN45001 have recently become the backbone of most national laboratory accreditation schemes or services.

For the successful introduction of a national laboratory accreditation scheme, government must initiate quality awareness campaigns to impress and convince industry that "quality counts". With central government financial assistance, the establishment of the infrastructure to support the use of quality techniques is essential.

Of the many components that make up this infrastructure, the national standards body and ISO play a key role with the publication of standards for the measurement and performance of quality techniques. Testing and calibration laboratories must be encouraged to make substantial contributions towards the development of the quality infrastructure. Standards for quality management systems, assessment of a manufacturer's quality system (ISO 9000), and verification of products by test must be developed. Quality consultants and specialists are to be registered and trained as assessors and developers of quality control techniques. A quality campaign within the national education system is a key component, and this infrastructure is completed with the establishment of national accreditation and certification bodies.

Since 1979, when the British Government decided to establish a national laboratory accreditation scheme to be complimentary with and incorporating the British Calibration Service (BCS), itself formed in 1961, the director of the National Physical Laboratory gave the laboratory community many opportunities to make their views known on all aspects relating to the development of a comprehensive set of policy documents now utilized by the National Measurement and Accreditation Service (NAMAS). Many aspects of the work of this team of volunteers raised difficult issues of which the following were of considerable concern to the laboratory manager:

1. Confidentiality of clients' information.
2. Direct and indirect costs of accreditation.

3. Frequency of assessment and surveillance.
4. Calibration intervals.
5. Assessor qualifications and training.
6. The use of proficiency testing programmes.
7. Responsibilities of the laboratory "technical manager."
8. Use of the national accreditation logo.
9. Directory of accredited laboratories (testing and calibration).

Of the above-mentioned list, there is still much work to be done on aspects of calibration, use of logos, assessor training and costs imposed on laboratories by accreditation schemes.

### 3 International Laboratory Accreditation Conference (ILAC)

It might be assumed that Walter G. Leight of the National Bureau of Standards (NBS) had, in 1986, ILAC in mind when he stated that "as the world gets smaller, the need for better ways to exchange information grows increasingly important." ILAC has since its advent in 1977 through the initiative of the U.S. and Danish governments, provided a forum for governments, accreditation practitioners, laboratory managers, standards bodies, and many international organisations to exchange ideas and experiences relating to the development of accreditation systems. The fact that more than 50 countries and 10 international organisations have participated in ILAC activities over the past 13 years is confirmation of the international interest in laboratory accreditation and the vital role the test report plays in international trade.

In 1981, when ILAC assembled in Mexico City, the International Union of Independent Laboratories (UILI) sent its first delegation. With over 800 member laboratories from 24 countries, a UILI delegation has attended all subsequent ILAC plenary sessions and many task force meetings, thus making available the views of the laboratory manager on many aspects of accreditation development. The forum provided by ILAC and its European cousin, the Western European Laboratory Accreditation Cooperation, (WELAC) will continue to be actively supported by UILI and the emerging national bodies dedicated to the development of their testing industry. It is the intention of the British Measurement & Testing Association (BMTA), founded in February 1990, to join the UK delegation to ILAC, thus mirroring the support given to the U.S. delegation to ILAC by the American Council of Independent Laboratories (ACIL). However, the costs associated with a decision of UILI to be more active in the workings of ILAC are currently beyond their financial resources. A way must be found around this issue for organisations like UILI have much to offer ILAC and its national delegations on such matters as proficiency testing, calibration intervals, the development of confidence in national accreditation schemes and training programmes to improve the technical competence of testing laboratories and those responsible for laboratory assessments.

## 4 The Demands of ISO/IEC Guide 25 - 1982

In 1986 Earl Hess of Lancaster Laboratories published a paper in which he called for the ideal national accreditation system to preserve the two essential values of technical soundness and practical efficiency. "The underlying philosophy of this national system," stated Hess, "is to unify, not fragment, to be efficient, not duplicate, to be all-inclusive, not myopic and to be demanding but reasonable." It is the opinion of the Stanger Group that, following well in excess of 1000 hours of assessments, the national accrediting bodies concerned are implementing the wishes expressed by Hess and many other laboratory managers.

ISO Guide 25 -1982, currently under review, sets down the guidelines for assessing the technical competence of testing laboratories. The document contains only five pages, whose contents are submitted under twelve headings. The brevity of the document is applauded, but beware any laboratory manager who assumes that the length of text in any way demonstrates that national laboratory accreditation is easy to achieve. To dispel such thoughts, headings adopted by ISO in their Guide 25 are used in this paper to collate just some of the demands on quality assurance policies of the testing laboratory and highlights a few examples of calibration and testing accreditation experiences of the Stanger Laboratories.

## 5 Scope and Field of Application

Experience has shown that national accreditation bodies have on occasion been slow to accept their obligation to assess laboratories for all tests or groups of tests that accreditation has been applied for, especially in the area of fieldwork. Recognising that many laboratories conduct the majority of their work away from their headquarters or central laboratory and that more and more laboratory clients, including government departments, are only accepting laboratories that are accredited to the national scheme, there is poor reaction time by some accreditation bodies to undertake assessments of testing, calibrating and sampling services undertaken either in the field or in temporary or satellite laboratory facilities. Pressure is also being applied by some laboratories for their materials R & D, materials failure investigation services and even opinions to be accredited. While the latter demands may never be realised, it has been found that a "greenfield" laboratory takes up to two years of preparation time prior to its first formal assessment and that accreditation bodies require up to three years to introduce a new field of tests to their range of assessment capabilities.

In time the voluntary mode that currently surrounds laboratory accreditation activities will change, especially in support of the European Single Market, to a mandatory status; thereafter, both laboratory and accreditation body will have to operate within much tighter time frames set by market forces.

## 6 Organisation

The decision of a laboratory owner to implement a policy of total quality management and to apply for laboratory accreditation normally takes 18 to 24 months before the first full assessment takes place. The most difficult part of accreditation is to obtain the will and active support of one's copartners or board of directors: "Why should we have other people telling us how to do our job?"; "Where will they find assessors who know as much as we do about this test programme?"; "We already have our procedures written so why do we have to have a quality control manual, and anyway do you want me to earn fees or spend weeks writing the manual and other registers?"

In support of the quality assurance objectives of the group, a quality control officer was appointed along with quality control representatives for each technical department that collectively met weekly as a committee under the chairmanship of the Managing Director. Manpower and funds were allocated for the production of all documentation and registers and in particular for the quality control manual. Management leadership with many years of experience in multiassessments as well as the enforcement of maintenance and calibration programmes provided the initial momentum. The submission of the formal application for accreditation confirmed to every staff member tht there was no turning back.

Accreditation demanded major changes in the chain of command within the company. Heads of departments found that their quality control representatives had access to the managing director via the quality control officer; this was a new situation. The major increase in documentation and preauthorisation required for many aspects of laboratory life were for some difficult to adjust to. However, after seven years as a nationally accredited laboratory we now have the means to verify the accuracy and quality of our reports.

## 7 Staff

The means at a laboratory's disposal to demonstrate that their staff are technically competent and that adequate supervision is in place are limited within the time normally provided for either an assessment or surveillance visit. While proficiency testing programmes have their place, an active in-house audit programme and a formal training programme for every member of staff are the proven basis of evidence for an assessor to review.

## 8 Security

The laboratory manager would wish to protect his client's information as diligently as a journalist does his source or informant. It is into this atmosphere of confidentiality that the laboratory assessor seeks access to clients' project files. For an effective assessment to be made, the assessor must have that access. However, it is the duty of the accrediting body to implement and be seen to implement procedures to ensure the maintenance of that bond of confidence expected by right of their clients. In this regard what proof is given to the laboratory by the accrediting body?

Adequate security of all artifacts is becoming more difficult every year, and the total computerisation of our business life adds a further dimension to the question of security.

## 9 Testing and Measuring Equipment

The requirements of quality assurance with regard to registration, utilisation, maintenance and calibration of testing and measuring equipment is seen by many managers to be the area of maximum effort required before the first assessment takes place. It would be inaccurate to imply that before the arrival of accreditation inadequate care was applied, but the records now required for, say, 800 sieves, alone introduces a major on-going cost.

The problems encountered by laboratories relating to calibration are legion. A greater awareness of the implications of the calibration requirements stated in paragraph 7.2 of ISO Guide 25 on the part of the laboratory is essential. Paragraph 7.2. states that "the programme of calibration of equipment shall be designed and operated so as to ensure that all measurements made by the laboratory are traceable to national standards of measurement....."In addition, the laboratory is required to specify in its equipment records the maximum period of time within which the next calibration of each item of equipment held is to be made." In some countries the government department or agency responsible for the custody of national reference standards is extensive, but in others support is almost non-existent. The laboratories require urgent help on calibration intervals.

## 10 Calibration

The commercial implications of calibration requirements are substantial. Test programmes have been withdrawn because calibration costs have made the test uneconomical. The author's laboratories are servicing over 2000 calibration certificates. Their decision has been to establish a BCS-approved in-house calibration service for load, tension and compression verification; calibration of concrete testing machines; force-measuring devices and external micrometers, dial gauges and pressure gauges. A wide range of calibrations outside of the BCS approval are also provided for in-house use. This facility provides calibration response times that all too often cannot be provided from other sources.

## 11 Test Methods and Procedures

It is laboratory policy that whenever there is ambiguity or uncertainty relating to a test method or procedure within a national standard and in every case where a test method is being used by the laboratory that has been compiled by parties other than national standards bodies, a technical work procedure (TWP) is used by the laboratory staff. The company has to date compiled over 200 TWPs with the greatest need coming from the chemical and engineering departments, the standards laboratory and all regional laboratories. All TWPs are reviewed annually, as are the series of technical guidance notes (TGN). To date twelve TGNs have been published by technical directors on such subjects as "Uncertainties of Measurement", "Writing Reports", Reissuing and Amendments of Reports" and "Handling, Sampling and Cutting Asbestos Fibre Materials".

The danger with this aspect of accreditation is the monitoring needed from our librarians to ensure that standards are kept up to date and that superseded standards, which are used at almost equal frequency, are clearly identified. The publication departments of most national standards bodies have, in recent years, significantly improved their services, but the price of standards has escalated.

## 12 Environmental Safety

The main developments adding substantial costs to the laboratory are in temperature monitoring and humdidity controls for the workplace and in temperature-controlled equipment. While most demands are while the test is in progress, it has been found more economical to install 24-hour monitoring systems. It is now possible to advise clients of environmental conditions for the past seven years, and the same facility is a mandatory requirement for those operating a calibration service from their standards laboratory.

## 13 Handling of Items and Their Storage

The basic routines relating to the reception and registering of samples received at the laboratory should cause few problems for an efficiently managed laboratory. However, problems commence when samples are dispatched by agencies on behalf of clients. It is on those occasions when up to three carriers have been involved in the transfer of samples from client to laboratory that records are difficult to compile.

At the end of the testing programme, many laboratories have major storage problems. Most clients want their samples retained for life. While a laboratory has a clear obligation to retain, in appropriate storage facilities, those samples that could be involved in litigation and samples that have produced suspect results, a firm policy has to be taken on the retention of routine samples. We have found that the initiation of a storage charge has had the right response from all but a few clients.

## 14 Test Reports

The subject of accreditation and test reports could be the subject of a paper in its own right. However, the main concerns of accreditation relate to the avoidance of the misuse of the accreditation logo coupled with problems associated with test reports that contain some data for which the laboratory is accredited and some data for which the laboratory is not accredited.

**15 Summary**

The development of quality assurance and national laboratory accreditation schemes continue to be challenged by a proportion of the laboratory community in most countries. Quality Assurance has many advantages for the laboratory manager and staff alike. The demands, however, of ISO/IEC 25 and EN45001 can only be achieved at considerable cost in terms of management time and finance. Pressure from the marketplace and the effects of quality campaigns by governments have made the testing industry take total quality management and laboratory accreditation seriously and recognise the value of both functions as valuable marketing assets.

The costs of accreditation on the laboratory are substantial, but the experience of the Stanger Laboratories supports the belief that national laboratory accreditation schemes are in the interest of the laboratory and their clients. The accrediting bodies mentioned in this paper have reduced the number of assessments per year. Accrediting bodies must have all relevant resources to accredit the total range of laboratory services provided by testing laboratories. All parties involved in laboratory accreditation must urgently find ways to significantly reduce the lead time before the first assessment of a laboratory and the introduction by an accrediting body of a new field of accreditation activity.

To attract the attention of our Customers at home and abroad has required the establishment of national accreditation systems to ensure that testing and certification bodies are consistent and reliable. These schemes are voluntary but evidence suggests that, to protect our existing UK based markets and take advantage of the Single Market, it will be pressure from the market place that will encourage more organisations to review their corporate policy on total quality management.

# 25 FORMATION ET MOTIVATION DES PERSONNELS (Training and motivation of staff)

M. VALLES
Centre d'Études et de Recherches de l'Industrie du Béton Manufacturé, Epernon, France

Résumé
L'auteur analyse les conditions nécessaires pour que la mise en place de l'assurance de la qualité dans un laboratoire d'essais contribue à la motivation des collaborateurs. Il passe en revue chacune de ces conditions et présente l'effet que l'on peut en espérer. Le rôle de la formation dans la motivation est analysé. Un programme de formation à trois niveaux est proposé.
Mots clefs : Laboratoire d'Essais, Assurance de la Qualité, Motivation, Formation.

---

Si la **formation** est une notion claire pour la plupart d'entre nous, il n'en est pas toujours de même pour la **motivation.**

Etre motivé, c'est quoi ? Peut-on motiver quelqu'un ? Naît-on motivé ou motivant ?

Pour nous, gens de laboratoire, ce concept qui rejoint le subconscient des individus est bien loin de nos préoccupations quotidiennes. Et pourtant, les femmes et les hommes de nos laboratoires sont, autant que les autres, sensibles au facteur motivation.

**Qu'est-ce qui motive nos collaborateurs ?**

Pour simplifier, admettons que la traduction "d'être motivé" est : **avoir envie de faire, de rechercher le progrès.**

Les raisons de la motivation résident pour grande partie dans l'inconscient. Les publications désormais classiques de Maslow et Herzberg nous éclairent sur les besoins de tout individu pour être motivé. La réponse successive à chacun des besoins rappelés dans la pyramide ci-après (en commençant par la base) contribue à la motivation.

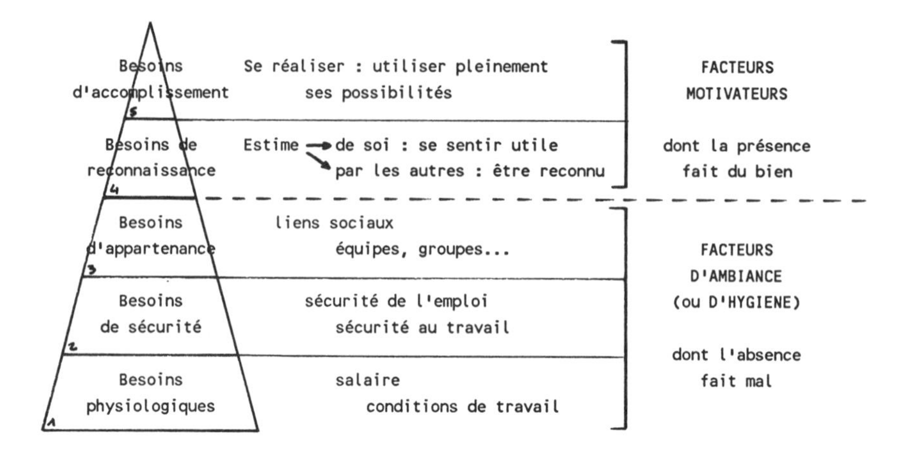


**Hiérarchie des besoins fondamentaux et facteurs de motivation de l'homme dans le travail (Maslow-Herzberg)**

**Quelle relation entre motivation et assurance de la qualité, dans un laboratoire d'essais ?**
Notre expérience, aussi bien en laboratoires que dans l'industrie, nous permet d'affirmer que la mise en place progressive de l'assurance de la qualité peut créer des conditions très favorables à l'épanouissement de la motivation des collaborateurs.

Au vu des besoins repérés dans la pyramide de Maslow, passons en revue ces conditions potentiellement favorables :

**Besoin 1 : Conditions d'organisation du travail**

SI lors de la mise en place de l'organisation qualité, le collaborateur peut s'exprimer, il proposera certainement des aménagements de poste qui, non seulement lui faciliteront la vie, mais entraîneront des gains sensibles de productivité.

**Besoin 2 : Sécurité**

SI l'organisation mise en place est telle que le collaborateur considère qu'il sait exactement ce qu'il a à faire, que ses limites sont claires, qu'il dispose de tous les renseignements nécessaires et qu'il comprend le pourquoi des choses..., il se sentira rassuré.

Il pourra alors se consacrer entièrement à sa tâche au sein de son **espace de liberté**, sans inquiétude sur les interprétations de l'encadrement, et son efficacité sera accrue.

SI le matériel est suivi en permanence, et mis hors circuit dès qu'il y a soupçon d'anomalie, le collaborateur se sentira en confiance et pourra mieux se concentrer.

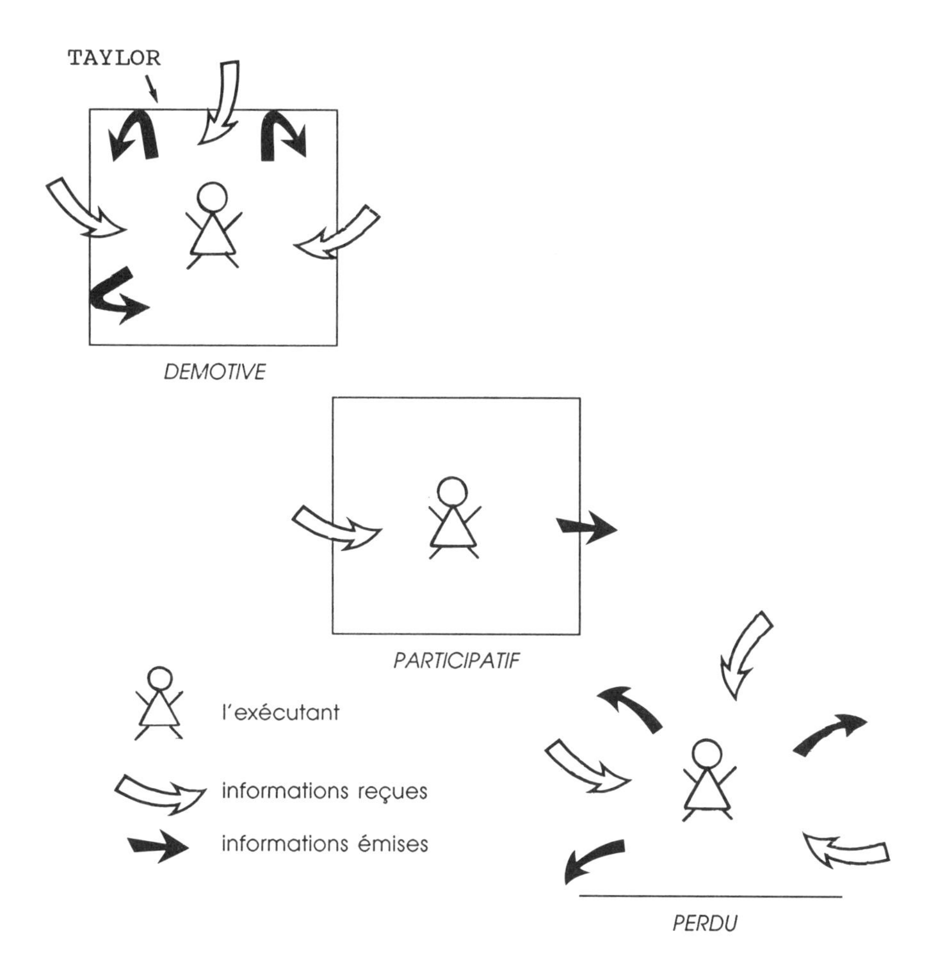


**L'espace de liberté**

**Besoin 3 : Appartenance**

SI l'action qualité comporte des groupes de travail destinés à organiser sa mise en place et à rechercher le progrès par l'analyse des anomalies, dérogations ou incidents, le collaborateur se sentira membre d'une équipe, sa participation sera accrue.

SI le laboratoire sollicite une accréditation officielle, l'obtention puis le maintien de cette accréditation constituent un challenge qui rassemblera tous les collaborateurs.

**Besoin 4 : Reconnaissance**

SI le collaborateur est associé à l'établissement de la partie de l'organisation qualité qui le concerne plus particulièrement (consignes, procédures, modes opératoires, feuilles de relevés...), il se sentira mieux reconnu.

SI les tâches à exécuter lui permettent d'être autre chose qu'une machine qui parle, grâce par exemple à la mise en place d'un auto-contrôle et par appel à l'avis sur des informations liées à l'expérience (allure de cassure, couleur, anomalies...), il se sentira plus responsable.

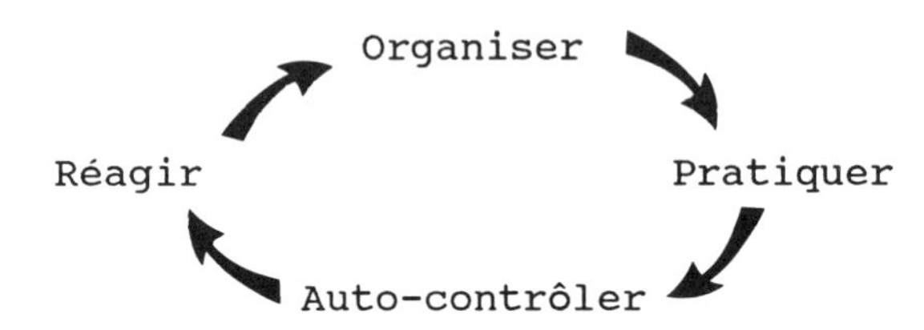


SI les tâches à exécuter comportent une part d'initiative contrôlée, selon le principe des feux de signalisation routière, le collaborateur se sentira encore plus utile.

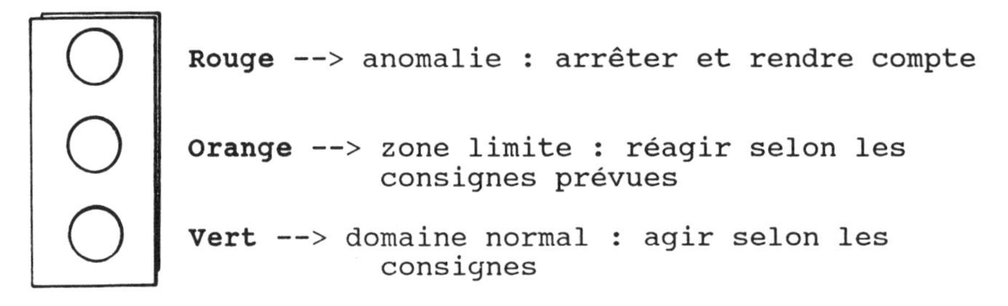


SI le collaborateur bénéficie d'une formation adéquate, qu'il sait faire partie d'un plan de formation qui lui est propre, il se sentira reconnu.

SI l'encadrement sait reconnaître les mérites...

**Besoin 5 : Accomplissement**

SI grâce à une meilleure délégation (rendue possible par les nouvelles sécurités en place), le collaborateur dispose d'une plus grande autonomie et d'un travail enrichi par de nouvelles responsabilités, il aura conscience de mieux utiliser ses possibilités.

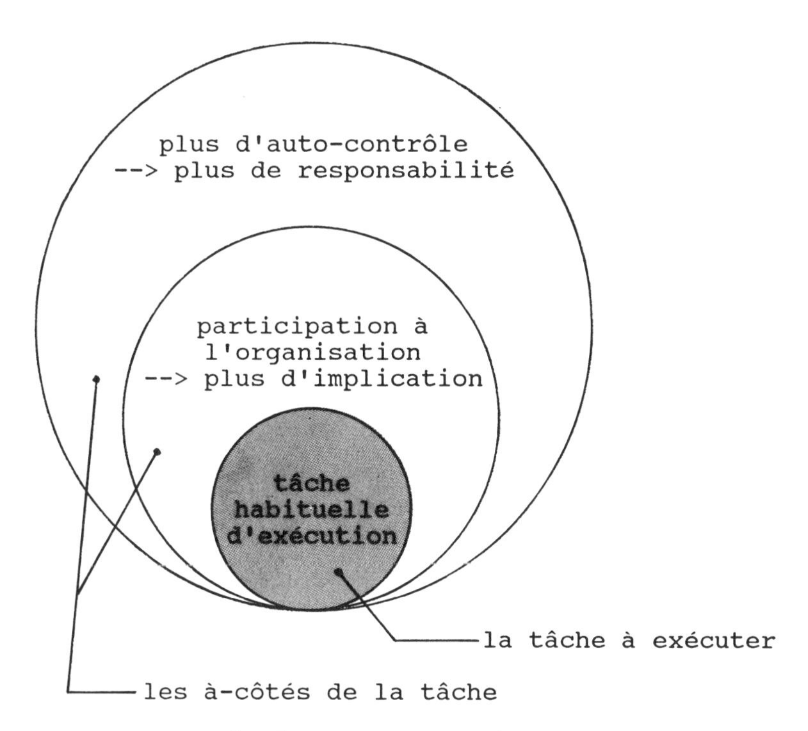


**L'enrichissement des tâches**

SI le collaborateur peut prendre en charge des actions de progrès dans le domaine qui le concerne, il estimera s'accomplir encore plus.

Ces précédents constituent beaucoup de SI parfaitement réalisables, mais il est des conditions **incontournables** que nous nous permettons de rappeler ici :

. **une Direction motivée et motivante,**
. **le droit (réaliste) à l'erreur,**
. **une formation adéquate.**

**La formation : un impératif**
La formation dans le cadre de l'action qualité joue un rôle considérable. Nous avons vu son importance au niveau de la satisfaction du Besoin 4. Mais en fait, une véritable stratégie de formation est nécessaire. En effet, de nombreuses notions sont peu familières aux intervenants ; citons par exemple :
- la combinaison des incertitudes lors du raccordement d'un instrument de mesure aux étalons de référence,

- les fréquences de maintenance et d'étalonnage des matériels,
- l'achat des matériels en termes de performances en adéquation avec le besoin,
- la traçabilité des essais,
- les limites juridiques d'un rapport d'essai...

En fonction de notre propre expérience en tant que laboratoire, mais aussi de formateur au sein du Réseau National d'Essais, nous proposons trois familles de programmes :

- **Cadres dirigeants**
  - les principes de l'assurance de la qualité
  - le dirigeant motivant
  - comment lancer une action qualité

- **Cadres moyens**
  - l'assurance de la qualité (principes et application aux laboratoires d'essais)
  - les audits internes
  - les performances des matériels d'essais (principes, incertitudes de mesure, étalonnages, maintenance...)
  - aspects juridiques liés aux essais
  - le cadre motivant
  - comment lancer et développer une action qualité

- **Techniciens**
  - pourquoi la qualité des essais
  - l'assurance de la qualité
  - étude du manuel qualité et des divers documents les concernant
  - la précision des essais
  - la pratique des matériels

En **conclusion**, nous pouvons affirmer que si la mise en place de l'assurance de la qualité est indispensable dans tous les laboratoires, les résultats à en attendre peuvent être très différents.

Sans réflexion préalable sur la motivation et la formation des collaborateurs, il s'agira d'un système de plus, plaqué sur une organisation existante.

Si ces deux facteurs sont convenablement pris en compte, les résultats dépasseront les espérances, tant pour l'adhésion des collaborateurs à ce projet, que pour l'accroissement de la productivité et de la satisfaction des clients.

Technical Instruction Manual
In which detailed instructions are given regarding the carrying out of technical activities, e.g test procedures; measuring procedures; check lists etc.
Quality Plans
The Quality Plans are documents which are applicable to define the quality system applicable to an specific job.
These plans are the documents used to define the Quality System applicable to the investigation programmes.

**3 Work procedures**

There follows a description of the general requirements established by the quality system in relation to investigation work.

**3.1 Evaluation of investigation proposals**
This aims to establish a course for internal petitions to carry out investigation programmes and to evaluate the proposals in terms of the objectives and policy of the Management of the company. Investigations assigned by clients are excluded from this process.

Petitions are channeled by way of the Division Directors who must in turn submit them for the approval of the Management Committee of which they are all members.

The documentation of these petitions must include the following information:

(a) Identification of the applicant.
(b) A description of the investigation work, indicating its scope and the expected results.
(c) Description of the necessary resources.
(d) Reason for the investigation. This consists of a description of the causes leading to the carrying out of a plan of investigation. The reasons may be grouped into the following concepts:

Professional Practice. Which requires the perfecting of test or control procedures.
Standardization. As the personnel belong to national and international standards commissions, they should detect obscure or insufficiently known aspects in the development of the standards which are clarified by the carrying out of a plan of investigation.
Personnel Qualification. Within the policy of quality, special emphasis is made on the professional qualifications of the technical staff. The carrying out of a doctoral thesis is promoted and the investigation work asociated is undertaken internally.

### 3.2 Plan of investigation

Once the proposal for investigation is accepted: The followed activity consists of the making up of a Plan of Investigation. The objective of this Plan is to define the following aspects:

(a) Quality Plan of the investigation programme
In this section the specific responsibilities relative to the investigation work are defined. The procedures and instructions of the general manuals to be applied are identified and the procedures specifically prepared for the work included. Besides, this document contains the applicable auditing programme. The Quality Plan is jointly prepared by the Research Testing Department and the Quality Assurance Department.

(b) The assignment of personnel with indications on whether additional training be necessary in order to carry out the work.

(c) Details of the equipment required, including the specifications for the purchase of new equipment if it be so required.

(d) Schedule of work.

(e) Budget.

Prior to the commencement of the work the plan is submitted for evaluation by a team of experts designated by the Management Committee with the object of checking the viability and suitability of the Plan, with regards to its proposed objectives. When it is considered necessary for the evaluation of final plans, pilot tests will be carried out prior to their approval.

All modifications to the plan require the previous evaluation and approval of the aforementioned team of experts.

### 3.3 Development of the work and recording of results

Development of the work will comply with the established plan as indicated in the previous point.

The analysis of the test results is jointly carried out by the Research Departments and by the technicians who requested the investigation. This analysis is reflected in a document which is subject to a further evaluation by the team of experts whose approval is required prior to its edition.

### 3.4 Application of results

Once the document of analysis of the results has been approved, and if these affect the development of the control activities, a Technical Instruction is prepared with its application to the works and is then included in the Technical Instruction Manual.

## 4 QUALITY ASSURANCE ACTIVITY DURING THE INVESTIGATION PROCESS

There follows a description of the most important aspects of Quality Assurance in relation to the process of the investigation work previously mentioned.

### 4.1 Verification of the Plan of Investigation and Results

The object of this is to establish a number of controls which ensure on one hand that the work was initiated in accordance with the established criteria and on the other that the results obtained are adequate and that they may be used for their proposed objective.

Given the level of specialization that this activity demands, this task is assigned to a team of experts from all the areas related to the investigation process which are subject to evaluation.

### 4.2 Process control

In order to control that the work is carried out in accordance with the Quality Plan established, the Quality Assurance Department establishes an auditing programme that covers all the activities from the carrying out of the investigation plan to the testing phase and onto the final edition of the results. In order to carry out the auditing programme the Quality Assurance Department relies on the advice of technicians who are specialized in the areas to be audited.

### 4.3 Equipment control

This aspect is of particular importance in any activity in a laboratory, and there is no reason why it should differ in essence in the case of investigation work. Investigation, however, does present the peculiarity that in many cases the definition of its characteristics and particularly its precision do not come under any regulations, instead it is the same investigator who must define them in terms of these technical requirements. The Investigation Plan must therefore include a section dealing with the equipment to be used and its characteristics. In the case of available equipment this will be in the form of an extract from the Laboratory's inventory, and where equipment has to be purchased it will require a specification of the same.

The suitability of the equipment to the plan is checked by means of an evaluation carried out by the team of experts and its use, calibration and maintenance by that of the auditing programme.

The equipment purchased specifically for investigation programmes is included in the laboratory inventory and its control with regards to its use, maintenance, verification and calibration is identical to that of equipment in normal use. A brief reference will be made, however, to the documents related to the control of equipment.

(a) Inventory of testing and measuring equipment.
(b) Equipment cards. This includes identification, characteristics and requirements and periods of maintenance, verification or calibration.
(c) Instructions of calibration.
(d) Calibration Diagram. This indicates in order the largest down to the smallest level, calibration and measuring equipment and gives information regarding the equipment used in the calibration of each equipment. This document is used for verifying that the measurements made in the laboratory are traceable to international standards of measurement.
(e) Reports or certificates of calibration. Indicating the equipment used for the calibration, the measures taken, the calculations necessary to determine the errors. Observations that are deemed necessary.
(f) Labels on the equipment indicating state and position for use.

Apart from these documents a procedure has been established for the purchase of testing and measuring equipment with the view to ensuring that the equipment acquired meets with the necessary requirements of their proposed usage, this procedure includes the carrying out of tests on delivery if it is seen fit.

### 4.4 Personnel

#### 4.4.1 Qualifications

As previously mentioned investigation work requires the assignment of personnel highly qualified in the areas to be covered by the investigation, the Investigation Plan should, as such, specify the personnel assigned and where necessary indicate the additional preparation that should be received prior to the carrying out of the work.

#### 4.4.2 Motivation

Motivation is a subject of particular importance and is very often forgotten in investigation work. The personnel carrying out this work often combine it with production work which due to external pressure usually requires most of their attention. The result is that investigation work which would be highly beneficial is not promoted or that the investigation work is carried out without the due attention.

Within the policy of quality established the carrying out of investigation work figures as one of the activities which improve the quality of the work as well as the qualifications of the personnel. It falls upon the managerial staff to stress the importance of these activities and to ensure that they are carried out with the same rigour and quality as the rest of the activities.

## 5 CONCLUSIONS

The main conclusions to be reached in all the aforementioned are that:

- Investigation work should be covered by the Quality System.
- Quality Plans are the way to adapt the Quality System to the peculiarities of the investigation activities.
- Everything that is established in a laboratory in its Quality System relative to the carrying out of auditing, the qualification of personnel, control of testing and measuring equipment, requirements and procedures of calibration of measuring equipment etc - is totally applicable to investigation work.

Finally it is important to emphasize one aspect mentioned at the beginning of this paper, that being the fact that quality systems do not limit the capacity of the investigator but rather systemize his or her work, errors are avoided and the investigator can justify irrefutably the validity of the results of the investigation.

# 27 LA VÉRACITÉ DES RÉSULTATS EXPÉRIMENTAUX: UNE CONDITION NÉCESSAIRE POUR UN LABORATOIRE DE RECHERCHE

(Accuracy of experimental results: a necessary condition for a research laboratory)

J. PAQUET
Centre Expérimental de Recherches et d'Études du Bâtiment et des Travaux Publics, Saint-Rémy-lès-Chevreuse, France

Résumé
Cet exposé indique la démarche critique qui doit être appliquée à tout essai conduisant à un résultat quantifié. Il importe en effet que ces mesures soient exportables, c'est-à-dire qu'elles soient reliées par une démarche logique aux concepts physiques de base. Des exemples sont donnés dans lesquels des pièges de l'instrumentation ou des insuffisances de connaissance théorique menacent la qualité des résultats.
Mots clés: Qualité, Instrumentation, Laboratoire de recherche et d'essais, Matériau, Essais non destructif.

## 1 Introduction

Par rapport aux autres communications de ce symposium, cet exposé se place sur un plan spécifique : en effet, je ne parlerai pas de l'organisation de la qualité à l'intérieur d'un Laboratoire de recherche. Je ne serais d'ailleurs pas le mieux placé pour le faire, puisque ce rôle est, au C.E.B.T.P., dévolu à Mr. SEGUIN, qui assume cette tâche avec compétence au sein de la Direction Technique et de la Qualité.

Je me présente ici en tant que producteur et utilisateur de mesures dans le domaine des études et recherches du BTP depuis plus de vingt années et particulièrement attaché à tirer quelques leçons de cette longue expérience. J'évoquerai plutôt le non-dit de la qualité, c'est-à-dire tout ce que l'on suppose acquis avant la mise en oeuvre d'un plan qualité proprement dit, et mon propos sera plus technique que politique.

Dans une conférence prononcée à l'ouverture du cycle de séminaires "Prospective de la mesure et de l'Instrumentation", intitulée "La Mesure, charnière entre Connaissance et Action" de P. GIACOMO, j'ai trouvé la formule d'introduction qui manquait à cet exposé: "La Mesure est l'art de consigner les résultats d'observation de manière qu'ils puissent être utilisés par autrui, en tout temps et en tout lieu".

J'ajouterai les deux réflexions suivantes :
- les résultats expérimentaux constituent la nourriture de base des chercheurs, et de la qualité de leur alimentation dépend la santé globale de la recherche.
- les résultats d'observation doivent être exprimés dans un langage international unique.

Nous allons, à travers un certain nombre d'exemples, montrer la difficulté d'application de ces principes pour un centre expérimental de recherche et d'études, et indiquer les différentes causes des manquements aux devoirs de qualité.

On peut citer dès maintenant :
- L'incompétence (cas exceptionnel bien sûr)
- Les pièges de l'instrumentation
- L'insuffisance d'analyse fondamentale.

Ces manquements au devoir de qualité peuvent causer des torts considérables à la communauté scientifique et technique. L'existence dans la littérature de résultats peu fiables peut troubler pendant plusieurs décades la bonne prise en compte d'un phénomène. En effet, même si la crédibilité de certaines publications est faible au moment de leur parution, elle peut s'accroître au fil des ans par le jeu des multiples références, alors même que les conditions de l'expérience ont été oubliées.

Le biais introduit par une expérimentation défaillante nécessite, pour être effacé, plusieurs expériences contradictoires.

Ceux qui ont la lourde responsabilité d'adapter la réglementation aux résultats d'expériences et certains assureurs sont confrontés à ce type de problème et ont essayé de quantifier la définition du "bon choix" face à des résultats expérimentaux dispersés ou évolutifs. L'approche bayesienne est souvent suggérée. Elle conduit à pondérer les résultats en fonction de leur certitude, dont l'appréciation va au-delà du simple calcul d'erreur et peut intégrer un facteur de confiance global.

## 2 Exemples de problèmes rencontrés

### 2.1 Exemple 1 - La bonne foi des expérimentateurs est mise en défaut par un piège de l'instrumentation

L'établissement d'un diagramme Effort-Déformation en régime cyclique est l'un des essais les plus courants de notre profession. On sait que, du résultat, on peut tirer l'énergie absorbée par le matériau, et par suite en caractériser l'amortissement (figure 1).

Supposons que pour une sollicitation à la fréquence de 1 Hz on observe un déphasage de 4 degrés dans le bon sens entre force et déformation. On s'empressera en général d'attribuer ce déphasage au matériau lui-même. Mais il suffit que dans la chaîne de mesure les deux voies aient respectivement des bandes passantes de 30 et 100 Hz, pour produire ce même résultat face à un matériau parfaitement élastique.

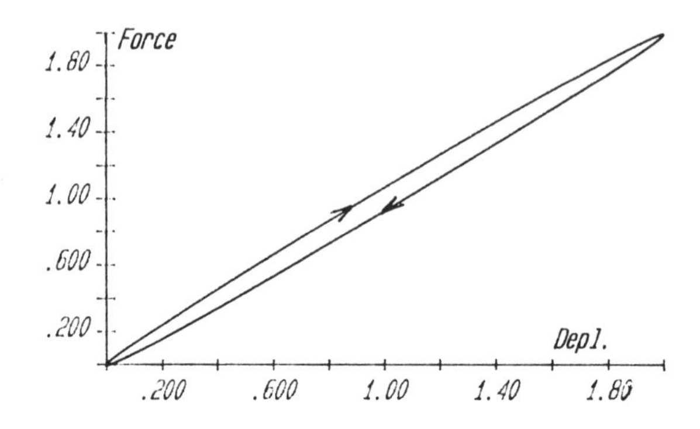


Fig.1. Diagramme Effort-Déformation

Il eut mieux valu alors que la boucle tournât à l'envers ce qui aurait permis de déjouer le piège.

**2.2 Exemple 2 - insuffisance des connaissances fondamentales ou comment mesurer une grandeur non définie : la déformation instantanée du béton**

Voici une notion qui fait partie des échanges courants entre laboratoires. Chacun de ces laboratoires a sans doute de ce concept une définition implicite, mais une analyse fine montre qu'il est impossible de la formaliser sans un minimum de précautions. Cette difficulté est liée au fluage à très court terme du béton, qui s'amorce pour des temps très courts inaccessibles à la mesure courante.

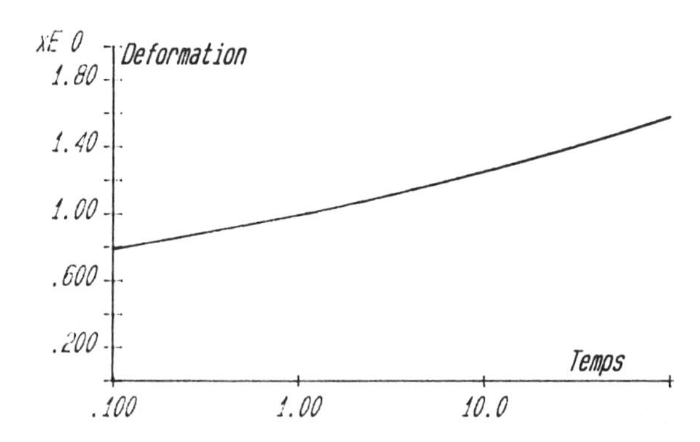


Fig. 2. Fluage à court terme du béton

## 2.3 Exemple 3 - Essai de flexion avec rupture d'un matériau fragile

L'apparition de matériaux ductiles tels que les bétons de fibres nous a amenés récemment à réexaminer les diagrammes obtenus sur des matériaux fragiles (figure 3)

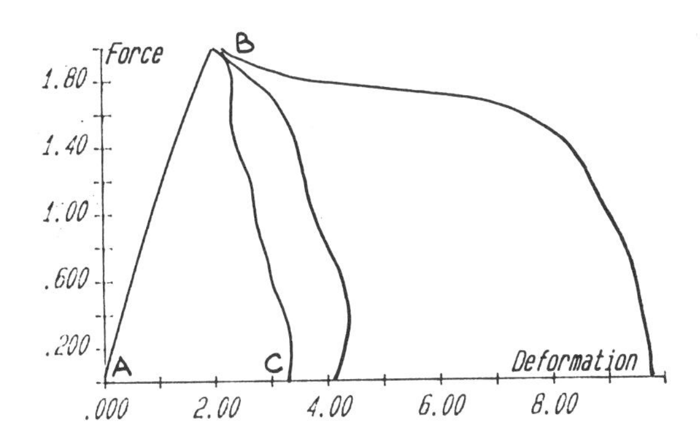


Fig. 3. Diagramme Effort-Déformation
Cas de matériaux fragiles et ductiles

Ces diagrammes comportent la phase AB de chargement élastique et la phase BC de rupture. Cette dernière, tracée par un enregistreur, peut sembler traduire les propriétés réelles du matériau. En fait, compte tenu de la vitesse du phénomène, ce trajet reflète plus les réponses transitoires des chaînes de mesures que le comportement intrinsèque du matériau.

## 2.4 Exemple 4 - Mauvaise adaptation de l'essai au problème posé : la recherche d'une non linéarité

L'essai classique conduisant à la détermination du module d'élasticité consiste à appliquer sur le spécimen une force croissante et à observer sa déformation. L'usage veut que tout écart par rapport à une droite soit interprété comme une non linéarité. Or, un comportement linéaire viscoélastique conduit au même résultat. Ici un essai parfaitement mené peut conduire à des conclusions erronées.

Bien sûr, ces maladresses ne conduisent pas à des catastrophes et globalement la technique s'en accommode.

Mais ce sont des facteurs qu'il est important de mieux maîtriser pour permettre l'application en toute conscience des procédures de qualité des essais au sens classique du terme.

**2.5 Exemple 5 - Faire de mauvaises mesures avec des appareils parfaitement calibrés**
Nous voulons ici attirer l'attention sur les problèmes spécifiques des mesures sur ouvrages, et les problèmes de mécanique des sols pour lesquels le couplage matériau-capteur est souvent à l'origine d'erreur importante.

**2.6 Exemple 6 - Insuffisance de diffusion des connaissances fondamentales ; définition non ambiguë de la vitesse de propagation dans les matériaux.**
Parmi tous les domaines accessibles à la mesure, les Essais Non Destructifs sont particulièrement vulnérables vis-à-vis des critères de qualité, car ces méthodes sont à la fois séduisantes et difficiles à traiter.

La définition non ambiguë de la vitesse de propagation des ondes élastiques (vitesse du son) et sa mesure dans les géomatériaux (béton, pierre, roche), constituent un exemple caractéristique.

La somme des documents existants dans ce domaine est impressionnante, à tel point qu'il est impossible à un non spécialiste abordant le sujet de se faire par lui-même une idée précise. Dans ce domaine, le CEBTP s'est déjà vu confier des travaux d'analyse et de tri pour attribuer des labels de qualité aux différentes publications bien qu'elles aient toutes franchi avec succès l'obstacle des comités de lecture.

L'impression globale qui se dégage de ces analyses est celle de l'incommunicabilité : en 1990, c'est-à-dire plus de 30 ans après la naissance de ces méthodes, les informations sur les vitesses de propagation dans nos matériaux ne peuvent être échangées car actuellement trop dépendantes des méthodes et des matériels mis en oeuvre.

Face à ce problème deux attitudes sont possibles:

1) Considérer que le paramètre vitesse n'est pas un but en soi mais que sa finalité est (par exemple) l'estimation de la résistance mécanique. Dans ce cas, chaque laboratoire va élaborer un "package" complet avec sa propre vitesse et sa propre corrélation vitesse-résistance.
2) Considérer que la vitesse est une notion physique porteuse d'information et admettre que la difficulté de sa définition, révélatrice de la complexité du matériau, est forcément enrichissante.

On peut dire malheureusement que c'est l'attitude 1 qui a prédominé, ce qui a stérilisé les échanges nationaux et internationaux dans ce domaine.

Il semble donc nécessaire de focaliser les efforts d'unification dans le domaine des END. D'autres raisons nous y poussent : une volonté générale de clarification, un prochain congrès de la RILEM sur ce thème et l'émergence de normes "expérimentales" dans le domaine de la vitesse du son et de la sclérométrie.

Sur ce dernier point, on peut dire qu'une normalisation avec des bases théoriques insuffisantes est particulièrement dangereuse car elle a tendance à privilégier l'accessoire au détriment de l'essentiel.

Notre opinion est qu'il serait préférable de procéder ici à des habilitations de matériel, plutôt que de vouloir imposer des caractéristiques métrologiques générales à des appareils forcément complexes. De même, la normalisation des procédures de calibration des appareils de mesure de vitesse du son nécessite une connaissance des phénomènes physiques, pas toujours accessible à l'utilisateur moyen.

## 3 Conclusion

Le tableau dressé ici peut paraître exagérément pessimiste. C'est vrai que nous l'avons voulu ainsi pour mieux mobiliser l'attention.

Nous retiendrons la fragilité du témoignage expérimental avec toutes ses composantes alors que les développements théoriques sont toujours vérifiables a posteriori.

Cette situation nécessite une vigilance particulière qui ne peut être que le résultat d'une action regroupant les compétences dans les domaines les plus divers.

## 4 Références

Giacomo, P. (1989) Extrait du Bulletin National de Métrologie (BNM) n°78

# 28 FIVE YEARS OF QUALITY ASSURANCE IN LABORATORIES ASSOCIATED WITH THE MANUFACTURE OF PORTLAND CEMENTS

P.J. JACKSON
Sampling and Scientific Services, Rugby, UK

Abstract
The role and the work load of testing laboratories at plants involved in the manufacture of Portland cements are discussed and a comparison is made of operations before and after independent assessment was implemented. In the case of U.K. cement making operations it is concluded that in its present form independent assessment has provided user and manufacturer with benefits which are only marginal relative to the additional effort and cost involved. Suggestions for improvements are made.
Keywords: Quality Assurance, Portland Cement, Testing Laboratories.

## 1 Introduction

The role of the testing laboratory on a cement manufacturing plant is as follows:

(a) Deciding which process variables to monitor.
(b) Ensuring that the monitoring is correctly carried out.
(c) Testing of samples supplied to the laboratory.
(d) Ensuring that the testing is carried out correctly.
(e) Using the test results to control the process and also where necessary to provide data for customer use.
(f) To assist the customer on the use of the product.

This paper deals with items (c), (d), (f) and part of (e).
It is also proposed to make a clear distinction between introducing and operating a quality system which complies with BS 5750 Part 2 (later EN 29002), and the quite separate matter of independent assessment in order to confirm to interested parties that the claim made to comply with EN 29002 (1) is valid.

2 The work carried out in the testing laboratory of a cement-making plant can be described under the following headings:

### 2.1 Raw Materials
In order to achieve continuity of operation and quality it is necessary to know both the chemical and the physical characteristics

of the quarried raw materials, the gypsum rock and the fuel. The former are assessed on the basis of borehole samples often tested many years ahead of usage. The gypsum rock and the fuel are normally sampled on arrival and before being passed to storage, as are the sand and the fly ash when these components are used.

Only broad specification limits are generally necessary for some aspects of the composition of these materials since compensating adjustments to proportions can and are made on an on-going basis in order to maintain a suitable feedstock to the kiln(s). Further, as the kiln feedstocks are completely transformed by thermal processing, rigid specification as intimated in EN 29002 is not necessary.

## 2.2 Clinker

Once the feedstock has been produced and introduced into the kiln system, the principle variable which affects product quality and which needs constant monitoring is the degree of combination achieved. This is normally assessed in terms of the free lime content (i.e. the amount of lime remaining uncombined) although quite often second order measurements such as the bulk density of a particular size fraction of clinker is used. If the firing process is inadequate then insufficient combination of the compounds will take place and the clinker will be deficient in $C_3S$ which is the main strength-giving component. Grossly inadequate combination will lead to unsoundness (a cement which expands in the hardened concrete or mortar). The test method(s) adopted here must be rapid in order to ensure that any unsatisfactory material is rejected.

## 2.3 Cement

As far as the ordinary Portland cements permitted in the U.K. during the period under discussion are concerned, there are two main testing needs at the grinding stage - notably to establish the amount of gypsum rock added, in order to comply with maximum $SO_3$ specification requirements and also that the cement shall be ground sufficiently fine to achieve the required strength level. Sampling and testing in the U.K. has normally been carried out at the grinding stage in order to reduce the time taken to apply any necessary process changes. It has not been the practice in the U.K. to regularly sample and test the product as loaded, but this may change if European Standards replace British Standards.

## 2.4 Product Service

Users from time to time experience difficulty in the use of all types of products and Portland cement is no exception. An important role of the testing laboratory is to assist the technical service representatives in the field to establish the reason(s) for a problem as a preliminary to suggesting a solution. This involves a knowledge of the product and of associated materials as well as different and sometimes more sophisticated test equipment. In the larger U.K. companies this work is often dealt with for all plants by one central testing laboratory.

## 3 Quality Assurance

### 3.1 Internally Assessed Quality Systems

Cement manufacture is a continuous operation and it is usual for the product to have been incorporated into a structure long before the results of the most critical (28 days) strength result are available. Manufacturers have therefore over the years developed quality systems which have proved exceedingly effective and which dealt with the important aspects of EN 29002 long before that document was ever conceived.

This activity has been further promoted by the existence in the U.K. for some 53 years, and over much of the period under discussion, of a Common Price Agreement (C.P.A.) which meant that manufacturers were restricted to competing on service and on quality. With the latter in mind, they had, through the Cement Makers Federation (C.M.F.) introduced on their own initiative a system of testing which had as its objective as near a common quality product as possible - this being consistent with the concept of a common price.

Firstly because of the known differences in the strength-giving properties of cements when tested in different laboratories (+/-8%) it was decided that each manufacturing unit should each month submit to a single reference laboratory a composite sample of its mainstream product and from the tests carried out in that laboratory and those carried out in the plant laboratory, a correction factor should be produced and known as a calibration multiplier. This multiplier was based on the running average of 9 pairs of test results, and updated monthly. It was applied to all the test results produced at the plant testing laboratory, such that the result obtained approximated to the figure which would have been obtained had the test work been carried out in the reference laboratory.

Secondly because a high proportion of concrete is produced at constant workability rather than at constant water/cement ratio, slump tests were also carried out on the composite samples at the reference laboratory and the industry average slump value calculated. The water/cement ratio in use in the plant laboratory was then changed from the 0.60 required in BS 4550 part 3 section 3.4 mix Cl (2) to that which was necessary to give the industry average slump level. This was normally within the range 0.58 to 0.63.

Thus for what were essentially commercial reasons the U.K. cement manufacturers took the steps required to improve the standard and also the traceability of their product testing.

In additon, through the circulation of samples within their own organisations they carried out an extensive programme of co-operative testing.

Compression testing machines were also calibrated in house before British Calibration Service came into being.

### 3.2 Externally Assessed Quality Assurance Systems

Following the U.K. Governments initiative to promote: 'Standards, quality and international competitiveness' (3) the U.K. cement industry through the Cement Makers Federation felt that they should assist by permitting their very satisfactory in-house testing procedures to be assessed by an external agency.

Discussions took place with the Quality Assurance Division of the British Standards Institution in order to establish if their Registered Firms Scheme (4) was appropriate in this case. As this approach involved a whole industry the 'Sector Scheme' approach had to be adopted. This meant setting up a committee chaired by a BSIQAD representative and comprising representatives from Government Departments, Specifiers, Users, Manufacturers and members of organisations vending products with which the cements in question are used (fly ashes etc.). The first step was for this committee to agree a Technical Schedule (later to be known as a 'Quality Assessment Schedule') (QAS) (5) which when used with EN 29002 provided a working basis specific to cement-making operations. Concurrent with the preparation of the QAS, work proceeded on the preparation of a model Quality Manual suitable as a framework within which to incorporate the procedures adopted during the manufacture at specific cement plants.

Because in the U.K. the quality of cement is normally monitored and controlled from the laboratory, the introduction and operation was carried through from that sector of the plant and the Works Chemist (or Technical Manager) in most cases became the 'Person Responsible for Quality Assurance' - (PRQA). The whole procedure was brought into being in May 1984 (6) when all U.K. manufacturing plants were registered.

## 4 Discussion

We are dealing with a situation where a well controlled industry has moved from an internal regulatory system to an externally monitored system based on EN 29002.

The balance sheet is assessed as follows:
It has been somewhat easier (in some cases) to replace test equipment which is getting towards the end of its useful life if it was brought to managements' attention that there was a risk of being criticised at the next assessment visit. It has also been helpful to spell out definitive responsibilities, although assessors have seemed singularly unwilling to establish if the level of authority afforded the individual is commensurate with the responsibility.

As far as disadvantages are concerned, there has been a significant increase in the amount of record keeping and instruction writing which had never before been found necessary and which has not brought with it any tangible benefits. Many matters which previously were committed quite adequately to memory have now to be recorded and the memory exercised in finding them if ever they are required. This has meant that personnel have been taken off such valuable items as process development to assist with this work.

Where complaints have been received, and have been accepted by the manufacturer as justified, too little time is devoted by the assessor(s) to establish if adequate action has been taken to prevent a recurrence.

However the most serious criticism to be made of external assessments is associated with the lack of support given to matters raised in internal audits - all too often these are ignored with the result that internal control is weakened 'Because the assessor has not followed them up'. This may be due to the continuing changes which take place in the personnel carrying out the assessments and the resulting lack of experience in the field in which they operate. They may be first class 'Systems people' but it is maintained that to be effective they must also have a thorough knowledge of an experience in the laboratory procedures themselves. The introduction of Technical Advisors is only a partial solution as they are only involved in a limited number of the assessments and the three or four hours each year in which they have to make assessments of the laboratory operation does not inspire confidence.

The real danger is the implication that because a plant has been assessed in this way, then the status of the laboratory is elevated. Clearly insufficient time is available for this purpose. The earlier versions of the QAS required manufacturers to relate their test results to a reference laboratory. This was an excellent concept which was in keeping with the need for traceability, but was later deleted. At the present time no independent sampling and testing of the products are called for.

## 5 Conclusions

It is estimated that external certification of the U.K. cement industry's testing laboratories has over the past five years cost at least M£1. The benefits achieved over this period are not considered to be consistent with this level of expenditure.

The assessment of the complexities of a modern cement plant to the requirements of EN 29002 demands a very high level of expertise in the assessor together with a not inconsiderable time in which to make a valid assessment. The writer's experience over the past five years suggests that this situation does not prevail, and that it would be more appropriate to the industry and those who use its products to adopt an assessment process based essentially on independent product sampling and testing and with a much reduced level of process assessment. Independent laboratory assessment should be separated from the general assessment and dealt with by a special body such as NAMAS.

## 6 References


(1) CEN 29002 'Quality systems - Model for quality assurance in production and installation'.

(2) British Standards Institution - Standard 4550 Part 3 section 3.4.

(3) British Government White Paper 1982 - 'Standards, quality and international competitiveness'.

(4) British Standards Institution - Quality Assurance P.O. Box 375 Milton Keynes MK14 6LL U.K..

(5) British Standards Institution - 'Quality Assessment Schedule' 2420/47.

(6) Evans, D. (1987) Independent quality assurance for the cement industry in Concrete January 18-20.

# 29 ISMES EXPERIENCE IN QUALITY ASSURANCE FOR TESTING

G.F. CAMPONUOVO
ISMES SpA, Bergamo, Italy

Abstract
ISMES activities deal with applied research, environmental and geological engineering, site investigation engineering, civil engineering, mechanical engineering, design and work management, software engineering and design, development and implementation of testing equipment, instrumentation and systems.

The paper presents ISMES' Quality System with particular emphasis on the different test activities: technological, structural and environmental; geotechnical, geomechanical and geophysical.

Beside laboratory tests, in-situ and off-shore activities and site laboratories are also considered ; details are given on both the general approach and philosophy, practical solutions and QA tools for a simple but effective Quality and Quality Assurance management of testing activities according to ISMES experience.

## 1 The ISMES Quality System

ISMES began systematically applying Quality Assurance (QA) methodology in 1982, in the context of nuclear and aeronautic sector tests, in which QA represented a contractual obligation.

The present ISMES' Quality Systems is equally applicable to the whole range of ISMES interventions, as well as to integrated QA/QC services, and allows all QA requirements to be met, whether referred to ISO 9000 series or any other standard of reference: the short-term objective is that of a systematic application to all ISMES activities.

In the context of ISMES operational practice, QA is understood in terms of enabling technicians to do things in an informed and well-documented way.

At the planning stage, need-analysis, the definition of quality criteria to be respected and the relative parameters and measurements, as well as the evaluation of skills and resources to be utilized, are among the aspects that technicians must decide upon: the QA structure's function is that of enabling them to carry out truly valid evaluations effectively and without loss of time.

At the operating stage, QA checks are carried out by actual colleagues, fully informed as to outline conditions and therefore in a position to dialogue concretely with the technician in charge without wasting time.

The company structure set aside for QA management is in fact made up of a central Unit, equipped with a secretary and operational on a full-time basis, and a QA Personnel group, responsible to the single Divisions or Laboratories and normally operational within the relative Unit, being available according to the needs of the order concerned.

Apart from having suitable QA training (analogous to that expected of personnel involved in audit activities), the person in charge of QA should be a good technician at a level appropriate to the functions (types of checks) to be undertaken.

With regard to Quality System documentation, given the considerable diversification of intervention types, standard QA Programs (QAP), specific for the main intervention sectors, serve to amplify the Company's General QA Manual (in which the general basic elements are explained).

The documentation of these standard sector QAPs is directly prepared by the technical officer responsible for the sector activities, in order to be effective: the structure follows the classic framework of a basic document (Manual) plus a set of generally applicable design or technical (operating) procedures, and, if appropriate, specific management procedures to complement the general ones of ISMES.

This documentation has been set out as generally valid basic documentation: personalization in terms of the needs of a specific order is determined on the basis of contractual specifications by means of an order document, in which reference is made to suitable procedures, plus other possible order procedures.

Whenever the needs of a particular order are not adequately met with by a standard sector QAP, and/or with respect to articulated and complex orders, a specific order QAP is created.

## 2 QA Operational tools

The operational instruments usually utilized to guarantee a simple but efficient handling of Quality-related problematics are:

**2.1 Check lists**
They represent the basic work instrument related to all evaluations made at both design and test stages: they are generally used for technical aspects as well as those concerning QA.

**2.2 Test Modulistic**
This is essential during the operating stage in those cases where activity management is lacking an adequate computer aid.

Besides paying due attention to identifying objects, actors, skill-interfaces and other possible annotations relating to instructions or acquisitions, the modules are structured in such a way as to guide the operator, using suitable blank spaces to be filled in, in order to concentrate on the important phases.

**2.3 Test Notebook**
This is the operative dossier (log) in which the person in charge collects or notes all important information (including potentially important information) and the operative documentation related to managing the internal interfaces during the preparation and execution phases of tests/activities, including recording, elaborating and accepting the results.

The Notebook, which is handled as an internal QA document, facilitates, as an instrument, the complete auditability of tests/activities, even without monitoring the process itself: the evidence supplied by its correct compilation guarantees, above all for the technician in charge, the way in which the tests/activities are (have been) carried out.

**2.4 Monitoring Dossier**
This is analogous to the Test Notebook, though related to the duties of the person in charge of QA: it guarantees the auditability of the checks carried out (inspections, monitoring, verification, QA analysis, possible corrective action) and the relative results.

**2.5 Computer assisted QA**
Procedure computerization, for both applied research and managing activities, whenever possible, is the most efficient way for operational realities to become "QA aligned". Of course the ideas as to the best way of operating and the details of the conceptual scheme to be translated into computer terms must be clear before, and the software employed should be of a sufficiently high quality: however the different operational phases as well as the quality system review are undoubtedly simplified.

Placing an emphasis on a high level of computerization in the different phases of the process is then fundamental to the various forms of ISMES intervention, and QA/QC methodologies have been applied systematically in software engineering.

Besides using standard products available on the market (for example "Project/II" for the management of design activities) ISMES can elaborate software products in different application fields, also of considerable dimensions (more than 100.000 statements): for example FIESTA for FEM stress analysis, EDDIS and MIDAS for problems related to dams, ISA for signal analysis; or sub-systems as IDAS. Also for expert systems utilization and development ISMES experience has been adequately consolidated.

As concerns test activities, ISMES software can be referred to:

- The effective management of the different test phases: control of test equipment (vibrating tables or general loading systems), data acquisition and validation
- The elaboration of test data: possibility to manage big amounts of data and the relative problems of accessibility, protection and safety; effective data elaboration, ensuring traceability and re-execution of procedures; support to man-machine interaction.

## 3 Examples of QA solutions for testing

### 3.1 Standard laboratory tests

Such tests are characterized by the fact that they are carried out on the basis of detailed technical standards: no need exists for new engineering or design contents with respect to specifications: it is merely a question of correctly carrying out what has been already established/requested.

The references for quality management in this case are obviously the ISO Guide 25 and the EN 45001.

In ISMES, tests of this type are carried out both on company premises, in the context of laboratories dealing with geotechnics, geomechanics, material tests and generally all technological tests related to construction processes, as well as in construction site laboratories (see 4.4).

### 3.2 Structural tests

These tests are characterized by the fact that often the specifications are not exhaustively detailed: this means that before "carrying out" a test it is necessary to "plan" it, either completely or partially: type and positioning of stress and restraints; meaningful parameters and relative measures; type and positioning of measuring instruments; and the elaboration of results.

A program-related document is prepared for such tests, in which a full definition of design content is given, together with the important stages of the testing process, in order to allow the reporting of hold/notification points. If necessary, this document is submitted to the Client.

### 3.3 Results requiring interpretation

In certain instances, although data has been obtained by qualified personnel following procedures known to be reliable and repeatable, such data does not in itself constitute an objectively meaningful response to the basic technological or scientific question of the test: this data must be in some way interpreted.

This type of situation is typical for nearly all geophysical investigations and in many instances of qualifying component and system tests, as well as tests on "as built". Similarly, the results of ND examinations must be interpreted, in the sense that, so as to determine a product's acceptability on the basis of reference standards, a defect typology must be attributed to each indication.

Should it prove necessary, the stage of result interpretation is fundamentally important and suitable procedures should define:

- The specialist's characteristics (experience, qualifications, etc.)
- The elements for which the interpretation must be programmed
- The criteria and modality of interpretation
- Documentation and/or certification associated with the interpretation and relative handling.

### 3.4 On-site laboratories

So as to be able to guarantee the monitoring of the entire construction process on large construction sites, the most efficient solution is often the setting-up directly on-site, of specialized laboratories for qualifying materials and procedures, and for a coordinated management of the various activities and data.

ISMES's first experience in the field of on-site laboratories was gained on the construction sites of ENEL's nuclear power stations, where QA was a contractual obligation: the solutions found were considered satisfactory and maintained as an operative standard even on other sites, when the activities involved were not deemed to be within a formal QA context.

Today ISMES supplies the following as standard services:

- On-site laboratories for material tests (geotechnical, bituminous conglomerates, concrete and ferrous materials; and chemical-physical tests): the laboratories operative at present are those at the ENEL power stations at Montalto, Brindisi, Fiumesanto,

Tavazzano and Pietrafitta

- Laboratories specialized in the coordinated and computerized management of ND tests and checks related to mechanical assembly, with a data bank for the filing and the real-time elaboration of important data: the laboratories at present operative are those on the ENEL sites at Montalto, Brindisi and Fiumesanto
- Laboratories for geotechnical monitoring, covering the development and manufacturing of geotechnical instruments, and management of data from measurements.

**3.5 In situ research and tests**

Besides paying due attention to aspects like operator qualification, a detailed procedurization of different stages and a correct registration and documentation management, a fundamental aspect of general ISMES QA philosophy concerning in situ research and tests is that of aiming at a high level of automation at the executive stage, so as to limit the root causes of test uncertainty.

Test apparatus are normally equipped with automatic advancement and registration systems, together with software enabling an accurate diagnosis to be carried out at each stage of testing: this permits both immediate intervention with regard to possible corrective action, as well as a correct interpretation at a post-result analysis stage.

Whenever possible, independent, fully-equipped mobile units are made available, allowing even objectively complex intervention to be handled as standard. In the oral presentation the examples of geotechnical and off-shore applications will be given.

At the planning stage, besides details on operational criteria, all elements that could directly or indirectly influence result quality in terms of aspects specific to the site of intervention, are identified in the programming documents: such information is of particular importance in those cases where there is no possibility of utilizing automated solutions, in order to guarantee the qualified operator suitable means and optimum outline conditions to allow him to fully express his own professionality.

## 4 Concluding remarks

In the operational reality of the different types of ISMES intervention, the role of human factor proves very often to be determinant. Besides sophisticated computer aids, it is then fundamental that QA is actually considered by managers and engineers, both single persons and teams, as the most effective way for defining and ensuring situations and boundary conditions such that the concrete objectives agreed upon can be attained at a minimum cost and by effectively exploiting the professional skills.

# 30 QUALITY ASSURANCE IN A RESEARCH LABORATORY – BRE's EXPERIENCE

B.O. HALL and T.W. PAYNE
Building Research Establishment, Watford, UK

Abstract
The Building Research Establishment (BRE) is the principal organisation in the United Kingdom carrying out research in building and construction and the prevention and control of fire. BRE is part of the UK Government's Department of the Environment (DOE). Its main role is to advise DOE and other Government Departments on technical aspects of buildings and fire and on related subjects, such as some aspects of environmental protection. The advice given is supported by a broadly-based research programme. Its findings and accumulated expertise are communicated to the building industry and building owners through publications, seminars and exhibitions, and through BRE's large input to British, European and International Standards and to Approved Documents under the English and Welsh Building Regulations. The Establishment's wide range of specialist skills and technical facilities are also made available to the construction industry and its suppliers and clients through BRE Technical Consultancy.

This paper considers the quality management requirements for an organisation of the type of BRE. It compares the quality management systems in place in BRE with the requirements of EN 29000 and EN 45000, and outlines the procedures BRE uses for its own quality management purposes, and indicates how it is developing third party assessment of the systems.

## 1 Introduction

The Building Research Establishment (BRE) is the principal organisation in the United Kingdom carrying out research in buildings and construction and the prevention and control of fire. BRE is an Executive Agency of the UK Government's Department of the Environment (DOE). Its main role is to advise DOE and other Government Departments on technical aspects of buildings and fire and on related subjects, such as some aspects of environmental protection. It is also developing this role, to an increasing extent, for industry. In order to keep its advice up-to-date BRE undertakes a broadly-based research programme, commissioned by Government and private industry clients, which ranges from studies of the basic properties of construction materials to investigations of specific building problems. The findings and accumulated experience are communicated to the building industry and building owners through publications, seminars and exhibitions, and through BRE's large input to British, European and International Standards and to Approved Documents under the English and Welsh Building Regulations. The Establishment's wide range of specialist skills and technical facilities are also made available to the construction industry and its suppliers and clients through BRE Technical Consultancy. This includes advice to professionals, designers and contractors on ways of assessing the quality of whole buildings.

BRE has a total of about 700 staff, approximately half of them being scientifically or professionally qualified in one or more disciplines; for example as chemists, physicists, mathematicians, geologists, engineers, architects, builders, or economists. They are supported by technical staff in an equally wide range of specialisms, including computer scientists, instrument and electronics engineers, information scientists, design draughtsmen, photographers, and administrative staff, and by extensive workshop facilities. This spread of expertise enables BRE to tackle the diverse and often complex investigations needed to solve the problems encountered in the field of building and construction at large.

BRE has at its head a Management Board comprising the Chief Executive, Deputy Chief Executive, 4 Directors and Secretary. The organisation of the Building Research Establishment (BRE) as a whole is depicted in Chart 1. This indicates the integrated structural organisation of the composite laboratories within Groups at BRE with responsibilities delegated through members of Management Board, each covering a range of related topics. The Chart also indicates the direct reporting link of the BRE Quality Manager to the Chief Executive.

Each Group at BRE consists of a number of Divisions concerned with particular classes of activity. Each Division generally comprises staff of a similar discipline and contains operational sections. Each Group has a Group Quality Representative.

## 2 Requirements of EN 29001 and BRE quality systems

It is of the first importance that the work carried out by BRE should be of appropriate quality. BRE has therefore evolved quality management systems over many years with this object in mind. While not specifically aimed at meeting the requirements of EN 29001 (Quality Systems, Part 1 Specification for Design/Development, Production, Installation and Servicing) the elements of compliance with such a system are already there and, as indicated below, are already formally in place in some areas. Following is an indication of how the elements of this Standard might be used for developing formal quality systems for an organisation such as BRE. The sub-heading numbers refer to the main clause numbers of Section 4, Quality systems requirements, in the Standard.

4.1 Management responsibility. This section in the Standard requires the management to define its quality policy and the responsibilities of staff, requires the organisation to recruit and train suitably qualified staff, to appoint a quality manager and to review quality systems periodically. Elements of this requirement are in place in BRE. Responsibilities and training requirements are developed both through formalised annual staff reporting and career development review procedures. BRE has appointed a quality manager whose role is set out in more detail below.

4.2 Quality system (documentation. A considerable amount of documentation exists incorporating the present informal and formal quality systems. Those parts of BRE activities which are covered by third party accreditation are fully documented. A considerable programme of extension of the formal quality system is being carried out.

4.3 Contract (project) review. This section in the Standard requires that projects should be clearly defined. BRE is essentially a contract research organisation for the largest part of its programme. Considerable effort is put into discussing and documenting the project requirements between BRE and its clients, both in Government and in the private sector.

4.4 Design (methodology) control. The methodologies used for research are very wide and varied. While it has been necessary to formalise those for which third party accreditation has been carried out, in other areas methodologies are more difficult to control formally. However, BRE has structured itself such that line management generally comprises sections and divisions with one or a small number of disciplines to ensure close control of professional standards. Over and above this, BRE has a policy of open publication of primary research results in journals with external referees, and uses Visitors who oversee the methodologies and resources used by BRE and report to the Chief Executive on the extent to which these are appropriate and to the highest scientific and technological standards.

4.5 Document control. This requires a system to ensure that the documents required by section 4.2 above are up to date and available. The system has been formalised at BRE in respect of those areas for which third party accreditation has been obtained. An overall document control system is in place, and gradually documented procedures which have not been brought into the third party accreditation scheme are being brought into compliance with this system.

4.6 Purchasing control. This section requires that all staff recruited, product and facilities purchased, and EMC contractors appointed are to specified requirements. BRE has stringent recruitment and purchasing controls laid down both through line management and its Finance Division. While these have not yet been brought formally into the BRE Quality System, little work would be required to do so.

4.7 Purchaser supplied product. (Care for products provided by clients (eg materials for test)). Procedures for this are laid down in documentation which has not yet been brought formally into the BRE Quality System except in those areas which have received third party accreditation.

4.8 Product identification and traceability. This section requires that all stages in the work are documented so that it is possible to trace easily who did what, to what, and when. Other than in those areas for which third party accreditation has been obtained, procedures in this respect at BRE depend principally upon the professional integrity of the research staff in research divisions. There are structures of laboratory notes, internal reports, client reports etc which comprise this documentation, but the procedure notes defining these structures, and the work processes involved, require further development.

4.9 Process control. This section requires that all activities are planned and controlled. In particular:

a) that they have documented work and test procedures,
b) that progress on work is monitored and reported as it proceeds,
c) that there are documented approval procedures,
d) that there are criteria for work which has been written down.

The research projects documentation, the development of which has been outlined above, contain within them descriptions and dates for output as well as descriptions of methodology. Most contract agreements with clients include agreements about progress reporting. In general there are quarterly meetings with clients for long term projects with half year formal reports. Approvals procedures for all reports or papers leaving BRE are documented and traceable. In those areas where third party accreditation has been carried out the procedures are more formalised than elsewhere.

4.10 Inspection and testing of incoming samples, products, facilities and EMC contractors; also inspection during action and completion of project. Procedures exist for this, although they are not completely formalised in all areas.

4.11 Inspection, measurement and testing equipment. This section requires the following:

a) Identify equipment, measurements and accuracy required.
b) Calibrate equipment.
c) Document calibration procedures.
d) Maintain equipment.
e) Mark equipment with calibration status.
f) Maintain calibration records.
g) Keep measuring environment so that equipment can operate within calibrated limits.

This is an area in which BRE has obtained third party accreditation for several aspects of its work. The way in which this is done is set out in more detail below, together with an indication of our future intentions.

4.12 Inspection and test status. To keep a record of the inspection and test status of every project. The procedures outlined in the comments against section 4.8 above ensure that such a record is maintained, although the third party accredited domain is small at present.

4.13 Control non-conforming product (project) to prevent inadvertent use. The end product of work at BRE is generally published papers or reports. However BRE occasionally produces computer software or patented inventions. The approvals procedures within BRE are both formalised and stringent.

4.14 Corrective action. This section requires documentation to be established and procedures monitored to ensure appropriate corrective action. Other than in areas in which BRE has obtained third party accreditation, procedures in this area are relatively informal at present, although it is generally possible to trace such actions through correspondence files, project records etc.

4.15 Delivery of results. This section would require documents in the system to be established and maintained for monitoring this process. BRE output largely comprises papers in technical and scientific journals. However BRE also has large publishing and advisory organisations which have their own quality systems including those related to delivery of physical output such as books and reports. These systems would require considerable formalisation before they could be claimed to comply with EN 29001.

4.16 Quality records. The BRE Quality Manager has established a system for the preparation, control and maintenance of quality records. This is fully operative in those areas for which BRE has obtained third party accreditation, and is actively being developed in other areas.

4.17 Quality audits. This requires checks to be carried out on the effectiveness of the quality system. This is formally undertaken in those areas where BRE has obtained third party accreditation. However, in other areas the effectiveness of the system is checked by such mechanisms as feedback from clients of their satisfaction, market research in terms of publications and other services, prior review of journal articles, and the establishment of advisory groups such as the BRE Visitors mentioned above.

4.18 Training needs of staff. This section requires the training needs of staff to be identified and supplied. As already mentioned, BRE has a system of annual review and career development for this purpose. BRE also has in-house training, and makes use of external training procedures both for mid-career development and for career development for incoming staff.

4.19 Servicing. BRE has an extensive organisation providing technical services including calibration and other workshop facilities. Where servicing forms a part of a project/test method which has received third party accreditation the requirements and processes are documented. It is an aim of BRE to extend such documentation for suitable cases as resources permit.

4.20. Statistical techniques. This section requires statistical techniques to be used where appropriate within quality systems. Much of the work within BRE is of such a varied nature that statistical techniques are not easily applicable. However, where they can be used, they are used and documented.

## 3 BRE quality control programme

The quality management system adopted by BRE is based upon a documented framework which defines a common purpose and aims to produce a common standard of documentation, where this standard is compatible with the subject matter. The framework is of a tiered structure to provide for maximum flexibility whilst at the same time minimising the extent of amendment activity when a need for change arises. The system also eases the function of change control.

The BRE Quality Control Manual (QCM) heads the quality management framework (the top tier). Below this level, and supporting the QCM, is a set of documented General Procedures (GPs), each separately and uniquely referenced. An individual GP addresses a particular aspect of BRE activity which may have relevance across the organisation as a whole. Such GPs, which include administrative activity, are followed by staff as they apply to their work. A third tier of documentation comprise Specific Procedures (SPs). SPs cover the detailed procedures associated with a particular work activity and are thus specific to that task and grouped accordingly. The quality management documentation does not, generally, repeat procedures that are documented elsewhere. Wherever possible references to such documents are given and appropriate copies of the references made available to the staff operating the procedure.

The QCM sets out the overall quality policy of BRE and covers in general terms the management and other organisations of the BRE and main functional elements of the quality policy. The terms of reference of the Management Board and of other key quality management staff, and lists of GPs and SPs, are included in a set of appendices. It is available for examination to all bonafide enquirers who have an interest in the BRE quality policy and its quality management system. GPs and SPs are not so readily available since they may include procedures which are (commercially) confidential to BRE or to a client of BRE. Therefore these documents require Management Board approval before being disclosed to third parties, to ensure the confidentiality of clients of BRE is fully respected.

The QCM, together with its supporting GPs and SPs, defines the Quality Control Programme (QCP) for BRE. As work activities are brought within the regime of this quality management system they become part of the QCP. Areas working within the QCP are subject to regular quality audits in accordance with a forward programme. Other quality audits are conducted as may be thought to be necessary. In addition the quality management system is subjected to a formal annual review; elements of the system are reviewed at other times as is seen to be necessary, either as a result of improvements in techniques, or following an audit, or following some shortcomings or customer/client complaint.

## 4 Accreditation

It is a policy of BRE to seek appropriate third party accreditation for suitable activities working within the QCP. Currently accreditation has been awarded to BRE by NAMAS (National Measurement Accreditation Service, part of the UK Government's Department of Trade and Industry (DTI)), for a range of tests on fresh and hardened concrete. The range of tests in this area is to be substantially extended. BRE are also working towards extending the schedule of NAMAS accreditation in several diverse areas of activity, such as biodeterioration of timber, window testing etc.

The award of NAMAS accreditation, which must of necessity include the relevant documentation (ie the quality management system framework) and quality management organisation, also indicates that BRE meet the requirements of ISO Guide 25, EN 45001, and BS 5781: 'Measurement and calibration systems', and may also be considered as meeting those requirements concerned with the adequacy of calibration or testing contained in the ISO 9000, EN 29000 and BS 5750 series of standards relating to quality assurance in manufacture and similar activities. Furthermore BRE NAMAS accredited areas may be considered competent to provide accredited calibration and testing services (dependent upon the schedule of accreditation) covered by the UK National Accreditation Council for Certification Bodies, NACCB. The comparisons given here derive their validity from the NAMAS Accreditation standard M10 which is compatible with the standards referred to above to the extent indicated. As the scope of its QPC widens BRE will seek formal third party accreditation to the appropriate parts of relevant EN series and other relevant standards for the quality management system as a whole.

## 5 The formal quality management system in action

The formal quality management described in Sections 3 and 4 above has been operative at BRE for about two years. Substantial changes in working practises had to be made in order to comply with the more formal requirements of the quality management system framework. These changes have led to a more orderly mode of operation and easy audit traceability through the test procedures and associated documents (eg specimen reception; identification marking, inspection, handling and storage; work sheets; job files; test result sheets; test reports etc etc). Working to documented procedures has also helped to highlight where beneficial changes can be made to improve the quality, efficiency and effectiveness of the operation, leading also to lower costs in some cases. There is a belief among some staff that the quality management system has generated a lot of 'paperwork'. Certainly this has been so initially when elements of the system were established. However, it has not been always true in the day-to-day application of documented procedures. What may seem to be additional day-by-day paperwork is in reality about the same amount, but which is produced systematically, formally identified (labelled) and property filed. Thus the paperwork produced can be seen collectively and thereby may be perceived to be greater in quantity.

Relevant SPs lay down set procedures for an activity and identify the (competent and adequately trained) staff who alone are permitted to undertake the work. The systematic recording of test results, in an indelible form, and the checking procedures which follow, coupled with the use of measuring and test facilities of known characteristics and associated uncertainties, help to reduce the incidence of error and assist in assessing overall uncertainties related to the activity.

GPs and SPs are written by sponsors, nominated by line management, in the area concerned where the detail, including technical detail, is fully understood. Those sponsors are also responsible for the review of their Procedures. Drafting quality management documentation demands discipline and is a time consuming and therefore costly process; the range of SPs required to cover a particular technical area can be extensive. The operation imposes a need for job analysis and for writing job specifications. It can therefore serve a useful purpose as an opportunity and exercise to carry out a formal systems analysis of the area under consideration with additional opportunities to review the system regularly (at review time).

Security procedures assist in ensuring confidentiality of test results, test reports and/or other matters. They also ensure that equipment is not tampered with, particularly during a test sequence. All equipment used within the QCP is uniquely identified; its location and the person responsible for its care is known and recorded.

Reference standards and/or transfer standards are treated in the same way and used only for calibration or check testing purposes, as may be relevant. These procedures ensure the safe keeping of such standards and the maintenance of their integrity.

Equipment calibration has always been a feature of the BRE activity, its associated costs being absorbed in the work. However, as part of the QCP a BRE Measurement System is under development. Within this system equipment calibration has become a systematised process, fully compatible with the UK National Measurement System to which traceability is established. Full records of equipment maintenance, calibration and associated costs are kept. From these records a history of in-service performance and running costs can be built up. Re-calibration periods are defined in relevant SPs and recorded in the measurement system computer based data-base; thus useful early warnings of calibration expiry dates can be given. The historical performance record helps users to redefine re-calibration periods as may be necessary. Such reviews may lead to extensions of periods of re-calibrations (and hence reduced costs) in cases of proven stability or, conversely, to more frequent re-calibrations for equipment which proves to be less stable during service. Analyses of the equipment history will also assist in developing performance criteria for equipment at the specification and selection stages in the purchasing procedures. This will help to ensure the purchase of equipment fit for the purpose, whilst at the same time securing value for money.

An example of current review activity is concerned with transferring responsibility from the BRE Quality Section to the BRE Library for the issue and registration of the quality management system documentation, including standards, and for the issuing procedures concerned with amendments for updating the quality documentation. This will place the activity in an area more suited to the purpose.

## 6 Conclusion

BRE already has in place most of the elements required for a quality system which would comply with ISO 9001. It is apparent that the standard could be applied to the research process as carried out by BRE. However, the effort required to formalise the system in this way would be considerable. Nevetheless BRE Management have taken the view that, as resources permit, BRE should move in this direction, and a BRE quality management system has been established with this eventual aim in view, and thus partial accreditation obtained in some key areas. The BRE quality management system imposes a formal discipline on all staff relating to the manner in which they carry out their work. Although initially the quality procedures are usually seen by staff to be irksome, they tend to become adopted as a way of working life as experience is gained. These features, in turn, instil a sense of confidence in the staff, and thereby in our customers. Effective quality management can also be expected to lead to better customer service and, hopefully, to greater customer satisfaction.

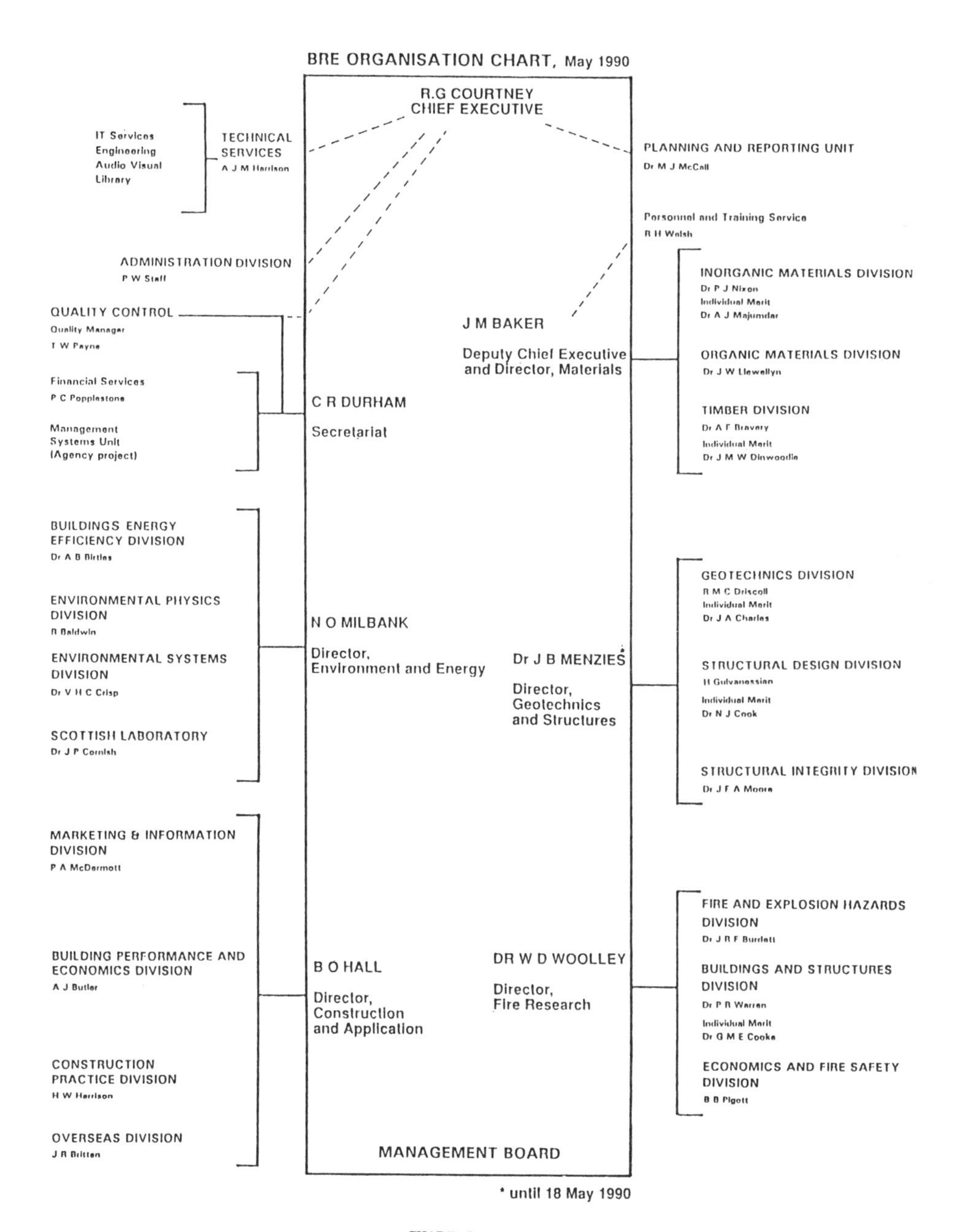
BRE ORGANISATION CHART, May 1990
R.G COURTNEY
CHIEF EXECUTIVE
IT Services
Engineering
Audio Visual
Library
TECHNICAL SERVICES
A J M Harrison
PLANNING AND REPORTING UNIT
Dr M J McCall
Personnel and Training Service
R H Walsh
ADMINISTRATION DIVISION
P W Staff
QUALITY CONTROL
Quality Manager
T W Payne
Financial Services
P C Popplestone
Management
Systems Unit
(Agency project)
C R DURHAM
Secretariat
J M BAKER
Deputy Chief Executive and Director, Materials
INORGANIC MATERIALS DIVISION
Dr P J Nixon
Individual Merit
Dr A J Majumdar
ORGANIC MATERIALS DIVISION
Dr J W Llewellyn
TIMBER DIVISION
Dr A F Bravery
Individual Merit
Dr J M W Dinwoodie
BUILDINGS ENERGY EFFICIENCY DIVISION
Dr A B Birtles
ENVIRONMENTAL PHYSICS DIVISION
R Baldwin
ENVIRONMENTAL SYSTEMS DIVISION
Dr V H C Crisp
SCOTTISH LABORATORY
Dr J P Cornish
N O MILBANK
Director, Environment and Energy
Dr J B MENZIES*
Director, Geotechnics and Structures
GEOTECHNICS DIVISION
R M C Driscoll
Individual Merit
Dr J A Charles
STRUCTURAL DESIGN DIVISION
H Gulvanessian
Individual Merit
Dr N J Cook
STRUCTURAL INTEGRITY DIVISION
Dr J F A Moore
MARKETING & INFORMATION DIVISION
P A McDermott
BUILDING PERFORMANCE AND ECONOMICS DIVISION
A J Butler
CONSTRUCTION PRACTICE DIVISION
H W Harrison
OVERSEAS DIVISION
J R Britten
B O HALL
Director, Construction and Application
DR W D WOOLLEY
Director, Fire Research
FIRE AND EXPLOSION HAZARDS DIVISION
Dr J R F Burdett
BUILDINGS AND STRUCTURES DIVISION
Dr P R Warren
Individual Merit
Dr G M E Cooke
ECONOMICS AND FIRE SAFETY DIVISION
B B Pigott
MANAGEMENT BOARD
* until 18 May 1990

CHART 1

# 31 ORGANISATION QUALITÉ DANS UN LABORATOIRE D'ESSAIS: LE CEBTP
## (Quality management in a testing laboratory: CEBTP)

M. SEGUIN
Centre Expérimental de Recherches et d'Études du Bâtiment et des Travaux Publics, Saint-Rémy lès Chevreuse, France

Aborder le problème de la qualité des prestations d'essais dans un laboratoire est un exercice difficile car il comporte des aspects multiples, complexes, se situant aussi bien au niveau des principes généraux qu'à celui des applications quotidiennes, aussi bien au niveau des questions de matériels que des problèmes humains. La qualité en effet doit être recherchée le plus possible en amont du résultat.

Après quelques réflexions sur l'apport du Réseau National d'Essais (RNE) et la présentation rapide de l'organisation qualité au C.E.B.T.P., j'évoquerai un certain nombre de points qui ont paru se dégager au cours du cheminement de la démarche qualité engagée au C.E.B.T.P.

Nul ne peut mettre en doute le fait que la qualité, même imparfaite, existait avant l'accréditation R.N.E.

L'engagement pris de la participation au réseau a conduit, d'abord, à une réflexion et une mobilisation puis à une formalisation.

Ce processus a été guidé, d'une certaine manière, par les exigences du RNE dont on pourrait dire qu'elles ont été l'"outil de drainage de la qualité". La rédaction du Manuel Qualité, par exemple, avec toutes les implications qu'il comporte, la remise en cause de certains éléments d'organisation, etc ont pu être "encadrées" par l'existence de cet outil.

En ce qui concerne l'organisation qualité au C.E.B.T.P., et sans entrer dans le détail, précisons que le responsable est le Directeur Technique et de la Qualité, chargé de la gestion de la qualité, assisté d'un délégué dont le rôle est de s'assurer de l'application des règles définies dans ce domaine et contenues, en particulier dans le Manuel Qualité.

Chacune des autres directions participe à l'obtention de la Qualité.

Dans chacune des unités techniques, il existe un Correspondant Qualité.

Par ailleurs, une section métrologie indépendante des services opérationnels est chargée des contrôles nécessaires.

## QUELQUES ASPECTS DE LA DEMARCHE QUALITE

### 1 MANUEL ET PLAN QUALITE

Le "Regard Qualité" au quotidien qui couvre l'ensemble des taches ou activités relatives à plusieurs types de prestations, est souvent, dans un premier temps, difficile à orienter : on ne sait pas très bien par où commencer. La continuité dans l'observation, l'échange des informations, l'analyse, la connaissance des contraintes réelles conduisent à une meilleure appréciation des situations et de leurs intéractions réciproques.

Par ailleurs, la prise en charge au niveau de chaque unité de son organisation propre en conformité avec le Manuel Qualité constitue une remise en cause extrèmement positive qui se formalise au moment de la rédaction du Plan Qualité. Ainsi, en retour, peut-on apprécier la validité du Manuel dans son application pratique.

## 2 REFERENTIEL ET AUDIT

Pour être efficace dans le domaine Qualité, il faut y être ordonné, disposer de certaines bases, de certaines références : c'est la technique du "référentiel" utilisé lors des audits.

Ces deux outils, "référentiel" et "audit" sont fondamentaux en particulier pour mesurer les dérives. Leurs objectifs peuvent, bien entendu, dépasser largement la démarche Qualité elle-même.

Ainsi ont-ils été employés au C.E.B.T.P. pour l'amélioration de la connaissance des diverses unités techniques.

Un référentiel comportant 8 chapitres et au total une centaine de questions a été mis au point ainsi que le support des réponses correspondantes.

Le repérage des questions est conçu pour faciliter un éventuel dépouillement informatique.

Ce document sert de base à un audit interne réalisé au sein de chaque unité en accord ou sur la demande même des responsables.

L'expérience acquise en ce domaine depuis sa mise en oeuvre permet de tirer certains enseignements :

Le dialogue et les échanges réalisés lors de l'audit sont très enrichissants.
L'analyse, même sommaire des réponses permet parfois de déceler des anomalies puis de proposer des corrections.
L'acceptation de cette procédure par un certain nombre de "pionniers" a tendance à faire "tâche d'huile" et permet d'élargir le champ d'application tout en restant dans le domaine du volontariat.
La diffusion de ces outils de la Qualité permet aux responsables, à différents niveaux, d'en apprécier le bien-fondé.

## 3 FORMATION - INFORMATION

La décision de candidature à l'Accréditaion entraîne un examen de la situation du laboratoire vis-à-vis des exigences du RNE. Parmi celles-ci, la recherche des textes de références, des normes, la rédaction de modes opératoires et de procédures d'essais entraîne une réflexion puis une action sur le plan des dispositions techniques à prendre pour s'assurer de la conformité lors de l'exécution des essais.

Les difficultés rencontrées, les solutions proposées, leurs limites, peuvent faire l'objet d'informations intéressant d'autres unités techniques d'où l'idée d'une diffusion périodique de notes techniques reprenant les éléments d'intérêt général, enrichis éventuellement par des compléments permettant de couvrir plus largement les questions soulevées.

Ainsi, le C.E.B.T.P. diffuse-t-il depuis plusieurs années des notes internes baptisées I.T.Q. (Informations Techniques Qualité) parmi lesquelles citons quelques thèmes :

Résistance à la compression.
La Qualité dans le déroulement de l'activité d'un laboratoire,
La Qualité et le matériel.

## 4 QUALITE ET MOTIVATION

Le Chemin de la Qualité au quotidien est long. Il demande persévérance et continuité.

Il ne peut être que volontariste et réclame, de la part de ceux qui tracent la voie, un effort constant pour convaincre. On pourrait comparer cette démarche à celle engagée pour l'introduction de la micro-informatique dans l'entreprise : l'efficacité devient réelle à partir du moment où la motivation de l'utilisateur est suffisante. Pour l'obtenir, un certain temps est nécessaire. On ne peut aller trop vite. Pour jouer du piano, il faut apprendre le solfège. Mais si cette première étape réussit, il y a rarement un retour en arrière.

## 5 LE JUSTE NECESSAIRE

Reprenons la définition de la Qualité donnée dans la norme Pr X 50-120, à savoir :

> "Ensemble des propriétés et caractéristiques d'un produit ou d'un service qui lui confèrent l'aptitude à satisfaire les besoins exprimés ou implicites".

Pour faire la preuve qu'un service, une prestation d'essais, est apte à satisfaire les besoins en question, il faut démontrer que le résultat a été obtenu conformément au processus prévu. D'où une nécessaire formalisation permettant, en amont du résultat obtenu, de trouver la trace des différentes étapes ayant conduit à ce résultat.

Ainsi apparaissent divers états, bordereaux, bons, etc, qui alourdisssent le fonctionnement. Pour en assurer le suivi, il faut créer des points de passage obligé où le contrôle peut être assuré.

Les dispositions systématiques ainsi adoptées doivent être bien comprises car elles limitent, parfois, l'initiative, et risquent dans certains cas d'allonger les délais et par conséquent, d'entraîner des difficultés avec le client.

Il s'agit donc de trouver l'équilibre souhaitable, le juste nécessaire sans perfectionnisme.

# OUVERTURES EXTERIEURES

## 1 LA REFERENCE "RNE" ET LA CLIENTELE

Les animateurs Qualité sont souvent confrontés à la question de savoir ce que cela rapporte à l'entreprise.

Dans le cadre des prestations d'essais, l'image de marque d'un laboratoire repose à la fois sur la rigueur et l'honnêteté des intervenants ainsi que sur l'expérience acquise. Par ailleurs, l'Accréditation RNE porte exclusivement sur une liste d'essais parfaitement définie au moyen de textes de référence (en général des normes). Enfin, la relation entre le Laboratoire et son Client ne se limite pas, en règle générale, à ces seuls essais. Les motivations conduisant au choix, par le Client, de tel ou tel laboratoire d'essais comportent plusieurs aspects. Par exemple, la confiance réciproque préexistante, la rapidité dans l'exécution et, bien entendu, les prix pratiqués. Ainsi, si le seul souci du Client est de faire réaliser des essais obligatoires, réglementaires, conformes aux exigences d'un cahier des charges auquel il est soumis, son choix se portera sur le laboratoire le moins disant, qu'il soit RNE ou non.

## 2 VERS LE CLIENT

La mise en oeuvre des exigences du RNE implique l'ensemble des acteurs liés à l'exécution des essais. La formation à la Qualité aux différents niveaux de responsabilité conduit à une réflexion individuelle et collective sur les méthodes et moyens pour y parvenir. De cette situation résulte, en général une ouverture plus large sur le plan professionnel.

# 32 COMMISSION OF THE TECHNICAL TESTING AND RESEARCH INSTITUTE FOR BUILDING (TZUS) IN THE CZECH BUILDING INDUSTRY

J. VESELÝ
TZUS, Prague, Czechoslovakia

Abstract
The Institute originated in 1953 as the Testing and Control Institute for Building. At present it is organized in the form of individual regional divisions with head office in Prague. Division 1 for Prague and Central Bohemia ensures preferentially the principal task of the Institute - to influence effectively the improvement of quality in the building industry. For this purpose it is provided with modern testing, measuring and computer technology. Apart from that the Division solves research problems, participates in the drafting of Czechoslovak standards, ensures experiments in the building industry, affords technical assistance to construction practice and participates in quality control for export. The knowledge attained in these activities is projected into the practice of Czechoslovak building industry, the work of design institutions, the programmes of higher education institutions and the Czechoslovak standards. Practical experience /see examples/ has proved the irreplaceable role of experiment in quality control.
Keywords: organization, purpose, equipment, application of experience.

## 1 Introduction

The new ways of Czechoslovak economy and the introduction of market mechanism and accentuated environmental protection will enhance the role of the Technical Testing and Research Institute for Building in Prague /TZUS/ in the Czechoslovak building industry the Institute has been playing for the past 35 years. The principal task of the Institute is quality control in all phases of construction terminating with the use of the work. In its performance the Institute links up with more than 60 years' tradition of Czechoslovak material testing. Objective testing of building materials and components and their joints as well as of whole structures both in modern laboratories and on construction sites by foremost specialists ensures reliably the safety, economy and long life of every construction work. The regional divisions of the Institute cover the whole territory of the Czech Republic. In its laboratories the Institute covers 24 specific branches and participates in the programme of comprehensive quality control and in the development of science and technology, affords technical assistance to construction practice, affords expertises and consultations, carries out testing and quality control of products and structures, such as bridges, tall buildings, TV towers, foundations of turbogenerator sets as well as the structures of nuclear power plants. The Division 1 in Prague employs 70 specialists and is provided with computers, mobile measuring laboratories for static and dynamic as well as thermal tests on construction sites and with automatic loading test control systems.

The Division is provided with a destructive testing bed, pulsators, a type UPS 1000 large capacity testing press of FRG make, generating the maximum force of 10 MN, modern measuring and assessment instrument sets of Hottinger Baldwin Messtechnik, Brüel Kjaer, Honeywell, a defectoscopic centre with the possibility of application of radiographic methods, magnetic and hardness testing, ultrasonic pulse methods, etc. incl. the respective computers. All that creates the good prerequisites for the high quality of work.

## 2 Experiment and Quality Control

### 2.1 Investigations of R.C. Structures

The development of testing methods in the investigations of R.C. structures should proceed towards speedy, simple and inexpensive methods which would

be applicable in and extend effective statistical processing of test results,
ascertain the properties of structures speedily without their destruction,
not be connected with high labour requirements.

The importance of these methods comes to the fore particularly in the field of the so-called technical diagnostics determining the technical state of the building, i.e. its ability to perform the required functions in the given conditions of use. After the determination /diagnosis/ of this state the causes of its origin /genesis/ are ascertained together with the prerequisites for further development /prognosis/.

### 2.1.1 Loading Tests

The Institute carries out classical static and dynamic loading tests, considering the determination of dynamic characteristics of building structure as a perspective method of investigation of their state. Generally speaking this method involves the determination of the natural frequency of the structure, its respective vibration modes and its damping. The excitation of vibrations of building structures can be performed by the sudden relief of the structure deformed by static load, by a sudden loading of the structure by a pulse /jet engine/ or by the excitation by stationary force or moment mechanical exciters with harmonically variable amplitudes.

Rotary exciters with directed force vector appear most advantageous. The Institute has six such vibration exciters of 600 kg and two of 50 kg in weight. In some cases this method is used also for the determination of the extent of damage or ageing of the structure. In the evaluation of results it is necessary to realize that a 10% variation of flexural rigidity manifests itself by a 5% variation of natural frequency.

### 2.1.2 Non-Destructive Testing

Non-destructive testing methods generally enable the testing of materials without their destruction or with the destruction which does not impair its functional characteristics. The Institute carries out current non-destructive tests of concrete, ascertainment of reinforcement position, radiological tests and tests of concrete structures by local damage methods. The extent of reinforcement corrosion is determined by the electrode potential method, the determination of pH of concrete is carried out by colorimetric indicators, the determination of the alkalinity of concrete by potentiometric method, chloride determination by potentiometric, semi-quantitative and other methods.

### 2.2 Knowledge Obtained from the Application of a Testing Machine with Defined Boundary Conditions to Compressed Component tests

Reliable comparison of accurate theoretical analysis with the actual behaviour of components in real experiments and the influence of variability of the conditions of production on its reliability are particularly important for vertical load-bearing structures, as their failure may result in the collapse of the load-bearing system of buildings with all gruesome consequences. For these tests the Institute uses a type UPS 1000 V universal four column testing press of the firm RaK-MFL from the Federal Republic of Germany. The crossbeam feed is infinitely variable with hydraulic locking in any position on smooth columns. The forces are measured directly by a force measuring sensor. The hinged support is ensured by hydrostatically mounted pressure plates with negligible friction even when maximum forces have been applied /the rotation of the pressure plate in all directions at $3^{o}$/. The regular distribution of the maximum force applied to wall elements is ensured by a rigid trussed girder of box cross section. The machine is programme-controlled by microprocessors, enabling e.g. the selection of the loading process in accordance with the force increment in time or deformation increment in time. From the control panel it is possible to operate direct computer inputs and outputs.

In the course of the relatively short application period of this testing equipment a number of interesting results have been obtained from the tests of both columns and wall elements. Perfect and definite assurance of boundary conditions /hinged mounting/ during the whole duration of the tests made it possible to discover a number of imperfections due to production and their influence on the bearing capacity of tested elements, such as the influence of the different strength of concrete along the cross section depth in the case of columns.

Test results have confirmed also the importance of accurate analysis of more complex load-bearing elements, such as wall elements with openings, which can bring about, apart from an improvement of the reliability of the whole structure, also better use of the material. In any case the installed equipment has proved beneficial both for the improvement of the quality of production and for the development of scientific knowledge not only in the field of concrete structures, but also - and particularly in the case of slender structures of other materials.

For instance, an analysis of test results of bar-shaped members has clearly proved that the locking of the hinge does not guarantee the condition of the defined static behaviour because the direction and point of force application which pass through the centre of the functional hinge are not known.In the very case of slender bars the elimination of the hinge changes the buckling length in such a way as influences their load-bearing capacity. The differences vary about 3% in the case of e.g. 300/300 mm columns 3 000 mm high, and between 15% and 26% in the case of slender walls 150 mm thick and 2 500 mm high in case of single side or double-side constraint as compared with the hinged support of the bar at the foot and in the head.

The perfect function of the hinge is of great importance in the tests of elements of such non-homogeneous materials as concrete. It is well known that in the manufacture of bar-shaped elements in horizontal position the working of concrete, dependent on the selected technology of production, results in non-homogeneity of concrete along the cross section depths. The concrete nearer the base of the mould has a higher strength than the concrete near the upper face, as a rule. These differences in concrete quality amount to one class or more, while the mean concrete strength in the cross section satisfies the design requirements. Irregular stress distribution in the cross section then results in an eccentric effect of the test sample on the test mechanism with all unfavourable consequences, such as the flexure of the column and the origin of horizontal forces affecting the piston of hydraulic cylinders. In such cases the failure usually occurs in those places in which the deformation ability of concrete has been exhausted. It is possible to state that in the tests of columns perfectly hinged at the foot and at the head the failure always occurred in the working /upper/ face of the column cast in horizontal position.

2.3 Cellular Concrete and Its Dynamic Characteristics
The specialists of the Institute are involved also in the research of new materials, among others also cellular concrete. Cellular concrete is favoured particularly in outside walls because of its very good thermal properties variability of assortment, easy workability, stability of volume and favourable relation between mechanical strenght and volume weight /density/. The static parameters of the material and of the components made of it are relatively well known, but the influence of dynamic parameters necessitated speedy elucidation.

It is coming to light with increasing clarity that these parameters play a decisive role in the applicability and life expectancy of cellular concrete components. In research, therefore, external dynamic forces /dynamic effects of wind, the effect of lifts, machines, genuine and technical seismicity, climatic factors/ as well as internal dynamic forces, particularly the natural frequency of transverse undamped vibrations, the dynamic modulus of elasticity, the Poisson's dynamic coefficient and the logarithmic damping decrement were monitored.

The specialists of the Institute thus contributed to the determination of the recoomended indicators for the assessment of the dynamic characteristics of cellular concrete, stating general relations for their computation incl. informative values obtained by measurements. These achievements concerning the behaviour of cellular concrete as a material in precast components under the effect of forces of variable character thus increased the knowledge of designers, builders and users about cellular concrete, cellular concrete building components, products and structures.

## 3 Conclusion

The brief survey of the mission and activities of the Technical Testing and Research Institute for Building, Prague, reveals the irreplaceable role played by testing in the Czechoslovak building industry. It is one of the ways to the constant improvement of the quality of Czechoslovak construction and the achievement of competitiveness not only on the home market, but also on European and world markets. For this purpose all modern methods available in the field must be applied.

# 33 AN EXPERIENCE IN COMPILING A LABORATORY QUALITY MANUAL

S. AVUDI
The Standards Institution of Israel, Tel Aviv, Israel

ABSTRACT

The laboratory of building materials which previously formed a part of a general construction laboratory has been running for 15 years as an officially recognised testing, certification, and approval laboratory of construction materials. It includes two sections dealing with structural and finishing materials (coatings, waterproofing, tiling, etc.). The Laboratory is staffed by 24 engineers and technicians.

The paper tells the story of how the decision to write a Q.M. was arrived at, conveyed to the staff and the steps taken to compile the manual with the cooperation of the staff.
Keywords Quality assurance, Quality manual, Building Materials.

## 1 BACKGROUND

This is the story of the compilation of a Quality Manual for the Building Materials Laboratory.

This Laboratory is one of five construction laboratories in the Building Division and one of an overall eleven laboratories of the Israel Standards Institute. It has been functioning as a statutory construction laboratory for over 60 years and has its established procedures, test methods and personnel with long service records.

When I retired from the position of Director of the Building Division, after 36 years' service in the Institute, the successor to this office asked me if I would help in drawing up a Quality Manual for the above laboratory. As this was one of the tasks that I had longed to do but somehow did not find the time for, I willingly accepted.

The reason that this laboratory was chosen to start with compiling a Quality Manual is that it is a compact laboratory, staffed by 24 persons doing mostly in-house testing, while most of the sister-laboratories performed on-site tests and boasted a much larger staff.

## 2 WINNING THE PERSONNEL

Usually when you start such a project, you come against diverse and complex reactions of the staff. These range from apprehension to

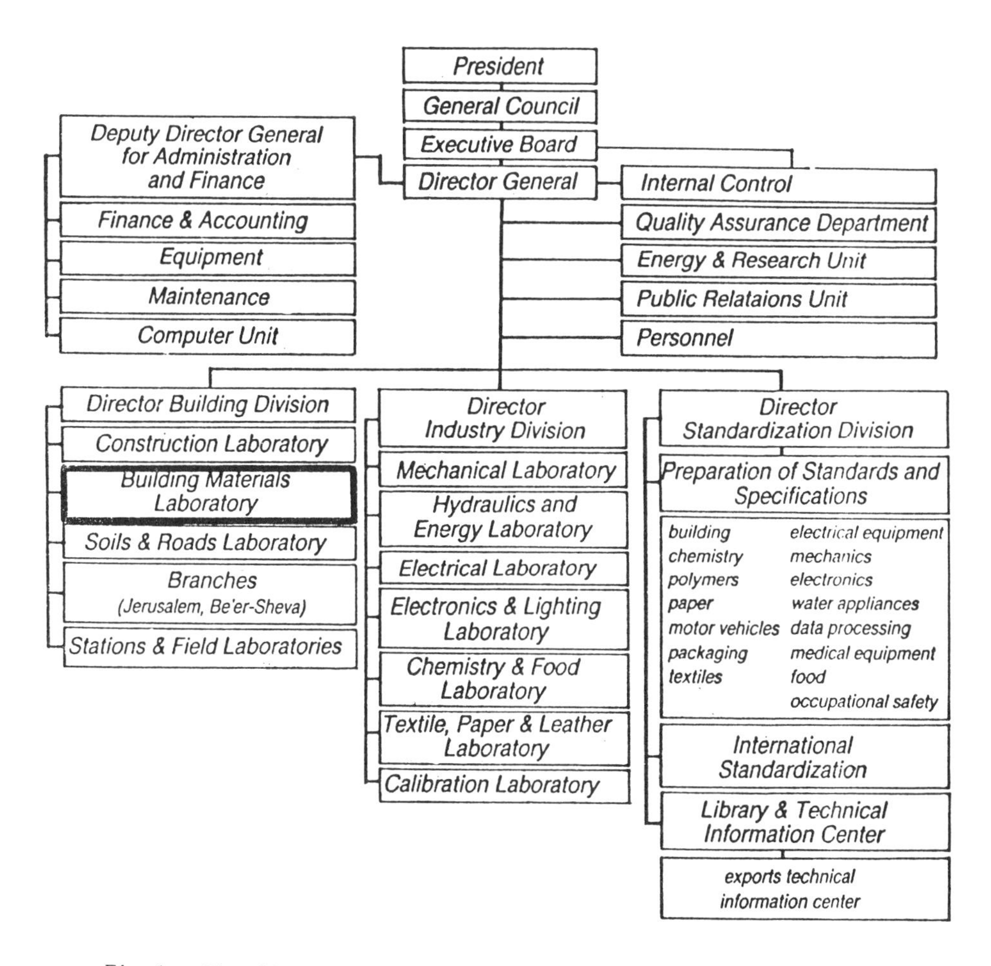


Fig.1. The Standards Institution of Israel Organization Chart

anxiety about how it would affect the person concerned. So we started enlisting enthusiasts. There are two levers that can enhance enthusiasm: (1) Pride - to be a member of a laboratory that can claim a Quality Manual, especially as a pioneer laboratory in the establishment; (2) that you are a participant or a co-author of the document.

We scheduled a meeting attended by the Director of the Building Division and the Head of the Building Materials Laboratory who was the dominant person on the list of enthusiasts to be enlisted. We convinced him that it was actually his project and that I only served as a coordinator.

Secondly, we set a meeting with the two Section Heads of the Laboratory and the Head of the Laboratory. Here the group dynamics worked to achieve proclaimed cooperation.

## 3 ORGANIZATIONAL CHART

Although there is an organizational chart for the Laboratory (see Fig. 2), the two sections are only generally defined and the distribution of tasks between them is not always consequential. We asked both Section Heads and the Head of the Laboratory each to fill a questionnaire which included definition of authority and responsibility and the listing of tasks and tests which fell under their jurisdiction. As was expected, we did not receive three similar answers to our questionnaire. In an attempt to find out how the discrepancy was caused, the Section Heads who boasted much longer service records than the new Head of the Laboratory, claimed past resolutions and precedents. We had however to take into account a balance of authority/responsibility definition for the Section Heads which would also be applicable and acceptable to other laboratories of the establishment.

## 4 EQUIPMENT

On the subject of equipment we had less disputes. Every Section Head was asked to submit a list of all items of equipment, stating make, vantage, precision and date of latest calibration. The only problem which arose was when we required a name list of persons authorized to use each item. We found that in some cases members of the staff who did not meet our definition of the required qualification to use a certain item were actually trained and allowed the use of such items. These were mostly staff personnel with a long service record in the Laboratory.

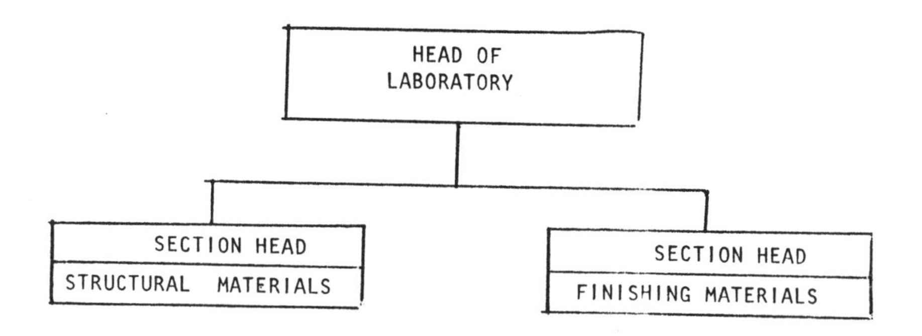


FIG. 2 - ORGANIZATIONAL CHART

## 5 TEST METHODS

We asked each Section Head to make a list of the tests performed with reference to the test method. Most of the testing is covered by an Israeli or ISO Standard. Where such a document does not exist, an ASTM, BS or DIN is used. However, there is a certain amount of testing made on casual or new materials that are not covered by approved specifications. For these materials an appropriate testing procedure is adopted by the Laboratory. We started to document these procedures and if possible to correlate them. It appeared that not infrequently some such material was submitted to and tested by one Section, while a similar material was submitted to and tested by the other Section, and not always were these procedures the same. It was stipulated that in all cases non-standard test programmes have to be approved by the Head of the Laboratory.

## 6 QUALITY CIRCLES

When we started to document test procedures, we initiated open discussions of every test, to which subscribed all staff members related to the performance of the test. Inadvertantly, many suggestions were laid across the table, pertaining to every phase of the test, from sampling to preparation of specimens, effectiveness of the test method, calculation and documentation. Every such contribution was examined, analysed, and, when found totally or partially worthy of condsideration, proper measures were taken for its adoption. So actually, although we did not aim to, we virtually ran quality circles in the Lab.

## 7 USE OF EXISTING MANUALS

Before we set out to compile the Quality Manual for this specific laboratory, we looked into existing ordinances of the Institute, and into manuals. When they were comprehensive, we adopted them fully. When they were partly applicable, we made a reference to them. Some of these existing manuals were: Safety Precautions in Laboratory Work, Documentations of Test Reports, Preservation of Samples After Test, Preservation of Test Records.

PART FOUR

# LABORATORY ACCREDITATION

# 34 LABORATORY ACCREDITATION WORLDWIDE AND PRESENTATION OF ILAC

A. BRYDEN
Laboratoire National d'Essais (LNE), Paris, France; ILAC Committee on Laboratory Operation and Management

## Introduction

Testing has been for many years an integral part of the process of building and controlling the quality and safety of products. Laboratory accreditation, meaning the formal recognition that a laboratory is competent to carry out specific tests or types of tests and implying the assessment of this capability, is a much more recent concept. Its development has grown mainly in the context of quality assurance and for the purpose of facilitating the acceptance of test results, especially across borders for international trade. It is remarkable that international standardization in this field has developed at the same pace as the accreditation activities themselves and they have in fact fed each other, so that today a lot of hope is placed in the multiplication of agreements between accreditation bodies to help solve some of the problems of technical barriers to trade.

This paper presents an overview of the international situation of laboratory accreditation and standardization activities in this field, as well as the more recent developments in Western Europe.

## 1 Quality of testing: a key factor for the economy

ISO Guide 2 defines testing as a "technical operation consisting of the determination of one or more characteristics of a given product, process or service according to a given procedure". Taking testing in its broad sense, that is covering also calibration and analysis, it is involved in almost all the sectors of the economy and many important issues, such as fairness of trade, environmental protection, safety of consumers and workers, energy conservation...

Reliable test and measurement data are needed in many instances, such as innovation and research, product certification and QA in production, private and public contracts, consumer protection and information, building and civil engineering and health services.

Making sure that testing services are available and of a suitable level of quality is therefore a key factor for the economy and a concern shared by all countries. During this century, testing activities have been considerably developed with the creation of many national standards and testing laboratories in the industrialised countries. Today, many developing countries are investing in this field, as a means of backing up industrial and economic development, so that the international community of testing has grown considerably in the past twenty years.

Testing is performed by laboratories having a wide variety of statuses, sizes and technical spectrum. In-house testing in industrial firms has also developed as part of the implementation of QA schemes.

As there are many users and prescribers of tests and many laboratories to provide testing services, one would think that letting the market law govern the situation would be a sensible solution. But very often, test data are the basis of important decisions that may affect the fairness of trade or such vital issues as health, safety or environmental protection. Moreover, not all testing activities are economically viable and public funding may be required, especially for capital costs, so that the practice of designating laboratories for performing specific tests has developed at the same pace as laboratory activities themselves. "Designation" is used here as covering broadly the procedure by which a government, public authority, certification body, or public or private organization designates one or more laboratories to perform specific tests for its own use or for the purpose of implementing regulations, standards or specifications in which it is involved.

Such designation is usually based on a number of motivations and criteria, such as ensuring the availability of testing services, ensuring homogeneity of the implementation of standards and regulations, independence of the laboratory or quality assurance of the tests performed. Quality is, or should be, one of the main factors of such designations, because there is a direct link between the quality of testing and the validity of the decisions taken on the basis of the results.

But the traditional definition of quality, namely the ability to satisfy the user's explicit or implicit needs, is difficult to apply to testing. Indeed the produce of a laboratory is a document, or even just a set of data, so that the client of a laboratory may be very satisfied with a faulty test result, as he may be very disappointed by a perfectly correct test report.

## 2 The concept of laboratory accreditation

It is because of the difficulty, but also of the necessity,

to evaluate the competence of testing laboratories that the concept of laboratory accreditation has developed over the years. It is defined by ISO-IEC Guide 2 as "the formal recognition that a testing laboratory is competent to carry out specific tests or types of tests". Laboratory accreditation is usually granted by an identified accreditation body, for tests specified in reference documents and after an initial on-site assessment of QA management and specific capability by qualified assessors. It is complemented by reassessment at periodic intervals and by proficiency testing or other forms of relevant auditing.

In the past ten years, laboratory accreditation has developed considerably in the world for two main reasons:

The necessity to identify and qualify the offer of testing services which are in growing demand in connection with the improvement of quality and safety.
It provides a very convenient mechanism to facilitate the international acceptance of test results, through bilateral or multilateral agreements between accreditation bodies.

## 3 International cooperation in testing

In the field of testing, international cooperation takes place at different levels:

For units of measurements and reference standards, the International Conference of Weights and Measures and its executive body, the Bureau International des Poids et Mesures, has been active for over a century to provide an organization to ensure international coherence of units and measurements.
In the field of legal metrology, which deals with measuring instruments connected with the fairness of trade and the implementation of regulations, the Organisation Internationale de Metrologie Legale is the international forum.
More recently, the need for international cooperation has appeared on the question of standard reference materials: ISO has a specific committee to deal with it - REMCO - and an international data base has been developed - COMAR.
Harmonization of test methods and test procedures is the field of international standardization, where ISO, IEC and numerous regional and national standards organizations have been active for many years: more than half of national and international standards are concerned with test methods.
When tests are used for the implementation of legislation, governmental authorities may get involved and this results in the implication of a number of

intergovernmental bodies, such as the U.N., the F.A.O. or the O.E.C.D.
Intercomparisons and round robin calibrations and measurements have for a long time been used to ensure international coherence of reference standards and measurements; in the same way, many intercomparative tests are organized on an international scale, especially in the field of materials testing and in such industries as the petroleum, steel, mining, textiles, paper or the chemical industries.

The mutual recognition of test results is of course facilitated by all this upstream cooperation. Many reasons may motivate the seeking of such recognition at the international level, such as measurements connected with the control of air or water pollution or multinational programmes of technical cooperation such as those in the aerospace or nuclear industries, but trade is undoubtedly the main incentive. It is no more necessary to underline how testing may create technical barriers to trade, even if it is designed to pursue other objectives such as safety or environmental protection. It is therefore natural that this issue should have been addressed by the GATT Agreement on Technical Barriers to Trade which came into force on January 1, 1980. It stipulates that all efforts should be made to encourage the acceptance of foreign test data in order to avoid unnecessary duplication of tests.

This is important to recognise that there are two levels at which mutual recognition of test results must be considered if it is to be efficient to remove barriers to trade:

The technical level, where agreements may be concluded between laboratories, certification agencies or accreditation bodies.
The acceptance level, which involves the prescribers and users of tests.

Whereas the first level implies mutual technical confidence, the second has legal implications which often have impeded the development of agreements, altough many examples and forms of such agreements are operational today.

Mutual recognition agreements between laboratory accreditation organizations clearly offers the most universal approach to the problem. That is why the concept of laboratory accreditation has been so popular and has spread so fast in the last decade.

## 4 ILAC: the International Conference on Laboratory Accreditation

International co-operation on laboratory accreditation as a

horizontal subject started in 1977 when Denmark and the USA took the initiative to convene in Copenhagen an International Conference on the mutual recognition of test results, which became the International Laboratory Accreditation Conference at its second meeting in Washington DC the next year.

The achievements of ILAC since then have been remarkable and well publicised. It is sufficient to give a few indicators to describe its success:

The publication of eight ISO-IEC Guides based on ILAC technical proposals, which form a complete set of reference documents for the implementation of quality programmes in laboratories and the operation of laboratory accreditation activities, see fig.1.
Nearly 50 counties representing the five continents and 10 international or regional organisations have taken part in ILAC conferences and task force meetings.
The distribution around the world of the venues of the conferences, fig.2.

ILAC has certainly directly influenced the creation of a number of new national accreditation programmes and the multiplication of bilateral agreements between accreditation bodies.

At its plenary meeting in Auckland (N.Z.), ILAC has adopted a new structure of work concentrated around 3 basic committees:

1 Testing and international trade
2 Practice of laboratory accreditation
3 Laboratory management and operation

ILAC, after having helped to constitute the basic standards for the operation of laboratory accreditation systems and bodies is now engaged in the development of the mechanisms to encourage bilateral and multilateral agreements and on the technical aspects of laboratory accreditation and QA management in laboratories. It has a major role to play to build confidence and facilitate exchange of experience in the testing community, as well as to encourage the development of national accreditation systems along similar lines.

Two important work areas have been initiated at the last ILAC meeting in 1988:

The revision of ISO-IEC guides 25 and 38, which deal with the criteria for the acceptance of laboratories, which is currently being undertaken in an ISO CASCO working group in conjunction with ILAC: these documents will cover both testing and calibration laboratories.
The coordination between these guides and the ISO 9000 series, with the aim to make the corresponding Q.A.

| | |
|---|---|
| Guide 25: | requirements for technical competence of testing laboratories (1982) |
| Guide 38: | requirements for the acceptance of testing laboratories (1983) |
| Guide 43: | proficiency testing (1984) |
| Guide 45: | presentation of test results (1985) |
| Guide 49: | quality manual for a testing laboratory (1986) |
| Guide 54: | recommendations for the acceptance of accreditation bodies (1988) |
| Guide 55: | recommendation for the operation of testing laboratory accreditation systems (1988) |

Fig. 1. INTERNATIONAL ISO-IEC GUIDES

| | |
|---|---|
| 1977: | Copenhagen - Denmark |
| 1978: | Washington - USA |
| 1979: | Sydney - Australia |
| 1980: | Paris - France |
| 1981: | Mexico City - Mexico |
| 1982: | Tokyo - Japan |
| 1983: | Prague - Czechoslovakia |
| 1984: | London - United Kingdom |
| 1986: | Tel-Aviv - Israel |
| 1988: | Auckland - New Zealand |
| 1990: | Torino - Italy |
| 1992: | Canada |

**Fig. 2. ILAC CHRONOLOGY**

requirements consistent so as to avoid the duplication of audits for laboratories.

## 5 A review of the national situations

At the national level, the first ILAC directory issued in 1981 which aimed at identifying national programmes for the designation of testing laboratories, contained hundreds of organizations and programmes. They were usually operating in specific sectors and according to quite diverse principles and criteria; quite probably very few operated against the principles contained in today's ISO-IEC relevant guides. They have not all disappeared today, by far, but a number of them have been adapted and either have been included in or are now relying on comprehensive national accreditation programmes.

The table in Figure 3 gives information on existing or recently created comprehensive national laboratory accreditation programmes. It has been drawn from an enquiry addressed to all the ILAC participants, as well as from the entries in the ILAC directory and information circulated at ILAC 88.

The main points which may be stressed by studying this table are as follows:

- The majority of accreditation bodies are public organizations or organizations with some direct involvement of governments. Only A2LA in the USA is entirely privately run, although it has some government contracts.
- There is a growing tendency to run calibration laboratory and testing laboratory accreditation systems in a co-ordinated way; several countries have already decided to merge these activities in the same organization.
- The field of accreditation covers practically the whole spectrum of testing activities.
- The creation of at least ten new accreditation bodies has been announced in 1988 and 1989.
- Some twenty bilateral agreements between accreditation bodies have been concluded so far; it is estimated that at least another dozen are currently under discussion.

A special mention of multilateral agreements should be made, the first example being the Nordic co-operation between laboratory accreditation bodies - NORDA - established in 1986. A multilateral agreement between the calibration services of the Netherlands, the UK, the Federal Republic of Germany, Italy, Finland, Sweden and France has been signed in 1989 to avoid the multiplication of bilateral assessments.

| COUNTRY | ACC-BODY | STATUS | | | FIELDS | | | | | | |
|---|---|---|---|---|---|---|---|---|---|---|---|
| | | Public | Semi-Private | Private | All | Calibration Only | All except Calibration | Year Esablished | No of Accredited Laboratories | Claimed Compliance w ISO-IEC Guides | No of Bilateral Agreements |
| AUSTRALIA | NATA | | | | * | | | 1946 | 1714 | * | 3 |
| AUSTRIA | OKD | * | | | | * | | 1983 | | | |
| P.R.CHINA | SACI | * | | | | | * | 1983 | 323 | | |
| | SBTS | * | | | * | | | 1984 | 75 | | |
| DENMARK | STP | * | | | * | | | 1973 | 106 | | 1 |
| EIRE | ILAB | | | | | | | | | * | |
| F.R.GERMANY | DKD | | * | | | * | | 1977 | 67 | | 3 |
| FINLAND | MSF | * | | | | * | | 1980 | | | |
| | TTK | * | | | * | | | 1988 | | | |
| FRANCE | BNM | * | | | | * | | 1969 | 130 | | 2 |
| | RNE | | * | | | | * | 1980 | 173 | * | 1 |
| HONG KONG | HKLAS | * | | | * | | | 1985 | 14 | * | |
| HUNGARY | MSZH | * | | | * | | | 1988 | 5 | | |
| INDIA | NCTCF | * | | | * | | | 1988 | | | |
| INDONESIA | PPMB | * | | | | | * | 1977 | 161 | | |
| ITALY | SIT | * | | | | * | | 1977 | 43 | | 2 |
| | SINAL | | * | | | | * | 1988 | | * | |
| MEXICO | DGN | * | | | | | * | 1982 | | | |
| NETHERLANDS | NKO | * | | | | * | | 1975 | 43 | | 1 |
| | STERLAB | | * | | | | | 1986 | 6 | * | |
| NEW ZEALAND | TELARC | * | | | * | | | 1973 | 285 | * | 3 |
| NORWAY | NOLA | * | | | * | | | 1988 | | | |
| POLAND | NCSP | * | | | | | * | 1983 | 30 | | |
| | NLMS | * | | | | * | | | | | |
| PORTUGAL | IPQ | * | | | * | | | 1979 | 18 | * | |
| SAUDI ARABIA | SASO | * | | | * | | | 1987 | 2 | * | |
| SINGAPORE | SINGLAS | | * | | * | | | 1986 | 12 | * | |
| SOUTH AFRICA | CSIR/NCS | * | | | | * | | | | | |
| SPAIN | RENLEI | * | | | | | * | 1981 | 56 | | |
| | RELE | | * | | * | | | 1986 | 6 | * | |
| SWEDEN | MPR | * | | | * | | | 1972 | 220 | | 3 |
| THAILAND | TLAS | * | | | | | * | 1988 | 1 | * | |
| TURKEY | TSE | * | | | | | * | 1987 | 15 | * | |
| U.K. | (BCS) | * | | | | * | | 1966 | | | |
| | (NATLAS) | * | | | | | * | 1981 | | | |
| | NAMAS | * | | | * | | | 1985 | 630 | * | 8 |
| U.S.A. | NAVLAP | * | | | | | * | 1978 | 200 | * | 4 |
| | A2LA | | | * | | | * | 1978 | 135 | * | |

Fig 3

## 6 The situation in western Europe

The objective of harmonizing the internal market in the European Economic Community by 1993 has resulted in a series of decisions relating to the harmonization of technical regulations and standards as well as to conformity assessment. Without going into details, the main features of the new global approach taken to deal with this issue are the following:

A mechanism of information on the development of new national standards leading to a reinforcement of the standardization work directly at the European level in the three regional standards institutes (CEN, CENELEC and ETSI).
The harmonization of technical regulations through European Directives setting "essential requirments" and the development of European Standards ("euronorms"-EN) to offer a preferential way to achieve and demonstrate compliance with these requirements.
The "modular approach" to conformity assessment which breaks down the various ways of demonstrating conformity with or without intervention of a third party (laboratory, certifying agency, certification of manufacturer's QA system...) into elementary modules (type test, QA models as described in ISO 9000...) and their combination to enable the affixing of the CE mark of conformity to EC Directives.
The global approach for developing the acceptance of test results and certificates through the development of mutual recognition agreements in the voluntary sector as well as a harmonized evaluation of third parties to be used for the implementation of EC directives ("notified bodies").

The main consequences of this overall policy on the development of laboratory accreditation in Western Europe are the following:

The EN 45000 series standards, which derive from the relevant ISO-IEC guides on laboratory assessment and accreditation will be the common base on which national accreditation schemes will develop.
Already 8 comprehensive national schemes are in operation within the EEC; Switzerland, Sweden and Finland are also developing national systems; the Federal Republic of Germany has taken the decision to create a system which, although it will probably be organised sectorially, will take the international standards as its reference.
In the field of testing, only one bilateral agreement has been signed between EEC countries, namely between NAMAS (UK) and RNE (France), but this may be explained

by the fact that the two systems are by far the most developed up to now. Other agreements are under discussion, as the other national schemes develop in terms of the number of accredited laboratories and the field covered by accreditation.
Testing will be one of the horizontal functions of the "European Organization for Testing and Certification" which has been set up by the signature on April 25, 1990 of a Memorandum of Understanding between the EC Commission, the EFTA Secretariat on one hand, CEN and CENELEC on the other; as a consequence, a structure of cooperation and coordination to provide in an organized way the input and representation of the testing community has been set up under the name of EUROLAB. This organization is based on national delegations representing the public and private testing and analytical laboratories of the countries belonging to the EEC or to EFTA and includes, in an observer capacity, all interested European Organizations, in particular WECC (Western European Calibration Coordination) and WELAC (Western European Laboratory Accreditation Cooperation). 16 countries signed the MOU creating EUROLAB on April 27, 1990.

## 7 Conclusion

The concept of laboratory accreditation has spread fast across the world in the course of the past decade. Apart from the national benefits which are attached to it, it has raised many hopes that it will effectively serve to facilitate international trade through the multiplication of bilateral or multilateral agreements of mutual recognition between accreditation bodies. This has been one of the main incentives for the creation of the more recent systems.

The efficiency of this mechanism depends on the degree of acceptance of the accreditation systems at the national level and since most of them are in the development stage, it will take time. But, in the long run, it should be a very powerful tool to enable international acceptance of test results.

In Western Europe, a clear option has been taken that it will be the basis for the development of the mutual acceptance of test results. It is therefore important that close international cooperation be maintained in this area.

# 35 THE PRACTICE OF LABORATORY ACCREDITATION

J.A. GILMOUR
National Association of Testing Authorities, Sydney, Australia

**Abstract**

While the term "laboratory accreditation" is of very recent origin, the concept of external (particularly second-party) evaluation of laboratory competence is probably as old as the industrial revolution. Indeed, formal systems have existed since the mid-1800s, most notably in the area of defence procurement.

The idea of a national third-party system to meet the needs of all users of laboratory services was developed in Australia immediately following the Second World War. Even then the activity was called laboratory registration and the expression "laboratory accreditation" was not in general use until the early 1970s.

Irrespective of terminology, the practice of assessing laboratories for technical competence is not new and it has evolved over many years to its present "state of the art". It is now in a process of formalisation and consolidation through the ILAC/ISO/IEC/CEN/CENELEC activities.

This paper describes this recent flurry of work to produce international standards (criteria) and guides on the practice and management of laboratory accreditation systems. It is a personal perspective so I might begin by reminding you that good reliable testing and measurement existed before laboratories were accredited and that the essential objective of the process is to give a wide range of users of test data confidence in that data - it is not to support yet another infrastructure bureaucracy.

**Definitions**

Like most activities and bodies of knowledge, laboratory accreditation has its own terminology which is to be found in ISO/IEC Guide 2.

The basic definition is that of the term itself:

> a formal recognition that a laboratory is competent to perform specific tests or specific types of tests.

The key words in that definition are

| | |
|---|---|
| "formal" | - there are predefined criteria to be met and procedural steps to be followed |
| "recognition" | - the accreditation is published |

"competent" - technical competence must be demonstrated

"specific" - the range of competence is defined in technical terms.

These are the elements which differentiate the activity from other processes by which laboratories are approved by a multiplicity of users for their own purposes.

The practice of laboratory accreditation also incorporates the features of on-site assessment, use of expert assessors, proficiency testing and surveillance of accredited laboratories - the definitions of these terms are contained in ISO/IEC
Guide 2 or in the other related ISO/IEC Guides.

**Laboratory Accreditation Systems**

There are two complementary documents, ISO/IEC Guides 54 and 55 and their derivatives, EN 45002 and EN 45003 which define the attributes expected of laboratory accreditation systems and of the bodies that operate them. These documents were largely developed from the descriptive material prepared by ILAC Task Force C between 1979 and 1984. This more descriptive discussion material was published by ISO in its 1986 publication "Laboratory Accreditation - Principles and Practice".

The objectives of this international standards work and the urgency associated with it are concerned with ensuring international harmonisation and international acceptance of test data as required by the GATT Standards Code and to meet the demands of the European Community's policy on testing and certification, the so-called "Global Approach".

The oldest, and still the largest, accreditation body is NATA in Australia; it predates any other comprehensive system by some twenty-five years. It, therefore, must be given much of the credit for pioneering the practice of laboratory accreditation.

Its major contributions to this subject have been its emphasis on peer assessment of technical competence of laboratories and developing and publishing criteria to be met by laboratories wishing to be accredited (the precursors to ISO Guide 25) which focus on these technical matters. It has also pioneered the codification of laboratory activities across all disciplines of science and technology with its concepts of fields of testing, classes of test and directories of laboratories.

The more recently developed Standards and Guides produced by the various standards bodies such as ISO/IEC, CEN/CENELEC and ASTM, accept those technic considerations so strongly advocated by NATA and superimpose a requirement for the adoption of sound quality management principles. This amalgamation of specific technical competence with a sound systems approach seems to be very valid although it imposes additional costs on the operation of the accreditation system and on laboratories. We must ensure, however, that the costs are contained within reasonable levels and neither should we lose sight of the overall objective of the activity which is to assure validity of test data. I sense a degree of overkill in some recent proposals being advocated in ISO and other fora.

Without laboriously quoting the various documents in current use in their totality, I intend merely to outline their content and to add some personal comment on current activity.

**Guides 54 & 55 (EN 45002 & EN 45003)**

These documents address the organisational requirements for bodies engaged in accreditation of testing laboratories and consider issues such as:

policy and decision making processes to ensure openness, fairness and integrity;

requirements for a quality system within the accreditation body;

procedures for the assessment and accreditation process;

provision for appeals;

availability of resources;

maintenance of records;

competence of assessors and their relationship to the accreditation body;

requirements for expressing details of accreditation;

publications.

The document EN 45003 contains a number of interesting innovations. For instance it requires:

*4 Organisation*

*The accreditation body shall:*

*(b) have the financial stability and resources required for the operation of an accreditation system. It should have and make available on request the means by which it receives its financial support;*

*(g) have policy and decision making processes to prevent any confusion between accreditation and product certification;*

*(h) be willing to take part in an exchange of information with other accreditation bodies to improve the quality of accreditation systems and to create confidence.*

*7 Sectoral committees*

*Sectoral committees shall have formal rules and structures.*

*14 Publications*

*(f) a list of accredited laboratories that identifies the scope of the accreditation granted.*

I suggest that there is room for debate about the meaning and implementation of some of the clauses in the document such as those quoted above and note they have little or no relevance to the assessment of competence of the testing laboratories to be accredited under the system. I do not disagree with the spirit of the clauses however, as they may be of interest in enhancing relationships between various bodies. I am not sure that they should be in a standard which may later be used in international negotiations.

For instance, the role of sectoral Committees varies markedly from one system to another. Some are purely advisory to the governing board or executive officers, while in other cases they have decision making authority. Is it always necessary to have "*formal rules and structures*?"

In NATA, the Registration Advisory Committees as they are known, are advisory only but they do have direct reporting links with the Council - the governing body. NATA also has many such committees with quite broad responsibilities, whereas those bodies more closely linked to Government, tend to have fewer committees with more limited remits.

Some of the differences between bodies are of an administrative nature, others, such as rigour of traceability chains, are technical. I suggest there is more scope for allowable variations in administrative matters than in technical questions.

While the primary purpose of laboratory accreditation bodies is to evaluate the competence of laboratories and their ability to produce reliable data, many of the existing organisations have the additional, but related, functions of developing standards and codes for good laboratory practice (the specific criteria publications referred to later in this paper) and providing advisory or counselling services to potential applicants for accreditation.

There are excellent historical reasons for these activities but there is also a potential for difficulties with conflict of interest for the accrediting body where it provides what amounts to a consulting service to its potential clients for accreditation.

It is also timely to note that a number of laboratory accreditation bodies are broadening their range of activities into areas which are related to their primary missions and which may enhance their traditional roles but which could lead to a some confusion between various functions.

Training of assessors leads naturally to training in laboratory management and quality systems for laboratory personnel. As ISO Guide 25 evolves and becomes essentially a combination of traditional technical matters and the systems concepts found in the calibration systems standards and the quality management standards such as the ISO 9000 series, it is not unreasonable for laboratory accreditation bodies to offer certification to such quality systems standards.

I also suggest that some bodies may also diversify into one or other more traditional product certification schemes.

This whole area of conformity assessment is being looked at afresh and I therefore suggest that edges between the various systems and concepts will continue to blur even further in the future.

It seems that the preferred option for international recognition of laboratories that is emerging is mutual recognition agreements between laboratory accreditation bodies and there is little doubt that such future agreements will be based on formal assessment for compliance with one or other of these international documents. Prior to the existence of these documents, mutual recognition agreements were based on quite subjective evaluations of each body. The essential question was "would I make the same accreditation decisions in the same situation?" Evaluation against these Guides or Standards may yield a different result in a mutual recognition negotiation, particularly if clauses such as those given above are mandatory.

NATA in Australia, TELARC in New Zealand (and perhaps one or two others) were created mainly to satisfy the internal needs of those countries. Almost all the more recently established accreditation bodies have as their primary purpose, the facilitation of international acceptance of test data to meet the demands of trade. There is, therefore, very good reason to adopt a rather internationally standardised approach. Naturally, local conditions must be considered and provided for in the international standards which form the basis of acceptance of the organisations and their systems. In the cases of Australia and New Zealand long established cultures have had to be changed to accommodate the new international approach and the transition has not been entirely painless.

**Technical Requirements**

Laboratory accreditation may be regarded as the end product of a process of evaluating laboratories for technical competence as distinct from a system for assessing compliance with a system standard such as for a calibration system or quality management system. It, therefore, requires criteria for technical matters. The generic document in this case is ISO/IEC Guide 25.

Guide 25 itself outlines the matters that all laboratories must consider to ensure competent operation but it is written in such general terms that before it can be used for a specific laboratory it must be interpreted in the context of the science or technology involved. For this reason laboratory accreditation bodies produce technology specific criteria. For example, NATA has specific criteria for all its technical fields of operation ranging from Acoustics and Vibration Measurement, Chemical and Mechanical Testing to Wool and Quality Assurance Systems. There are twelve separate publications in all, and each, with the exception of our wool surveillance program in which laboratories play a relatively small part, is in harmony with Guide 25 but use language and calibration requirements that relate to the specific field.

Other comprehensive laboratory accreditation systems have similar arrangements.

There is no doubt that, with some notable exceptions, most laboratory managers and scientific personnel recognise the value of the new concepts of quality assurance and the adoption of formal systems of quality assurance. This will have increasing impact on criteria used by laboratory accreditation bodies and the current revision of ISO/IEC Guide 25 certainly takes the new philosophies into account.

## Applications for Accreditation

Applications by laboratories for accreditation result from either a requirement imposed by a client or proprietor, or because the laboratory management itself perceives some value from the recognition resulting from accreditation or from the external audit provided by the assessment. Nevertheless, for the staff of any laboratory, assessment is a relatively traumatic event and there is almost always considerable pre-application discussion between the laboratory and the accreditation body.

Most accreditation bodies prefer that applicants for accreditation are well prepared for the assessment and that a formal application is only made after all possible preparations have been completed.

## The Assessment Process

The assessment process comprises a number of reasonably well defined steps which are widely adopted throughout the practice of laboratory accreditation.

### Application

The assessment process begins with the filing of an application form and, usually, complementary documentation such as laboratory quality manuals and such questionnaires as required by the accreditation body. This documentation is evaluated by the accreditation body and forms the basis for the briefing of the assessment team.

### Appointment of Assessment Team

Assessment teams usually comprise individuals competent in laboratory quality assurance together with technical experts selected on the basis of the type of work being performed in the laboratory. Teams would never be less than two, and may be as many as four or more depending on the nature of the laboratory. Assessors must be skilled in the assessment procedures and policies of the accreditation body.

### On-site Visit

The heart of the assessment is the on-site inspection. It will rarely occupy less than one day and may extend to a number of days.

The assessment team will focus on the requirements of the accreditation body and will audit the laboratory against those requirements. The team will have discussions with laboratory management and staff, witness tests being performed, review procedures (both technical and administrative) and inspect accommodation and facilities.

### Exit Interview

At the conclusion of the visit it is normal, but not universal, practice to conduct an oral review of the visit and often present some form of interim report.

### Review of information

Following the on-site visit and any subsequent follow-up action required by either the accreditation body or the laboratory, all information and documentation is reviewed for compliance with the requirements.

### Decision Making

Formal decision making processes vary between accreditation bodies. In some cases the decision is a purely executive matter for the staff while in others there are external review processes.

## Proficiency Testing

Proficiency testing is the use of interlaboratory tests to evaluate or confirm competence of a laboratory or a group of laboratories. There are many form of such tests which are described in ISO/IEC Guide 43. In one form or other they are playing an increasingly important role in the overall assessment process.

Many of the advocates for their extended use have ignored their cost. They are expensive to operate and to be of most value they must be repeated regularly - at even more cost. For some tests and materials, however, it is not feasible to conduct them at all. They should be seen as supplements to the other techniques used in assessing laboratories, not as substitutes.

There is also considerable scope for international cooperation in this area.

There are those who dismiss altogether the need for proficiency testing as an ingredient of accreditation. I suppose it depends on whether we regard accreditation as being a statement that laboratories comply with defined published criteria or whether we consider it to be a stronger statement of demonstrated competence. I suggest that these are not the same things and that if we are looking for proven competence, then some concrete evidence of performance is essential.

Our experience with proficiency testing leads us to the unequivocal conclusion that it is necessary for the overall credibility of the accreditation scheme. In addition, proficiency testing programs have a number of side-benefits which greatly add to their value.

Proficiency testing programs provide objective data which basically support the more subjective elements of the assessment process and there is a regular flow of surprises, some welcome, some not. For instance:

(i) not all standard methods are as standard as we would like;

(ii) not all published precision data is reliable;

(iii) some tests are very equipment dependent;

(iv) sometimes there are regional biases;

(v) laboratory competence is not related to ownership;

(vi) laboratories conducting tests frequently will usually perform better than those which perform them rarely.

**Surveillance of Accredited Laboratories**

For mature laboratory accreditation systems, surveillance of accredited laboratories accounts for eighty percent or more of their activities. There will always be some new laboratories seeking accreditation, but surveillance is the routine business. Mr Summerfield will discuss the practices used in this area and I do not wish to pre-empt what he may wish to say.

**Other Activities**

The above is therefore a brief description of what might be strictly called the practice of accreditation, but work in laboratory accreditation is much more diverse than that. The following activities are essential to the effective operation of a laboratory accreditation system.

**(a) Criteria Development**

While general criteria such as ISO/IEC Guide 25 retain their currency over extended periods of time, specific criteria for individual tests require constant monitoring, new test procedures are developed and new areas of testing are brought within the area of accreditation. The detailed working documents of accreditation bodies are therefore under constant review and development.

In some systems much of this work may be carried out by others such as standards writing bodies, industry associations or the like. In an organisation such as NATA, criteria development and review is an integral part of our activities and probably accounts for two to five percent of resources.

**(b) Directories**

One of the essential functions of a laboratory accreditation body is to make available information on accredited laboratories to users of test data. The ISO/IEC Guides 54 & 55 specify a level of detail which should be used to express the scope of accreditation - products, test methods or other descriptors - which defines the boundaries of competence. This is not the same as describing the capability of the laboratory for use by others. This can become quite a complex subject and there is no universal agreement on it.

Some organisations, including RNE in France and NATA, take the view that a comprehensive directory is most useful and we attempt to use language which is both precise and yet understandable by users of the directory. ASTM Committee E 36 is currently studying this problem with a view to internationalising language to promote international transparency.

**(c) Training and Education**

Clearly, despite efforts to establish and define criteria, the accreditation process still relies on the judgment of people - the assessors, advisers and staff - and if variation is to be reduced, all these individuals require training and continuing education. This can be provided either externally or internally or both. Either way it is a continuing activity for a laboratory accreditation body.

## Conclusion

This paper outlines the current practices used in accreditation of laboratories and a number of activities peripheral to, but necessary to support, that activity.

With the immense political pressure within the European Community for each Member State to establish national laboratory accreditation systems which are to be operationally transparent and harmonised, the development of the various international guides and European Standards has also applied pressure to systems in other parts of the world. It is not always clear, however, that identification of laboratory competence is really the ultimate goal. Let us hope that laboratory accreditation in particular and conformity assessment in general do not in themselves become additional trade barriers. These are questions which will have to be resolved in the near future.

In the practice of accreditation there is, however, considerable international agreement and there are opportunities for cooperation in proficiency testing and developing a thesaurus of terms to facilitate international acceptance of test data and access to specialist laboratories in foreign countries.

Testing will inevitably become even more important in trade than it is now. Competent laboratory accreditation provides an excellent mechanism to remove it from the political agenda provided it is not oversold and its limitations are appreciated. There is no substitute for building confidence from the ground up and this requires continuing personal interactions at all levels. Accreditation practitioners have a remarkable record of cooperation and achievement in this area. May that tradition continue.

# 36 THE SURVEILLANCE OF ACCREDITED LABORATORIES

J.A. ROGERS
National Measurement Accreditation Service, NPL,
Teddington, UK

Abstract
An increasing number of countries are establishing national, independent, third-party accreditation bodies which use internationally agreed requirements to assess the competence of calibration and testing laboratories. As the number of accredited laboratories grows the procedures used by these bodies to check that the accredited laboratories are maintaining the required standards becomes increasingly important. This paper outlines the areas that must be assessed and monitored by the accreditation body to give the users of accredited laboratories confidence that such laboratories will achieve the required standards at all times.

## 1 Introduction

One of the main reasons why laboratory accreditation is necessary is that clients need to be confident that a laboratory will perform tests consistently from one day to another and that the tests will be carried out in accordance with their requirements. Accreditation cannot be awarded on the basis of reputation alone and therefore all accreditation bodies of any standing carry out an assessment of the laboratory's operational practices before issuing accreditation. This one-off exercise is designed to establish whether the laboratory has a documented quality system in place, whether the system is appropriate and effective, and whether staff in the laboratory are following the instructions laid down in the system at all times.

Most laboratory accreditation bodies have developed formalised requirements for laboratories to adhere to and so this initial assessment also involves checking that the laboratory's quality system is in compliance with these requirements. Since the majority of laboratory accreditation bodies have developed criteria based upon ISO Guide 25, "General requirements for the technical competence of testing laboratories", the criteria are generally very similar and one could therefore expect laboratories in different accreditation schemes to achieve similar standards. In practice this may not always be so, partly because the methods they adopt to assess a laboratory's compliance with their requirements differ significantly. Some accreditation bodies will be more rigorous than others in the application of their requirements some place greater emphasis on particular features that they consider important.

It is not the purpose of this paper to draw comparisons but to describe the measures that all accreditation bodies should take to ensure that their accredited laboratories maintain the appropriate standards at all times.

## 2 The Form of Surveillance Programmes

The approach taken by different accreditation bodies to the task of monitoring accredited laboratories varies throughout the world but all the effective accreditation schemes rely upon regular visits to establish whether laboratories are continuing to comply with their requirements.

The nature of these surveillance visits and their frequency varies from one accreditation scheme to another. The visits generally take the form of a fairly comprehensive assessment when the interval between visits is set at 2 years and this assessment is usually similar in scope to that carried out prior to accreditation. Where shorter visit intervals have been adopted the accreditation bodies tend to carry out partial assessments with a more comprehensive assessment at say 3 or 4 yearly intervals.

The combination of interval and type of assessment chosen by each accreditation body will depend upon a number of factors such as the type of testing involved, the financial resources available, the level of confidence that has been established in the accredited laboratories and the availability of other measures of performance such as proficiency testing or interlaboratory comparisons.

The accreditation body will generally settle on a fixed interval for the bulk of its surveillance visits but will reserve the right to shorten this interval should it consider that certain key areas of testing need special precautions. It is for the accreditation body and any other accreditation body that it wishes to be recognised by to establish whether the surveillance programme chosen ensures that the necessary standards are attained at all times in all areas of testing.

The administrative procedures adopted by the accreditation body will also have a bearing on the effectiveness of surveillance programmes and it is essential that any departures from laid-down requirements identified at any time are acted upon if the accreditation body is to be certain that the laboratory is maintaining the required standards. For example, there should be mechanisms in place to ensure that any non compliances with the accreditation requirements identified during visits are discharged within a very short time and not allowed to remain partially or wholly discharged until the next visit. Similarly the accreditation body should ensure that when a laboratory does not achieve the expected minimum performance in proficiency tests immediate action is taken to identify and, where appropriate, rectify the cause of inadequate performance.

Accreditation bodies should have in place mechanisms that allow them to suspend or totally withdraw a laboratory's accreditation should it fail to demonstrate at a surveillance visit that it is capable of or prepared to maintain the necessary standards.

## 3 Timing of the First Visit

All surveillance visits whether they take the form of partial or full assessments should be designed to establish continuing compliance with the requirements of the accreditation scheme. This means that the accreditation body must satisfy itself that the operating practices adopted by the laboratory since receiving accreditation are still in line with the laboratory's documented quality system and with the accreditation requirements. Too often laboratories drift back rapidly into bad practices after initial accreditation and, therefore, many laboratory accreditation schemes only leave a short interval of say 6 months before carrying out the first surveillance visit to make sure that the laboratory is meeting the required standards and, more importantly, shows a clear understanding of the requirements and how to satisfy them.

## 4 Planning Surveillance Visits

If surveillance visits are to be effective they should be planned in advance and form part of an overall programme that ensures that every aspect of the laboratory's operations that has a bearing on the quality of its testing is examined over a fixed period of time.

Responsibility for the planning of surveillance programmes usually rests with the accreditation body's technical staff but often the details of each visit will be drawn up in consultation with the expert, external assessor or, where a team is involved, with the team leader or lead assessor. While these experts will identify areas of the laboratory's activities that must be examined they may also identify areas of potential difficulty or weakness which should be covered during surveillance visits. Thus, although the main purpose of each surveillance visit will be to check that the laboratory is complying with the requirements of the accreditation scheme the accreditation body may also tailor the surveillance programme, and therefore individual visits, to ensure that the weaker or potentially weaker aspects of the laboratory's activities are examined closely.

It is essential, therefore, that the laboratory is made aware of the form of surveillance visits and has made any preparations that may be required to demonstrate its capability in any particular area during the visit.

It is also essential that any assessors used by the accreditation body are not only fully familiar with the detailed requirements of the accreditation scheme but are also experts in the area of testing that they are assessing and are trained in the techniques of assessment.

## 5 Conduct of the Surveillance Visit

The manner in which the surveillance visit is conducted by the assessor or assessors will vary from scheme to scheme but the visit will usually include an opening meeting with laboratory management at which the purpose of the visit is explained, the assessment, and then a final meeting at the end of the day at which the outcome of the visit is relayed to the laboratory. Some accreditation bodies identify any non-compliances with the accreditation criteria on standard forms and pass these over to the laboratory at the end of the day so that the laboratory can propose a corrective action in writing. Other accreditation bodies prefer to make comprehensive notes during the visit, summarise their findings orally to the management and provide a detailed written report after they have left the laboratory and have had time to receive the assessors reports.

Whichever mechanism is used the laboratory management should be left with a clear picture of how well they are performing and what actions they must take straight away if they are to retain accreditation and maintain the required testing performance.

## 6 Areas to be Surveyed

Since some surveillance visits take the form of full assessments and others only partial assessments or slices through the laboratory's activities it is not possible to define a typical surveillance visit. However, all the elements of the laboratory's activities that are covered during initial assessment should be covered by the overall surveillance programme and the assessors should make sure that this is achieved. Some accreditation bodies provide assessors with check-lists for this purpose.

A surveillance visit carried out as a full assessment will involve the actions described in the following sections. A partial assessment type surveillance visit might typically involve less in-depth study of the quality system and/or the monitoring of a smaller section of the testing activities.

### 6.1 Quality system - documentation

The first task for the assessor or assessment team is to check that the quality manual is up-to-date, reflects current practices and any changes that may have occurred in the laboratory's organisational arrangements, facilities, ownership or testing activities.

If the manual is not up-to-date this will alert the team to the need to check that the management is carrying out periodic reviews of the quality system and that the arrangements for control of all documentation are satisfactory. Further checks of the latter can be made as the team tour the laboratory and establish whether or not staff have access to all the necessary documentation and have up-to-date copies especially of test procedures and standard specifications.

### 6.2 Quality system - audits

Having established the state of the quality system documentation the assessment team leader will then usually ask to see the records of any internal audits performed since the last visit.

The team leader will be checking to see that all the activities and areas of the quality system that should be covered by internal audits are indeed being audited by someone independent of the area being audited and that the audit programme has not slipped. He will pay particular attention to the audit records sheets to see what type of non-compliance is being found, what areas are failing to meet the requirements and how often the same or a similar type of problem is occurring. He will check very carefully that appropriate corrective actions have been proposed, staff nominated to carry out these actions and that the Quality Manager has checked that they have been done correctly and within a satisfactory timescale.

The audit records reveal not only how effective the quality arrangements are but also how seriously the laboratory treats quality. The individual audit records provide pointers to areas that the assessment team need to examine in depth to ensure during surveillance that not only has the corrective action been carried out but that the root cause of the problem has been identified and that there is little likelihood of a recurrence of the problem.

The internal audit records must show that staff have been watched while performing tests or measurements to check that they are following laid-down procedures and not applying short-cuts or adopting inadequate procedures.

### 6.3 Quality system - reviews

The team should then ask to see the minutes of the last meeting at which the quality system was reviewed to check that a review had indeed been conducted, that all the appropriate senior staff were involved and that the outcome of the meeting has been acted upon. The review should take into account any changes in the organisation, staffing, range of testing, workload etc that may have a bearing upon the quality arrangements adopted, together with the results of any audits performed and any complaints received from clients.

### 6.4 Quality system - general

While touring the laboratory the assessment team leader should be paying particular attention to establishing that staff are aware of what is in the quality manual and associated documentation, have a clear understanding of their duties and responsibilities and, in the case of the quality manager. have the necessary authority and access to the highest level of management to ensure that the quality arrangements are implemented at all times. The assessment team should also check that the staff are not subjected to any pressures or inducements that may affect the quality of their work and in particular any influences that may arise from the commercial or production management whose objectives may conflict with acceptable practices in the testing area.

Other elements of the quality system that the assessment team will need to check are still working satisfactorily may be examined by individual technical assessors but the team leader should be forming an overview particularly of those systems that cut across sections or departments.

#### 6.4.1 Sample handling

The team leader should check that there is still a sample handling procedure that ensures that all samples are individually and uniquely identified at all times especially when they move from one department to another and that there are clear rules for the storage, handling, bonding and disposal of all test samples.

#### 6.4.2 Measurement and calibration system

In the same way the team leader should check that the centrally coordinated arrangements for the maintenance and calibration of all test and measuring equipment including ancillary items are effective. He should also check that calibrations are being performed at the appropriate intervals and over the range used for the accredited tests. He should ask to see the results of any calibrations performed in-house including evidence that the operator concerned has been trained, is following documented procedures, and is using reference standards that are calibrated and that the calibrations are traceable to national standards. Finally, he should check that the calibration uncertainty is appropriate to the test measurement to be made.

#### 6.4.3 Test certificates and reports

To check that all test data is being correctly reported the team leader should randomly sample the laboratory's central test reports file. Reports and certificates should be examined, compared with original observations, checked for content as required by the test specification, and checked for accuracy, pagination and unique identification. The assessor should also confirm that the reports have been checked and signed by an authorised officer and that they do not contain misleading statements. Individual technical assessors should check samples of reports in their area of expertise

### 6.5 Technical Assessment

After the main elements of the quality system have been examined (or at the same time if there is a team) the technical experts on the assessment team should go to the testing areas and assess the technical competence of the staff, the suitability of the equipment and facilities, the availability and content of test and calibration procedures, the procedures for handling, storage and disposal of test items, the methods for recording and reporting results and any other aspect of the laboratory's activities that may affect the quality of testing.

#### 6.5.1 Testing Capability

In particular, the assessors should ask laboratory staff to demonstrate their competence in testing by observing tests performed on clients samples. The assessors should check that the testing staff understand what they are trying to do and that they have adequate documented instructions available for operating the equipment and to enable them to perform the test consistently. If the assessors have any doubts about the competence of any of the staff they should ask the laboratory management to show them the training records for the staff concerned and the criteria used to adjudge when such staff can be authorised to perform specific tests and/or operate specific items of testing equipment.

#### 6.5.2 Test and measuring equipment

For each accredited test that the laboratory performs the assessors should check that the laboratory has all the equipment required for the test and that it complies with the requirements of the test specification. In the case of equipment that involves computers or software the laboratory should be asked to provide evidence that it has checked that all test output data is what it should be.

The assessors should also check that all major items of test equipment and any ancillary items having a bearing on the quality of the test results are calibrated and the calibrations evidenced by current certificates (preferably certificates issued by accredited calibration laboratories) showing traceability of the measurements to national standards. Where the laboratory uses reference materials the assessors should check that they are certified reference materials and that they are traceable to national or international reference materials or standards or measurement.

In examining these certificates and the calibration records generally the assessors should be checking to see that the uncertainty of measurements are appropriate to the uncertainties the laboratory claims for its testing and that all calibrations are being performed at intervals in keeping with these requirements and any requirements stated in test specifications.

#### 6.5.3 Test methods and procedures

While watching tests being performed the assessors should compare the procedures adopted by the testing staff with those stated in the published specifications and/or in-house procedural documents. Should the testing staff deviate from the stated procedures this fact should be recorded after discussing the deviation with the staff concerned. The assessors should take note of any observations made as part of the testing process and should point out any additional information or data that the testing staff should be recording to ensure that the test can be repeated at a later date, that the requirements of the testing specification have been met and that the sources of any errors can be identified. Checks should be made of worksheets and workbooks by the assessors to establish that records are consistently made at all times and that any alterations to observed data or calculations have been made in such a way that all the original observations can still be identified.

Where appropriate, the laboratory should be asked to show that it is making regular use of certified reference materials to "calibrate" methods or instrumental techniques and that it is employing quality control checks to certify the quality of its test data.

The assessors should encourage laboratories to participate in proficiency testing programmes and where they have done so should ask to see the results and obtain explanations for any apparently inadequate performances. The assessors should make sure that the laboratory takes any actions necessary to improve their performance.

#### 6.5.4 Environment and confidentiality

Finally, while the assessors are touring the laboratory facility they should check that the staff still have sufficient space and light to perform the tests and that the testing environment still meets any requirements of the test specifications. They should also check that the arrangements to prevent unauthorised access to the laboratory and to its testing and measuring equipment and the arrangements to ensure that clients' information, reports etc are held securely and in confidence are still operating effectively.

#### Summary

If all the actions that have been described are carried out as part of a formal surveillance programme over a defined period of time and if the laboratory has carried out any corrective actions that have been identified then the accreditation body can be confident that it is maintaining the right standards.

The role of the assessment team is the same whether visiting a construction materials testing laboratory or a laboratory testing medicines and the same types of departure from the accreditation body's requirements tend to be identified.

One of the most common observations assessors make during surveillance visits is that staff are not working strictly to the accredited test methods. The reasons for the departures vary from inadequate training or lack of understanding to the desire to take short cuts or introduce variations to the method. Such findings by assessors will alert them to the need to examine the laboratory's auditing arrangements very closely since the laboratory may not be devoting sufficient resources to such activities.

At surveillance visits assessors also find that some of the systems have been allowed to drift away from the standards they set themselves at initial assessment. For example, it is often noticed that standards are not maintained when a higher than average throughput is suddenly required.

This is often seen in concrete testing where the laboratory fails to recognise the dangers from overloading its curing tanks or that consideration should be given to increasing the frequency of calibration of its testing machines when its throughput increases significantly.

If the laboratory has designed its quality system properly and has the necessary resources to implement its quality arrangements it should be able to anticipate or identify such dangers rapidly and not allow the standards to waver. During surveillance visits assessors should point out the long-term effects of departures from their quality system and make sure that any corrective actions sort out the root cause of the departures.

Surveillance visits are not only therefore the mechanism by which accreditation bodies confirm that accreditation should be maintained they are also an opportunity to assist the laboratory in developing a more effective quality system that fully meets their needs.

# 37 DEVELOPMENT OF ASTM STANDARDS FOR LABORATORY ACCREDITATION AND OTHER ACCREDITATION ACTIVITIES IN THE UNITED STATES

J.W. LOCKE
American Association for Laboratory Accreditation,
Gaithersburg, Maryland, USA

Abstract
ASTM Committee E-36 has been aggressively developing standards in the area of laboratory accreditation. Some of these standards have lead to similar ISO standards, some have lagged behind. It is the Committee's commitment to consider all relevant standards and develop those needed for use in domestic and international trade and for regulatory purposes. The work of this committee and its commitment for the future is described in this paper. Also described is a brief summary of the state of the art in accreditation in the United States, based on the 2nd National Forum for Laboratory Accreditation sponsored by A2LA and held on May 1, 1990.
Keywords: Laboratory Accreditation, Standards, Criteria, USA Programs.

## 1 Introduction

The American Society of Testing and Materials (ASTM) and its Committee E-36 on Laboratory and Inspection Agency Evaluation and Accreditation (LIAEA) present this summary of our recent accomplishments in development of laboratory accreditation standards and assure you of our commitment to work with RILEM and ILAC in the continued development of important standards to help improve the quality of our testing laboratories. ASTM Committee membership includes interested participants from all countries and E-36 has received some very able contributions form Canada and Australia.

Perhaps the best way to summarize this activity is to present the long range plan we have adopted, identifying those standards that have been developed to date and those which are being worked on.

## 2 Goal #1: Develop Standards.

The primary goal of the Committee is the development of standards to provide criteria, definitions, and other guidance for assessment, evaluation, and accreditation for testing laboratories and inspection agencies and for the operation of laboratory accreditation and evaluation systems and bodies.

**2.1 Objective 1. Maintenance of Current Standards**

The committee shall maintain the standards already developed by subjecting them to an annual review in accordance with E-36 By-Laws. All standards under E-36 jurisdiction shall be balloted for reapproval, revision, or withdrawal according to ASTM regulations (at least every five years). Strategies for 1990 are:

Conduct an annual review of all current standards.

Summarize the status of existing standards on a regular basis. The current status is as follows:

ASTM E 329 "Standard Practice for Use in the Evaluation of Testing and Inspection Agencies as Used in Commerce". This standard, originally published in 1967, was revised and approved in March, 1990. There is nothing like it in ILAC. It deals with laboratories testing and inspecting construction materials, composites, and construction practices and references other ASTM standard practices for laboratories: C1077 on Concrete and Concrete Aggregates; C1093 on Unit Masonry; D 3666 on Bituminous Paving Materials; D 3740 on Soil and Rock as used in Engineering Design and Construction, and E 543 on Qualification of Nondestructive Testing Agencies.

ASTM D 548 "Standard Practice for Preparation of Criteria for use in the Evaluation of Testing Laboratories". A revision is on the May 1990 society ballot. This revision brings E 548 in line with ISO Guide 25.

ASTM D 994 "Standard Guide for Laboratory Accreditation Systems". This standard has been reaffirmed as it is in the spring of 1990. It was the basis for ISO Guides 54 and 55, but relates more directly to Guide 55 on Accreditation Systems. The Committee saw no need for a Guide 54 on Accreditation Bodies, since most key provisions are described under systems.

ASTM D 1187 "Standard Terminology Relating to Laboratory Accreditation". This standard was reapproved in October, 1989.

ASTM D 1224 "Standard Guide for Categorizing Fields of Testing for Laboratory Accreditation Purposes". This standard was approved in December, 1987 and is scheduled for review in 1992. It is based on the Australian (NATA) structure identifying scopes of accreditation. The Committee is working on a document for use in standardizing directory listings for accredited laboratories and the outcome of this work may directly affect this standard.

ASTM D 1301 "Standard Guide for Development and Operation of Laboratory Proficiency Testing Programs. This standard was approved in September, 1989 and is scheduled for review in 1993. It is much advanced over ISO Guide 43 on proficiency testing because it actually presents criteria for evaluating proficiency testing programs.

ASTM D 1322 "Standard Guide for the Selection, Training, and Evaluation of Assessors for Laboratory Accreditation". This standard was approved in March, 1990 and there is nothing like it in ILAC, although much of the material was discussed there.

ASTM D 1323 "Standard Guide for Evaluating Laboratory Measurement Practices and the Statistical Analysis of the Resulting Data". This standard was approved in December, 1989. It focuses on the concept of statistical measurement control (SMC) and how the availability of statistical data may be taken into consideration in laboratory accreditation assessments. There is nothing like it in ILAC.

**2.2 Objective 2. Determination of Need for New Standards**
It is the duty of the E-36 Committee to be sensitive to the needs for laboratory accreditation standards in United States industry and government laboratories and to monitor all accreditation standards written in the national and international arena which might affect markets of U.S. industry. Strategies for 1990 are:

Review standards, guides, and draft documents of international groups such as CEN/CENELEC, ILAC, ISO/IEC, ISO CASCO, OECD, RILEM and ANSI/ASQC to determine those areas in which no relevant ASTM standard exists.

Make assignment of responsibility to the appropriate sub-committee to study these areas and identify where need for an ASTM standard exists.

**2.3 Objective 3. Complete Standards Currently in Development Stage.**
These include:

Guide for Surveillance of Accredited Laboratories.
Guide for Accreditation of Site Laboratories.
Guidelines for Establishment and Review of Mutual Recognition Agreements.
Guidelines for Preparation of Cooled Manuals for Laboratories. This would relate to ISO Guide 49 on the same subject.
Guide to Presentation of Test Results. This would relate to ISO Guide 45 on the same subject.

**2.4 Objective 4. Plan the Development of Standards for which a Need is Seen but which are not in Progress.** Some suggested areas are:

Standard on availability, use, and development of certified reference materials for laboratories.
Calibration and maintenance of test measuring instruments in laboratories.
Guide to organization of technical seminars.
Generic guidelines on sampling and sample retention for laboratories.
Guide on the traceability of samples through the laboratory system.
Identification of test areas

Guide for handling laboratory and accreditation system complaints
Guide for the development of specific criteria for use in evaluating testing laboratories.
Use of computer acquired data.
Guide to the use of laboratory comparison programs.

## 3 Goal #2. Organization Development.

Maintain the E-36 Organizational Structure and Promotional Activities for Committee Growth.

**3.1 Objective 1. Maintain Membership and Recruitment** of New Members. Strategies for 1990 are:

Develop and distribute a flyer about the work of the Committee.
Encourage personal contact between members and prospective members.
Invite the executive committees of all other ASTM main Committees to appoint a member as liaison to E-36.
Conduct a letter survey of membership to find areas of improvement.
Develop interesting agenda for Committee meetings and publicize if possible.
Inform Committee members of interesting developments by letter invitation.

**3.2 Objective 2. Stimulate Participation in E-36 Activities.** A strategy for 1990 is:

Contact by letter each member who has not participated during the past two years through attendance at a meeting or voting by mail and encourage them to take an active role.

**3.3 Objective 3. Provide Training and Recognition.** Strategies for 1990 are:

Develop a system for integrating new members into the system quickly.
Develop a system for leadership training for Committee responsibilities.
Present appropriate tokens of recognition for outstanding service, such as ASTM Award of Merit.

**3.4 Objective 4. Review the Scope, By-laws and Long Range Plan.** The Executive Committee should conduct a periodic review of the scope and the by-laws of the Committee. The long range plan of E-36 should be up-dated annually at the beginning of the year. Strategies for 1990 are:

Schedule the Scope and By-laws for future review, inspection and revision. The Scope and By-laws were last revised at the end of 1989.
Prepare the E-36 Long Range Plan for adopting in 1990. Up-date it annually thereafter.

**4 Goal #3. Improve Committee Efficiency.**

Make Efficient Use of all Resources Available to the Committee and Inform the Public of E-36 Activities and Encourage the Use of E-36 Standards.

**4.1 Objective 1. Cooperate with ASTM Staff.** Strategies for 1990 are:

Cooperate fully with the assigned staff person by communication with him in a timely manner in matters such as minutes, balloting, and problems needing staff assistance.

Develop a plan for exchange between the Committee and other ASTM Committees having an interest in accreditation standards.

Actively seek to co-sponsor seminars and like functions with other ASTM Committees.

**4.2 Objective 2. Explore the Use of Non-ASTM Sources Which Could be Helpful to the Committee in Accomplishment of Its Goals.** Strategies for 1990 are:

Investigate the interest of other non-profit organizations in accreditation standards and see if cooperative efforts could effect time and labor savings.

Plan a more effective use of government contacts and seek to involve more government employees in the activities of the Committee.

Make a concerted effort to get more industry involvement in the standards writing process.

**4.3 Objective 3. Inform and Educate the Public about the Importance of Accreditation Standards to National and International Trade and Encourage the Use of E-36 Standards.** Strategies for 1990 are:

Assist in the work of the ad-hoc subcommittee formed to promote cooperation between E-36 and other ASTM Committees engaged in producing specific accreditation standards.

Encourage ASTM Committees who have not written specific accreditation standards to do so if deemed needed.

Seek effective ways of giving information to the general public about Committee goals and activities.

This is a very ambitious plan for E-36, but the Committee is proud of its accomplishment in the past few years and is prepared to move ahead rapidly in this area of standardization. The committee recognizes RILEM and ILAC as groups which develop basic "pre standards" and want to assure members of both groups of the need for their continuing input.

**5 Second National Forum on Laboratory Accreditation.** This was held on April 30 - May 1, 1990.

The keynote speaker, **Donald R. Johnson, Director, Technology Services, National Institute of Standards and Technology (NIST),** spoke of great eras of change (1492, 1776) and indicated that now, 200 years after the U.S. integration of markets, Europe is preparing to unite her "common markets". This move to market unification will also have a profound effect on the way people in Europe are governed, and the concept, EC-92, is expected to go well beyond the 330 million EC community and may encompass the 800 million people in the European market place. He pointed out that eventually, accredited laboratories in the EC will be able to certify that products meet Euronorms and are thus fit for sale in Europe. Dr. Johnson indicated in order to keep U.S. trading opportunities open in the EC we will have to do product testing in Europe, or develop an acceptable, technically-equivalent system of accredited testing laboratories here. He noted that the U.S. Secretary of State James Baker and the Vice President of the EC Commission, Dr. Filippo Maria Pandolfi, two of our worlds foremost statesmen, are reaching positive agreement on the major points related to very difficult trade problems and wonders why there can't be cooperation in the United States in the field of testing. He quoted Joseph O'Grady, President of ASTM, "For the free flow of trade, equivalent standards, test data, and certification procedures must be mutually accepted and reciprocal."

The larger problem will fall on those companies without plants or even officers in Europe. Since these are the companies growing rapidly in the United States, a solution is needed soon. Dr. Johnson concluded his keynote remarks with the hope that we can begin dialogue at one firm point of agreement -- "...that we must work together to develop a laboratory accreditation framework that will see us beyond 1992 and into the twenty-first century." The building blocks of such a system will have to include the following elements:

Technically Strong Foundation
Systematic Organization
Public/Private Cooperation
Private Sector Administration
Periodic Evaluation
Official Recognition

One possible role for the government could be to referee the technical validation process so that official recognition can be given to accredited laboratories. A "USA" mark with the backing of the full force of the United States government will be our best hope of indicating those products certified to privately developed U.S. standards, or the ability of specific labs to certify, in this country, conformance to Euronorms.

A brief summary of procedures of a number of accreditation systems in the United States and Canada provided some idea of the breadth of approaches to accreditation and the problems which will have to be faced in arriving at a cooperative agreement.

**5.1 A2LA.** The American Association for Laboratory Accreditation program was summarized, noting that it is was based on international criteria, has accredited 220 laboratories in a variety of fields of testing and has achieved international recognition by accrediting laboratories in Canada and by signing a Memorandum of Understanding with the Hong Kong Laboratory Accreditation Scheme, an agency of the Hong Kong government.

**5.2 NIDA.** The National Institute on Drug Abuse program is a very sensitive program, formalized in the Federal Register, Monday, April 11, 1988, "Mandatory Guidelines for Federal Workplace Drug Testing Programs". The system has accredited 48 labs; 100 are in process of accreditation; proficiency testing is a very integral part of the program.

**5.3 NRTL.** The Occupational Safety and Health (OSHA) Nationally Recognized Testing Laboratory (NRTL) system accredits laboratories for testing, listing, and labelling of products used in the workplace, based on regulations in the Code of Federal Regulation (CFR) 29 CFR 1910.7 dated May, 1988. Two NRTLs were recognized for five years in the regulation, three more have been accredited, and several others are in process. No fees are charged; most of the accreditations are in the electrical/electronics field.

**5.4 RCRA.** The U.S. Environmental Protection Agency (EPA) Resource Conservation and Recovery Act (RCRA) program and its relationship with the states and the laboratories which they authorize was described. The needs for verifying reference materials commonly used, for highly experienced and educated auditors, and for the efforts of the program to arrive at a common understanding related to instrumentation, QA programs, facilities, and documentation were described.

**5.5 SCC.** The Standards Council of Canada (SCC) system was summarized. The program was open to laboratories in the United States as well as Canada as a result of the Canada/USA Free Trade Agreement. This program includes an ability to accredit calibration laboratories, in cooperation with the Canadian National Research Laboratory.

**5.6 NVLAP.** The National Voluntary Accreditation Program operated by NIST was described. This program is set out in 15 CFR Part 7. It is a self-supporting program and has accredited 1174 laboratories through 1990; 949 of these in the field of asbestos where accreditation is required to test asbestos in U.S. EPA's Asbestos Hazard Emergency Response Act (AHERA) program.

**5.6 FSIS.** U.S. Department of Agriculture, Food Safety and Inspection Service (FSIS) system to accredit meat and poultry laboratories was described. Methods of the Association of Official Analytical Chemists (AOAC) and check samples are the foundation for this program. This is an example of a very formal federal program.

**5.7 FDA.** The U.S. Food and Drug Administration (FDA) consumer protection and regulatory mission was described. This is an example of very informal government involvement through Good Laboratory Practice regulations for toxicology laboratories.

**5.8 NJDEP.** The Office of Quality Assurance, New Jersey Department of Environmental Protection system was described. The system has accredited 270 drinking water laboratories, 60 A-280 (Special New Jersey drinking water requirements), 470 waste water laboratories, 6 radiological laboratories and 18 bioassay laboratories. The system is well developed and planning growth; there is no reciprocity with other states.

**5.9 CDHS.** Environmental Laboratory Accreditation Program, California Department of Health Services, and how it has evolved from 1951 to its present stage of consolidation in 1988 was described. This system recognizes 23 fields of testing in the environmental arena. They have no reciprocity with other states but are interested in exploring the possibility.

**5.10 ICBO.** The International Conference of Building Officials (ICBO) recognizes test data when approved by the ICBO Evaluation Service. This is a very flexible system, recognizing laboratories which provide test data on given tests or more formally by listing testing agencies. Evaluation reports on building products, structural assemblies, etc. are prepared to determine compliance with design or code standards, and these reports are used by enforcement agencies as demonstration of compliance to standards.

**5.11 GE.** The General Electric Aircraft Engines system is intended for suppliers to GE. The main thrust is on comparison of round robin data. Audits are done on each laboratory to determine technical knowledge and to establish a working relationship with the laboratory.

**5.12 NADCAP.** The Society of Automotive Engineers recently developed the National Aerospace & Defense Contractors Certification Program (NADCAP). This system is built on Department of Defense assessment type procedures. It recognizes laboratories, such as nondestructive testing facilities, and suppliers, as in the case of sealants. This program cooperates with a similar one run to accredit fastener suppliers by the American Society for Mechanical Engineers (ASME).

**5.13 SHELL.** Shell Development Company uses environmental laboratories and has developed a procedure to qualify the laboratories used. The need for a national system to accredit environmental laboratories was stressed.

**5.14 WLCP.** U.S. EPA Drinking Water Laboratory Certification Program is set up under statutory requirements. It was pointed out that one source of unified criteria is legislation. EPA's program is in response to legislation which establishes a uniform criteria to be used by states which perform the accreditations. Reciprocity was possible in Florida, Georgia, Illinois, Michigan, Minnesota, Missouri, North Dakota, North Carolina, South Carolina, Tennessee, Washington and Wisconsin. Reciprocity is not permitted in California, Kansas, Massachusetts, New Hampshire, Ohio, Oklahoma Pennsylvania, and Utah. EPA encourages reciprocity and mutual recognition among systems.

**5.15 General Standards Picture in the United States.** Consultant Helen Davis compared the standards situations in the United States and Europe and noted the willingness of U.S. companies to shift to EC requirements because it is easier and it is mandated. This bodes ill for the U.S. enterprise which created the American Standardization System. Many ISO standards in the laboratory accreditation area trace their genesis to U.S. standards and the basis exits for operating U.S. systems every bit as adequate as the EC systems now being developed. Specific note was taken of the work of ASTM Committee E36 on Laboratory Accreditation which has developed a number of standards which were the basis for ISO standards.

**5.16 The American National Standards Institute.** George Willingmyre presented the ANSI view of happenings in Europe and the commitment of ANSI to initiate contacts with its counterpart in Europe, CEN/CENELEC was explained. Four models for international cooperation were presented for consideration:

1. National Treatment: all applicants accepted (including foreign) (Most U.S. systems operated this way).

1A Modified National Treatment: Adding quid quo pro trade considerations.

2. Reciprocal Treatment: Access given to laboratories in other countries if access is available in those countries. EC countries now operate this way.
3. International Model: A National body recommends various accredited labs who are then inspected by an international accreditation body; if accepted, a laboratory would be recognized in all participating countries.
4. Bilateral Models: Agreement between principals; UL and BSI for example.

The EC has two distinct spheres of operation: 1) the regulatory sphere, and 2) the voluntary sphere; the models used in each case are likely to be different. Emphasis should be placed on obtaining notification capability for U.S. laboratories.

**5.17 Government.** John Donaldson, NIST pointed out the problems associated with accrediting laboratory accreditation systems. It was suggested that people in the United States do not really understand the number and quality of U.S. accreditation bodies so it would be difficult to convince Europeans of their acceptability. NIST is developing a new directory of U.S. systems to be available in the fall which might be of help clarify the scene. Criteria already exist which could be used to make decisions about the quality of these systems. Government would probably have to be involved if a coordinated system is to be recognized in Europe.

**5.18 A2LA Ad Hoc Committee on National Accreditation Policy.** The results of a questionnaire which had been distributed by the A2LA Ad Hoc Committee on National Laboratory Accreditation Policy was summarized. Specifically, the committee addressed the idea of establishing a policy for accrediting agencies. Most respondents saw the need for some type of system in order to gain international recognition. Key program issues dealt with auditor qualifications and staff/organization power checks. Most saw such a system working on a set fee basis. There was no agreement on whether the coordinating system could be operated by ANSI or whether there needed to be government involvement.

The Chairman of the Ad Hoc Committee, spoke eloquently of the need for government involvement in any coordination system so that recognition by government entities would be facilitated.

**5.19 ANSI Certification Role.** The ANSI role in coordinating U.S. standards bodies was explained and the recognition ANSI receives in Europe because it is the formal U.S. representative to the ISO and IEC was noted. ANSI is the U.S. source for international documents. The ANSI program to accredit product certifiers was described. ANSI is acting on a request received from A2LA to establish a program to accredit laboratory accreditation systems. ANSI does not now have a program, but a special task force of the Certification Committee is being established to: 1) develop the criteria for accrediting such systems to be approved by the ANSI Board of Directors; and 2) to review the A2LA program in detail to see if it meets the criteria established.

**5.20 General Picture in Europe. Mr. Alan Bryden, Director General, Laboratoire National d"Essais (LNE)**, the featured speaker, presented a summary of the current status of relevant EC 92 activities. The history of the EC program, arriving at "The Global Approach" and the European Organization for Testing and Certification (EOTC), was described. The institutions in Europe which were established to develop cooperation among all interested parties were established:

EUROMET (Metrology Laboratories)
Western European Calibration Cooperation (WECC) (Calibration laboratory accreditation systems)
Western European Laboratory Accreditation Cooperation (WELAC) (Laboratory accreditation systems)
EUROLAB (representative from public and private testing laboratories -- Alan Bryden was elected Chairman)

There are two distinct areas: 1) voluntary; contractual relationships between suppliers and clients; and 2) mandatory, safety and health requirements. A gray area is in the testing verification by government needed to conduct trade. The Global Approach will encourage mutual recognition of programs in the various countries rather than establishing one European Community certification mark. The European Organization for Testing and Certification (EOTC) has been established as an EC body with CEN/CENELEC (the European standards organizations) acting as secretariat, to coordinate testing and certification activities. The European Quality System (EQS) has been established to coordinate quality system registration activities in the various countries. EOTC actions will be based on the Euronorms EN 45000 while EQS is based on Euronorms EN 29000. These Euronorms are based on ISO standards and guides and will be updated as the ISO documents are updated. These organizational efforts will mean more transparency, non discriminatory access to the market, and more flexibility in implementation. It was pointed out that there is a need for Memoranda of Understanding in the areas of laboratory accreditation, product certification and quality system registration. Specific focal points in the United States are needed to address these European counterparts.

**5.21 Chester N. Grant, Chairman of A2LA**, briefly summarized the conference, indicating that the variety of activities seemed to warrant the need for continuing this Forum next year. Hopefully, at that time, there will be a report on success in coordination between systems will have been achieved this year.

The forum focused on EC 92 product testing issues and environmental testing issues because they represented the two greatest areas of concern in laboratory accreditation activities in the United States. The issues focussed on quite different concerns during the forum and attendees appeared to focus on one or the other. The audience was asked if they should be separated into two forums in the future. By a vote of about four to one, the participants preferred that any future forum should address both issues at the same time.

# 38 LABORATORY ACCREDITATION BY THE STANDARDS COUNCIL OF CANADA

M. ARCHAMBAULT
National Accreditation Program for Testing Organizations (NAPTO), Standards Council of Canada, Ottawa, Canada

Abstract
In 1973 the Standards Council of Canada (SCC) established the National Standards System (NSS) as a framework for coordinating voluntary standardization in Canada. The national accreditation program for testing organizations (NAPTO) is a critical component of the NSS. This paper reviews current SCC testing and -since 1984- calibration laboratory accreditation operations. It notes the importance of cooperation between the National Research Council of Canada and SCC to establish a national network of accredited calibration laboratories, and concludes with a review of the resulting benefits of NAPTO accreditation.
Keywords: Accreditation, Assessment, Laboratories, NAPTO, Standards Council of Canada.

## 1 Introduction

### 1.1 Need for Laboratory Accreditation

There are many reasons for setting up a laboratory accreditation program. For the testing or calibration laboratory, accreditation is an independent evaluation of the quality of its operations; it gives corporate management assurance that the laboratory is operating in a sound technical and administrative fashion, and it provides extensive publicity for the laboratory through publication of its capabilities in the accreditor's directory.

Many large manufacturers have their own laboratory accreditation programs to ensure that goods and services provided by their suppliers will meet their needs. Some corporations have set up programs to satisfy the requirements of off-shore customers. Some federal government departments run their own programs to ensure that manufacturers and service organizations meet the requirements of specific legislation or contracts. Provincial government agencies also run accreditation programs for laboratories that test products to provincial procurement specifications. There are also non-governmental, non-commercial groups, such as hospital associations or consumer organizations, which have set up laboratory accreditation programs. Again, such programs are aimed at ensuring that a uniform, quality service is provided.

In the United States, there are hundreds of formal and informal accreditation programs for testing laboratories, each established for specific reasons. Most are private programs which have been set up to evaluate and accredit the wide range of testing facilities in the U.S. Often there is a great deal of similarity of criteria and requirements between the systems, resulting in frequent duplication of assessment activities that wastes increasingly scarce resources. Although Canada has multiple laboratory accreditation programs, the average laboratory is not faced with as many audits or assessments as its U.S. counterpart.

**1.2 The Standards Council of Canada Program**

The Standards Council of Canada is a federal Crown corporation which was established by an Act of Parliament in 1970. SCC operates the National Standards System (NSS) which is a federation of independent organizations engaged in voluntary standardization. The main role of the Council is to establish for the NSS the policies and procedures for accrediting NSS members.

Nationally, SCC activities comprise programs for the accreditation of standards-writing, certification, and testing organizations. Internationally, SCC is the Canadian member body on the International Organization for Standardization (ISO), sponsors the Canadian National Committee of the International Electrotechnical Commission (IEC) and represents Canada at the International Laboratory Accreditation Conference (ILAC).

Our National Accreditation Program for Testing Organizations (NAPTO) began operating in 1980. This program is open to qualified laboratories in any field of testing. By qualified we mean:

Competent in the fields of testing practiced.
Staff, facilities and procedures meet SCC requirements.

Size of the organization is not a criterion; competence is the prime consideration.

SCC has accredited laboratories in such diverse fields as environmental, concrete, oceanographic, petroleum, electrical, Non Destructive Testing, and armaments testing. An applicant must identify the specific tests or measurement capabilities for which it wants accreditation; SCC then must verify the organization's competence in the activities for which accreditation is sought.

## 2 NAPTO Program Operations

### 2.1 Program Overview

NAPTO operates in a manner similar to other major accreditation programs such as NATA (Australia) or NAMAS (U.K.). Once an applicant has listed the specific tests for which accreditation is sought, as part of its detailed application, a team of assessors is assembled, with expertise covering the stated area(s) of testing. The team, including an SCC staff member as team leader, then carries out an on-site assessment. The assessment is a two-fold process: verification of compliance with generic criteria and demonstration of technical competence.

NAPTO operations are structured to help ensure the integrity and quality of the program and to make sure that evaluations are carried out in a consistent manner. This provides a maximum level of confidence in the program for both accredited laboratories and the users of their services.

### 2.2 Effect of the Free Trade Agreement

With the signing of the Canada/USA Free Trade Agreement, both parties to the Agreement are obliged to accept applications from laboratories located in the other country. SCC has already received applications from 4 U.S. laboratories; these are being processed along with those of 29 Canadian applicants.

### 2.3 Accreditation Standards

The present accreditation criteria and requirements, published in SCC document **Accreditation Criteria and Requirements for Testing Organizations** (CAN-P-4B), are much improved over the original set. This document is compatible with the international laboratory accreditation criteria in ISO/IEC Guide 25 and the European Normes EN 45,000 Series. CAN-P-4 is a generic document in that its criteria and requirements must be met by each applicant for accreditation, regardless of size or field of testing. Because it is now clearer and more concise, candidates have less trouble preparing applications and these are more consistent. A "fill-in-the-blanks" application form has been developed, available in both hard copy and electronic (floppy disk) form, that requires the applicant to present its information in the, format of CAN-P-4.

By its very nature, an accreditation program is a dynamic, changing process. The SCC laboratory accreditation program criteria and procedures are regularly reviewed and updated to meet the needs of changing technology and revision of international standards. In addition, improvements may be made as a result of comments and suggestions from those involved at all stages of the accreditation process - SCC staff, laboratory staff, assessors, and members of the technical, advisory and executive committees.

**2.4 Accreditation Process**

The NAPTO accreditation process covers 5 basic steps. Once an application is received, (1) it is reviewed by SCC staff for completeness and any additional necessary information is obtained from the applicant; then (2) a suitable assessment team is assembled to (3) carry out the on-site assessment of the applicant's facilities. The purpose of an assessment is to confirm compliance with the SCC generic requirements and to verify the applicant's ability to do the tests for which accreditation is sought. Once any non-compliance item(s) identified during the assessment have been resolved, (4) an assessment report is prepared and (5) reviewed by technical and advisory committees before final accreditation is granted.

NAPTO assessors get a solid grounding in the SCC criteria, requirements, and procedures as well as basic instruction on assessment techniques. Training includes viewing an SCC-produced video entitled **Lab Bench** that describes the accreditation process with emphasis on the assessment process. A formal assessor training program has recently been initiated.

The assessors selected for an assessment must be acceptable to all parties concerned. They participate in the following activities during an assessment: (1) review of the applicant's operating procedures; (2) inspection of the physical facilities of the laboratory; (3) confirmation of appropriate calibrated testing equipment; and (4) verification of the ability of the applicant laboratory's staff to perform the tests for which accreditation is sought, through demonstration of a selection of tests representative of its scope of testing. Based on the summation of these activities, a decision is made on the applicant's accreditibility.

**3 NAPTO Program Costs**

Not all accreditation programs charge a fee to applicant laboratories; some users of goods and services set up no-fee programs specifically for their suppliers of testing or calibration services. For example, the Canadian Department of National Defence (DND) has a "Recognition" program for calibration laboratories. DND charges no fees, but as a condition of Recognition, laboratories must agree to provide services to the Department or to its military contractors.

Most other national laboratory accreditation programs charge a fee, ranging from a minimal basic fee to complete cost recovery. SCC originally charged flat application and annual fees of $500 to each applicant, regardless of its size or scope of testing / calibration capabilities. In January 1988, a revised fee structure was introduced. It included an application fee of $850, an annual renewal fee of $750 and lesser fees for expansion of the scope of accreditation. In 1989, government action to increase fiscal responsibility led to budgetary constriction at the Council that resulted in further fee increases. The present accreditation fee is $2000 with an annual fee of $1,200. These remain flat fees with no consideration of the scope of accredited activities; however, this will change as NAPTO moves to a cost recovery policy.

## 4 Calibration Laboratories

To date, SCC has accredited 7 calibration laboratories. The assessment of calibration laboratories is similar to the two-fold process for testing laboratories: (1) verification of compliance with generic criteria and (2) verification of technical competence. Step 1 is the same for all applicant laboratories. For calibration laboratories, step 2 involves a measurement exchange process. To verify its uncertainty claims for a stated parameter and range of measurement, the applicant makes specified measurements on metrological artifacts characterized and supplied by a reference laboratory. The reference lab then analyzes the applicant's results to determine if it can indeed measure as accurately as it claims.

The first 5 applicant calibration laboratories were accredited using the Standards Laboratory of the Department of National Defence as the reference lab. This arrangement was possible because the 5 applicants applied to both SCC and DND for accreditation of their measurement capabilities.

Because other applicants are not interested in applying for DND recognition, a formal agreement was signed with the Laboratory for Basic Standards of the National Research Council of Canada (NRC). This agreement established the NRC Calibration Laboratory Assessment Service (CLAS), performing reference laboratory functions similar to those of the Department of National Defence. The first calibration laboratory to be processed under this partnership was recently granted accreditation. At present, there are 14 applications from calibration laboratories being processed, and several others in preparation. The foundation of a Canadian Calibration Network has been laid.

## 5 Conclusion

### 5.1 NAPTO Program Status

NAPTO is growing steadily. At present there are 69 accredited laboratories with more than 33 applications for accreditation at various stages in the process. It is conservatively estimated that by 1995, there will be about 500 accredited laboratories in the NAPTO system.

As a result of this success, users of testing services are increasingly prepared to choose SCC-accredited laboratories to provide testing or calibration services. Arrangements for the acceptance of laboratories accredited by SCC have been signed between SCC and several large users of testing services, such as the federal Department of Consumer and Corporate Affairs and the Ministry of Transportation in the Province of Québec. Discussions are currently underway aimed to finalize similar agreements between SCC and other major users of testing services.

### 5.2 Major Benefits

SCC has developed the NAPTO program requirements and procedures in harmony with ISO/IEC, EN and ILAC guidelines on laboratory accreditation. A major goal of NAPTO now being achieved is to have SCC-accredited testing and calibration laboratories recognized and accepted by Canadian industry, government agencies and other users of laboratory services, thereby reducing duplicate assessments and audits that waste limited resources.

The SCC program compares favourably with national programs around the world. We are confident that as the importance of international trade continues to increase, so too will the need for the NAPTO national network of competent testing and calibration laboratories able to meet the demands of our expanding economy.

# 39 IMPACT OF ACCREDITATION ON LABORATORY QUALITY ASSURANCE

R.L. GLADHILL
National Institute of Standards and Technology (NIST), Gaithersburg, Maryland, USA

Abstract
Since 1978 the United States National Institute of Standards and Technology (NIST) (formerly the National Bureau of Standards (NBS)) has accredited testing laboratories through its National Voluntary Laboratory Accreditation Program (NVLAP). NVLAP uses on-site assessment, proficiency testing, and other means in order to evaluate a laboratory to determine suitability for accreditation. In particular, this evaluation emphasizes a thorough review and assessment of an applicant laboratory's quality assurance system.

Over the years, NVLAP has uncovered and helped to resolve significant problems in quality assurance systems and procedures employed by applicant laboratories. These problems, NVLAP's approach to solving them, and the resultant implementation of corrections are discussed.
Keywords: Accreditation, Criteria, Testing, Laboratory, Proficiency Testing, Quality Systems, Quality Assurance, Quality Control.

## Background

The U.S. Department of Commerce established the National Voluntary Laboratory Accreditation Program (NVLAP) in 1976 to provide national recognition, through accreditation, to qualified testing laboratories. NVLAP is administered by the National Institute of Standards and Technology (NIST) (formerly the National Bureau of Standards (NBS)). NVLAP provides an independent, third party evaluation of testing laboratories in accordance with well-defined, pre-established written criteria.

A primary goal of this program is to upgrade the quality of testing services offered by participating laboratories. Another major function is to provide assistance in the improvement of test methodologies and standards and improved testing laboratory practices.

NVLAP offers accreditation in specific technical disciplines only after a demonstrated need has been established. The determination of need may be based on the impact on public health and safety, regulation by a government agency, petition by laboratory users or laboratories themselves. Before establishing a specific accreditation program, necessary and sufficient standards and technically sound criteria must be available so that fully objective evaluations of applicant laboratories can be performed.

The decision to offer accreditation is based upon the needs of laboratories, the number of laboratories that might participate, professional organizations, public and private organizations concerned with commerce, trade, government regulations, public law, and public health and safety.

Qualified laboratories are granted accreditation for a quite specific scope of activities. They are issued a Certificate of accreditation and a Scope of Accreditation that define exactly the tests or types of tests for which a laboratory has demonstrated its capability and has been granted accreditation.

NVLAP began accrediting laboratories in 1978. Today accreditation is offered in thirteen technical disciplines, including construction testing services, asbestos fiber analysis, computer applications, personnel radiation dosimetry, and thermal insulation materials. The program is growing rapidly and currently serves 1200 laboratories, both domestic and foreign.

## Accreditation Criteria

The U.S. Code of Federal Regulations (15 CFR part 7) define the authority of NVLAP to grant accreditation, the procedures to be followed, and the general criteria to be applied. The criteria, based on the quality systems concept, define in general terms what a laboratory must have in place with regard to laboratory staff, facilities, equipment, calibration, procedures, quality assurance, quality control practices, recordkeeping and reporting.

Laboratories are evaluated against specific criteria for specific technical areas, derived from the general criteria and developed through a public consensus process. NVLAP actively encourages participation in the development of accreditation criteria from technical experts including those in government, private industry, standards and professional societies. Opportunities are provided at every stage of the criteria development process for comment and technical input from any interested party. The criteria are subject to periodic revision as the result of experience and to ensure harmonization with national and international standards and guidelines for laboratory accreditation systems such as ASTM Standard Practices E548, E329, C1077, D3666, and D3740, ISO International Standards 9000 and 9004, and ISO Guides 25 and 38.

The NVLAP criteria emphasize that a laboratory must have an effective quality system, including appropriate elements of both quality assurance and quality control. NVLAP's definition of "quality system" is consistent with the ISO definition, that is; "The organizational structure, responsibilities, procedures, processes and resources for implementing quality management."

The NVLAP criteria relating to quality systems require development and documentation of quality procedures compiled in a manual (or equivalent) which describes the quality assurance practices and quality control measures performed. In order to satisfy the criteria a laboratory must establish for its own unique needs and scope of testing services:

- qualitative and quantitative measures of testing performance;
- regular and routine implementation of those measures;
- statistical and analytical evaluations of performance;
- feedback mechanisms;
- a process for implementing corrective actions as necessary; and
- active revision and enhancement of the procedures in the light of experience and knowledge gained in the laboratory.

Written procedures must address: (a) an overall quality policy for the laboratory to efficiently control the administrative activities it conducts; (b) the initial training and methods to ensure the on-going competency of the staff; (c) methods of document control; (d) methods of sample control; (e) methods of data validation; (f) quality assurance for each function of the laboratory; (g) quality assurance practices and procedures for each test or type of test; (h) the use of internal audits, inter-laboratory testing, and reference materials; and (i) procedures for dealing with complaints. An individual staff member must be assigned to be responsible for periodic review and update of all quality procedures.

NVLAP provides applicant laboratories with a written outline and oral discussion of subjects that should be addressed in their quality manual. Detailed examples of manuals are not provided because the main purpose of the exercise is to encourage a laboratory to develop documentation unique to its own operations.

Laboratory Quality System Evaluation

The evaluation of a laboratory's ability to meet the NVLAP requirements and to perform the testing services for which accreditation was requested is based on the general and specific criteria. For a given testing discipline, the accreditation criteria are fully documented in a NVLAP Handbook developed for that discipline. The Handbook is available upon request to all interested parties. A Laboratory inquiring about accreditation is provided with the Handbook so that it may be fully aware of accreditation requirements prior to application and subsequent evaluation.

Evaluation for conformance to the NVLAP criteria is accomplished through on-site assessment of the laboratory, participation by the laboratory in proficiency testing programs, and technical expert review of assessment findings and laboratory responses to identified technical deficiencies.

NVLAP experience has been consistent in the sense that most laboratories (regardless of the technical discipline in which they are being evaluated for accreditation) initially have woefully inadequate quality systems. Very few laboratories have overall quality policies or procedures, and many laboratory managers do not understand the concept of quality systems.

Some laboratories choose not to apply for accreditation because of the perceived costs of establishing and maintaining quality systems. They apparently do not recognize the long term business and financial benefits of quality systems, not the least of which is protection of their interests in litigation by being able to produce adequate documentation of test procedures, process control and test results.

The observations presented below are based on the evaluations of 1200 individual laboratories, some assessed several times over a period of twelve years. They reflect the quality system deficiencies most commonly noted by NVLAP assessors.

It is important to note that the most valuable result of the entire evaluation process is the correction, by the laboratory, of all the deficiencies uncovered.

Documentation

The most common quality system deficiency, is by far, lack of documentation. Documentation is meager, at best, in almost every laboratory visited for the first time. Few laboratories have written operational procedures or laboratory test procedures. Adequate quality manuals are rarely found; many laboratory manuals are developed only because required by government contract. In some cases, manuals are written to satisfy a specific contractual requirement and are shelved after completion of work for that client.

Although most laboratories have some sort of organizational structure, few have any formal documentation describing that structure. Staff members are assigned duties orally and are expected to carry them out; no written position/ duties descriptions are available for review; the inter-relationships of staff members are often unclear.

Very few laboratories have written procedures for training, retraining, or determining the competence of staff members. The usual explanation is that they conduct on-the-job training, but cannot identify the principles or subjects covered in the training. Task check-off sheets are almost unheard of.

Many laboratories have minimal (or no) written procedures describing how they conduct specific testing activities. Many rely solely on the fact that they have a copy of a consensus method, such as an ASTM Standard Method, with no additional information on quality assurance or quality control procedures to be used when the method is performed in their laboratory.

For example, in construction laboratories, as part of the quality assurance in making concrete cylinders, procedures should be in place to monitor the condition of incoming molds for dimensional stability and water leakage. Capping compounds should be periodically checked for strength and mixture. Procedures should also be in place for other key variables requiring monitoring.

Laboratory Equipment

Another common quality system deficiency involves the condition of laboratory equipment. Many laboratories lack routine maintenance or calibration programs for their test equipment. Calibration or maintenance that is performed is usually not documented, and no records are available later to determine what (if anything) was done, or when.

Items such as slump cones or sieves are often found to be in very bad condition; the test results developed when using them are suspect. Some laboratories were found to keep a special slump cone or set of sieves to be used only for inspections, whereas those used in everyday testing were severely damaged.

In order to satisfy the NVLAP criteria, a laboratory must have a record-keeping system to provide evidence that all equipment has been periodically maintained according to a formal schedule and was in proper working order during testing. The laboratory must also have a system for periodic calibration of all applicable equipment, and ensure that if the equipment goes out of calibration it can determine the approximate time when that occurred. All calibrations must be traceable to appropriate, nationally accepted, standards.

### Sample Control

Sample identity and control is often found to be haphazard. Although all laboratories have some sort of sample identification scheme, sample control is often improper and inadequate through the various stages of sample receipt, log in, preparation, environmental conditioning, analysis, and storage.

For example, many concrete samples are not properly protected at the jobsite, or are not taken back to the laboratory in a timely manner for proper aging and conditioning. Sample conditioning parameters, e.g., temperature, humidity, time, are frequently not controlled properly or not recorded for future reference. Lack of any temperature/humidity recording device on a laboratory moist room is a common deficiency.

A deficiency commonly found in asbestos laboratories is sample contamination. This occurs from lack of proper cleanliness in the laboratory, or carelessness in keeping samples properly sealed to prevent cross-contamination.

The NVLAP assessors are alerted to look for sample control problems which are normally unique to each laboratory. The assessor evaluates the problems and makes suggestions to the laboratory for an adequate solution. However, the laboratory must notify NVLAP of any proposed solution, and must include the solution in the quality documentation.

### Quality Assurance/quality control

Very few laboratories regularly or routinely perform self-initiated and self-directed internal or external audits of their measurement processes. Quality control testing is rarely performed. Except as required by a contract specification or a regulation, few laboratories conduct blind testing.

In order to satisfy the NVLAP quality system criteria, laboratories must develop and institute review procedures for all laboratory activities. Laboratories are encouraged to tailor their quality assurance procedures to the nature, scope, and function of the laboratory. The quality assurance manager must design an appropriate and effective mix of internal testing, inter-laboratory testing, blind sample analysis, blank analyses, and participation in round-robins and other managed proficiency testing programs to assure quality testing. When appropriate, well characterized known samples (quality control samples) should be placed into the test cycle to check the test procedure and equipment.

As a quality control function, laboratories should use control charts and other statistical techniques to determine if the test results are drifting, to identify the cause of such trends, and to monitor individual technicians to determine their performance.

For example, as a quality control function, NVLAP requires concrete testing laboratories to participate in a within-lab proficiency testing program. They are required to maintain a statistical analysis of the concrete cylinders which are broken during routine testing operations. This analysis is patterned after ACI 214.

Labs use these analyses in different ways. They can watch for trends to determine if and how the test results are drifting then look for the cause. Sometimes the concrete may be bad, other times the testing may be bad, or the trend may be attributed to a new technician, poor procedural techniques, or some other variable. Some labs use the technique to monitor individual technicians to evaluate their performance.

Laboratories are also required to participate in reference sample programs so that they can measure their performance against other laboratories that perform the same tests.

## Impact Of NVLAP Accreditation Process

The NVLAP evaluation and accreditation process is, in most cases, an educational process for participating laboratories. The laboratories benefit greatly from the knowledge and constructive advice provided by the peer assessors. If the assessors feel that they cannot adequately address the problems uncovered at the laboratory in the time allotted for the assessment, they will advise the laboratory how or where to get the needed assistance.

Due to the emphasis placed on quality systems, most laboratories need to initiate new activities and install new procedures or upgrade existing procedures to comply with the criteria requirements. This often requires a substantial effort on the part of the staff, as well as a financial outlay.

Almost without exception, participating laboratories come to view the quality systems that they devise for their own unique needs and conditions to be well worth the efforts and costs. The overwhelming experience to date is that, after going through the evaluation process that leads to initial accreditation, laboratories institutionalize and continue to maintain and improve their quality systems. Quality system deficiencies noted in subsequent evaluations of accredited laboratories are usually minimal and generally pertain to changing needs or a misunderstanding of the original concepts.

Laboratories report back to NVLAP that as a result of the quality systems being put into action, that staff now knows not only what to do, and how to do it, but they now know what the most important events in the testing cycle are, and the impact on the results. They also report a significant change in the status quo, a whole new mentality is present to review, change and update procedures.

## Conclusion

In the NVLAP experience, most testing laboratories have minimal, if any, quality systems when they first apply for accreditation. However, a laboratory cannot achieve accreditation until it has a quality system in place and all deficiencies have been satisfactorily corrected or resolved.

Thus for a laboratory to become accredited, it is forced to establish a suitable quality system, containing all of the quality assurance/quality control measures needed to provide quality testing services in the technical area for which it is to be accredited. This often requires that a laboratory's quality system and documentation and procedures undergo several rounds of revision and review before approval. Once accredited, a laboratory must sustain the rigor of its quality systems due to periodic proficiency testing and on-site assessment required to maintain the accreditation.

Once educated as to the benefits of establishing and maintaining a quality system, beyond that of achieving accreditation, laboratories respond positively and develop appropriate systems. Even though there may be a moderate initial monetary cost, laboratories realize that quality systems are integral and necessary functions. Having and maintaining a quality system serves as an objective basis not only for achieving accreditation, but for claiming competence and reliability in litigation, in dealing with government regulators, satisfying contract specifications and in meeting the competition for testing contracts.

The feedback that NVLAP receives from participating laboratories is overwhelmingly positive. Laboratories are thoroughly pleased with the techniques established, the efficiency of operation, the new awareness of staff members of their jobs, and their overall ability to do a better job.

An independent survey of NVLAP-accredited asbestos laboratories was conducted by the National Asbestos Council's Sampling and Analytical Committee, PLM/PCM Subcommittee. The report stated "More than 40% of NVLAP participants saw the program as a boon to external credibility and a solid move to enhance internal quality assurance. Many laboratories that have been operating for five years or more without such advice now have a benchmark against which to monitor performance".

## References

American Concrete Institute, (1978) **ACI Standard 214 Recommended Practice for Evaluation of Strength Test Results of Concrete.**

Gladhill, R.L., (1989) **NVLAP Program Handbook, Construction Testing Services, NISTIR 89-4039.**

Laubenthal, T.G. and Weber, M.F. (1990) Work Practices of Laboratories analyzing asbestos-related samples by PCM AND PLM. **American Environmental Laboratory**, Vol.2, No.2, April 1990.

**Code of Federal Regulations**, Title 15, Part 7

# 40 ORGANIZATION OF THE SPANISH NETWORK OF TESTING LABORATORIES (RELE): CONSTRUCTION SECTORIAL COMMITTEE

J.C. MAMPASO
Ministry of Industry and Energy, Madrid, Spain
M. LOPEZ-BLAZQUEZ and A. ARRANZ
Ministry of Public Works, Madrid, Spain
C. ANDRADE
Institute E. Torroja, Madrid, Spain

Abstract
The "Red española de Laboratorios de Ensayo", RELE, was created on August the 5th, 1986 under the auspices of the Ministry of Industry and Energy. It was organized as a non-lucrative private organization which operates throughout Spain and it is open to any laboratory which observes the practices and procedures established by the 45000 Series European Norms.
RELE has as main objectives: a) to promote the accreditation of testing laboratories in accordance with the internationally recognized criteria, b) to contribute to ensure a quality service in laboratories and c) to cooperate with international bodies that seek similar objectives. In order to achieve these objectives, RELE is structured along the following lines: a General Assembly, the Head Council, the Board, the Sectorial Committees (five at present: Construction, Mechanics, Chemistry, Electrical and Electronics), a General Secretariat and a bank of Technical Evaluators.

**INTRODUCTION**

Due to the important development of the construction Market inside Spain during the 60's, the Ministry of Public Works established an Accreditation Procedure, in order to recognize the expertise of Laboratories, mainly for the control of site works. Afterwards the Ministry of Industry and other Ministerial Departments established similar procedures in order to recognize the competence of Laboratories to test products and materials in different fields.

However, the perspective of a single Market in 1993 promoted by the EEC, decided the Spanish Government to uniform the different existing laboratory networks in order to adapt national procedures to follow European Directives.

The "Red Española de Laboratorios de Ensayo", RELE, was created on August the 5th, 1986 under the auspices of the Ministry of Industry and Energy. It was organized as a non-lucrative private organization which operates throughout Spain and it is open to any laboratory which observes the practices and procedures established by the 45000 Series European Norms.

**OBJECTIVES AND ORGANIZATION**

Rele has the following objectives:

- To promote, direct and coordinate throughout Spain the accreditation of test laboratories in accordance with internationally recognized criteria.
- To create a Spanish network of test laboratories.
- To provide the necessary information of development, introduce and maintain quality systems, and in this way contribute to ensure a quality service in laboratories.
- To strengthen testing quality in Spain and to make the results known.
- To cooperate with international bodies that similar objectives.
- And in general, to coordinate and to promote all those activities and techniques of interest in the field of testing.

In order to achieve these objectives, Rele is structured along the following lines:

- A General Assembly : The governing body of the association, and made up by the founding members, associates and other members.

- The Head Council : It represents the General Assembly in day-to-day affairs.

- The Boards of Directors: It manages the association.

- Secretariat : It is in charge of administration and decision-making

- Sectorial Committes : Basically made up by representatives of the different laboratories. They are highly technical working groups that are in charge of carrying out Rele activities in the areas of research and advisory work. Up to the moment, five committees have been constituted:

. Construction.
. Mechanics.
. Chemistry.
. Electrical.
. And Electronic and Computer Systems.

- Technical Evaluators : Quality experts and technicians who carry out inspection and valuation actions , principally based on Rele's accreditation activities.

**PROCEDURE OF ACCREDITATION.**

The Rele System for the Accreditation of Test Laboratories lays down those prerequisites that must be met by laboratories to assure the quality of their services, so that they can be valid both in Spain and abroad.

The general prerequisites are layed out in the Rele's technical documents. They are complemented by other specific technical documents drawn up by the sectoral committees of the Rele and included in the accreditation programmes.

Rele Accreditation can refer to one or more tests, or types of tests, that are defined with the highest degree of precision. Each Rele accreditation will correspond to a "technical unit", which should be understood as a combination of perfectly defined technical and human means.

**An application for accreditation.**

When a laboratory expreses its interest in obtaining accreditation, the Rele Secretariat sends it the following documentation:

- The procedure to follow, together with a list of rights and obligations of the laboratory after it obtains accreditation.

- An application form and a questionaire before the evaluation process begins so that they can be filled out.

- A collection of Rele technical documents.

The description of the accreditation application is important, since no limitations exist once the methods and procedures necessary are clearly defined. The initial questionaire is also important, since it represents an autoevaluation by the laboratory in which its organizational and technical characteristics are evaluated, together with its system of quality control, If the questionaire is not filled out to the satisfaction of the Rele Secretariat, the accreditation process is not begun. The questionaire is considered as a valid declaration of the interested laboratory since it will be the basis for the evaluation that will be undertaken and it will serve as the estimate for the expenses that will have to be paid by the laboratory for Rele's work.

The information received by the Rele Secretariat, both during the application process and during the accreditation process will be considered and treated as confidential. Once received by the Rele Secretariat, the official laboratory application is analyzed, and if passed upon, will be sent to the appropriate sectoral committee. The sectoral committee will draw up the Accreditation Programme which will include the prerequisites, norms and methodology that must be met by the laboratory.

## Designation of the group auditor

The Sectorial Committee will designate a group of auditors who must make an evaluation at the laboratory. This group must be made up by at least one expert in quality control techniques and another expert in the testing field in which the laboratory is specialized, in accordance with Rele document number four.

The quality experts will belong to the Rele Secretariat, while the technical experts will be selected from among those who are included in the so-called "experts bank". The laboratory making the application will be previously informed of the names of the experts who will undertake the evaluation.

## Undertaking the audit

The auditor group will verify the laboratory's capacity to satisfy the accreditation demands in accordance with the check-lists prepared for this purpose and drawn up in each area of technical expertise on the basis of the criteria and actions established in the Rele document number four.

On the scheduled date, the evaluation visit begins with an initial meeting in which the members of the audit and the audited groups will be identified. The auditing program will be outlined according to both parties.

The audit will pay special attention to the following areas during the period in which the testing is undertaken:

- Organization.
- Efficiency and expertise of the personnel.
- Testing equipment.
- Environmental conditions.
- Testing methods and procedures.
- The storing, handling and identification of the testing apparatus, materials or elements.
- Documentation and records.

Once the evaluation visit is concluded, the audit group will hold a final meeting with the persons responsable for the laboratorys testing facilities in order to inform them of the main conclusions they have reached. The auditor group's report which is drawn up according to the Rele document number four will be delivered to the corresponding Sectorial Committee.

A copy of this report will be sent to the laboratory that made the application so that once it has been analyzed, the laboratory can inform Rele of whether it agrees with the document. The laboratory may also make observations regarding the report in question.

## Decision for accreditation

On the basis of the information, the appropiate sectoral committee will draw up a proposal for a decision of accreditation which will be sent by the Secretariat to the Head Council of the Association which will make the final decision on accreditation.

If the decision is favorable, the appropiate accreditation will be granted to the laboratory so that it can undertake the pertinent tests. A certificate of Accreditation will also be granted that will be valid for three years.

The accredited laboratory will be considered a full member of the Rele and will have the rights and obligations established in the association's statutes.

The Rele will also control the work of the accredited laboratories during the period of accreditation so that it can verify that the conditions which led to the original accreditation are maintained.

A laboratory that has already been accredited by Rele may apply for its accreditation to be enlarged so that it can include other testing procedures. If this is the case, the accreditation process we have previously described will be followed, although the auditors will limit themselves only to those new test processes seeking accreditation.

The Rele System to Accredit Test Laboratories is a

basic process which meets the needs of the European Economic Community in its testing and certifications policies, contributing to a better quality for Spanish products.

## ACTIVITIES OF THE CONSTRUCTION SECTORIAL COMMITTEE

During these first years of constitution in which any laboratory has an Accreditation, the Construction Sectorial Committee has had as members the most relevant laboratories related with this field. An average of 25 members monthly were meeting in order to develop the philosophy of RELE in the Construction field and discussing the procedures, for a better integration of Spanish Laboratories in the European market.

Several specialized courses prepared by RELE staff have been attended by the members. The Committee was chaired by the Inst. Eduardo Torroja, being Vice-president a representative of the Ministry of Industry and Energy and acting as Secretary a representative of the Ministry of Public Works.

The Accreditation Programs approved during these 4 years of activity dealt with the following products: cement, ceramic floor tiles, ceramic materials (bricks, blocs, tiles, etc.), stock of cocks, asphaltic concretes, bituminous materials, gypsum and related prefabricated products, natural stones, stainless steel pipes, steel pipes, reinforcing and prestressing bars, modular metallic chimneys, transit rods, thermal isolating materials, hydraulic and aerial limestone, and armors. 10 others are now in the way to be approved.

Four laboratories in the field of ceramic products and cement testing are already accredited and more than 20 have asked to be audited.

Finally, most of the members of this Sectorial Committee are representatives of RELE in the different committees of Certification (Quality Marks) in order to assure a correct coordination of the overall system (Standardization -Certification - Testing).

## AKNOWLEDGEMENTS

The authors are grateful to the President, Manager and Secretary of RELE for the information provided in order to prepare this communication.

PART FIVE

# TRANSNATIONAL RECOGNITION OF TEST RESULTS

# 41 TRANSNATIONAL RECOGNITION OF TEST RESULTS – AN INTRODUCTION

G. OOSTERLOO
STERLAB, Delft, The Netherlands

Preceding speakers have explained and made clear why and in what way the quality of the function of laboratories in general and more specifically in the field of construction materials and structures, can be improved. Attention has been given to:

- Standardisation of test methods
- Quality systems
- Test results
- Comparability of test results

In GATT negotiations, it is accepted that, when test reports are required, laboratories of exporting and importing countries should have equal opportunities, provided of course the laboratory is of recognised quality. The same principle is accepted for the free market in the EC after 1992.

It is therefore evident that there is an urgent need to create a structure that will make it possible to accept test results over frontiers: duplication of testing is then not necessary any longer; shipping between supplier and receiver of goods will be much faster and more efficient and technical barriers to trade can be demolished.

Sometimes the question is raised whether this would mean less work for laboratories. The answer is that it can be expected that testing will not lessen but increase. I would refer to:

- The growing number of products coming each year onto the market.
- The shorter lifespan of those articles.
- The impact of logistics and lot sizes, (just in time).
- The growing demand for test reports in view of product liability.
- New fields of activity such as safety, health and environmental problems.

On top of that there is no excuse to keep unnecessary paperwork or laboratory activities going, as a means of reducing unemployment.

It is evident that mutual recognition of laboratories and their test results based on assumed, or better, on assessed quality, is a long and complicated procedure. Moreover, those mutual recognitions must be accepted and sanctioned by trade partners and authorities, depending on the kinds of tests and subjects.

The essence of a mutual recognition agreement is that it implies that a test carried out in these laboratories will give the same or at least comparable results and that there is confidence that this is not only so in a special case on a certain date, but for a whole field of activities now and in the future, say for one to two years. In this respect I refer to the quality system in the laboratories.

The most objective base for mutual recognition is through third party assessment, in other words an accreditation body.

The platform on which accreditation bodies meet is ILAC. Worldwide there are two-yearly meetings at which the criteria, procedures and regulations under which accreditation bodies operate are discussed and agreed upon. The programme of those meetings is discussion to try to harmonize the papers prepared by working committees in the previous two years and to formulate new commissions for the next two years. The subjects have been: the criteria for accreditation worked out in ISO standards, surveillance visits, selection and training of assessors, reference materials, calibration, proficiency tests, in short all the subjects of preceding speakers.

As already mentioned, in GATT negotiations it is accepted that in the case of mandatory tests, or other cases where tests are required, those tests should be carried out either in the importing or exporting country. In this matter protection should not play a role. A condition is of course that the relevant laboratories should have the proper qualification. This is a gratifying statement, but I fear full implementation will be a long and difficult procedure.

In the EC, where a free market will be opened after 1992, it is of immediate urgency. Tariff barriers will disappear, but it may not be impossible that technical barriers will be raised, unless adequate measures are taken. I will now review the role of testing and laboratories in this context.

During the last three years EC and EFTA members of ILAC meetings have, along the lines set out by ILAC, discussed the way to come to mutual recognition of accredited laboratories and their test results. The policy of the EC is that such recognitions should be on a voluntary basis. With 1992 coming near, there was a clear need for a structure that will make mutual recognition possible along the lines of discussions in ILAC, and on December 6th 1989 a WELAC Memorandum of Understanding was signed between most accreditation bodies in EC and EFTA. WELAC stands for Western European Laboratory Accreditation Cooperation, and this MOU is intended to lead to bilateral or multilateral recognition of laboratories. I will now discuss the essence of this agreement.

*Participation in Welac is open to one Testing Accreditation Body from each of the states of the European Communities and of the European Free Trade Association to build up and maintain mutual confidence and to promote agreements between National Testing Laboratory Accreditation:*

*to open and maintain channels for a continuous flow of technical knowledge relevant to laboratory accreditation between the National Bodies and with other relevant bodies such as EOTC, BCR, WECC, ILAC laboratory organisations and accreditation bodies and certification bodies;*

*to establish activities aiming at a multilateral agreement based on mutual confidence stating the equivalence of the operation of the Member Bodies and declaring their commitment to promote and work for mutual acceptance of Test reports and Certificates issued by Laboratories accredited by them;*

*to develop a close collaboration between Member Bodies to promote the recognition of testing in line with the requirements of the EN 45000 series of standards;*

*to promote agreement between National Laboratory Accreditation Bodies and the international acceptance of those agreements;*

*to promote the international acceptance of Test reports and Certificates issued by accredited Laboratories;*
*to promote the exchange of information between Member Bodies;*

*to encourage visits between Member Bodies and cross-participation of experts in Laboratory assessments;*
*to co-operate in training of assessors and staff;*

*to organise international proficiency testing between accredited Laboratories in collaboration with organisations where relevant;*

*to organise expert meetings in special fields of accreditation;*

*to operate an evaluation and re-evaluation programme based on periodical assessment visits, to those Member Bodies which participate in that programme, by a team of experts in which each Member Body has the opportunity to mandate an expert and which makes a statement concerning the compliance of the valuated Bodies with relevant international standards or the WELAC guidelines and their suitability to enter into a multilateral agreement with the other Member Bodies.*

This Memorandum of Understanding is of an exclusively recommendatory nature. It does not create any binding legal effect on Member Bodies. In some countries laboratory accreditation is mature, some countries are some years in operation, some countries have just started.

RNE (France) and NAMAS (UK) had signed a MOU of mutual recognition before WELAC had phrased its objectives, aims and tasks. Essentially however it covers these aspects. Recently a MOU was signed between NAMAS (UK) and STERLAB (the Netherlands) along the lines set out by WELAC. These MOU's are based on full confidence that each accreditation body will come to similar conclusions in assessing a laboratory in either country. This confidence can not be based solely on reading and comparing relevant documents, regulations, criteria and discussions. This was done only to decide whether mutual recognition had a chance. In the same way as an assessment in a laboratory is carried out, the real confidence is based on observation of the assessment operation in the field. Objects of observation were:

the preliminary visit to the laboratory,
the actual assessment visit,
a surveillance visit, and
reading the reports and conclusions written after the visits.

This is a time consuming procedure. Before reaching the MOU each organisation had to spend approximately 30 to 40 man days. Each year there will be a surveillance visit to see whether the MOU can be prolonged. This will cost yearly 10 man days. We have to multiply these figures by 17, as 18 EC and EFTA countries are involved. Then all the relevant paper work must be translated to be accessible for other parties. In most cases a interpreter is needed when visiting a laboratory.

So the structure is ready and all accreditation bodies have expressed their intention to come to agreements, but the practical difficulties to overcome are enormous. The MOU's between UK and France and between UK and the Netherlands, and the experience gained can serve as examples for future agreements.

One of the first activities within WELAC will be to find acceptable solutions. One can think of multilateral MOU's by working with assessment teams consisting of more than one accreditation body. Cross participation of accreditation bodies in assessing laboratories can reduce the surveillance costs.

Personally I expect that within a year four to five bilateral or multilateral agreements can be signed, all under the prevailing condition of full confidence and voluntariness.

My conclusion is that there is a lot of work to be done: it will be difficult and there is a long way to go before complete mutual recognition is reached. But on the other hand all signatories of the WELAC MOU realize the importance and urgency in view of the free market after 1992 and have spoken out their intention to give full support.

I am confident that the structure and procedure now set up will lead ultimately to a worldwide network of multilateral agreements.

# 42 NETWORKS, COOPERATION AND CONDITIONS IN WESTERN EUROPE

A. PREVIDI
Commission of the European Communities (CEC), Brussels, Belgium

## 1 THE OBJECTIVES AND PRINCIPLES OF THE GLOBAL APPROACH

On 5 July 1989 the Commission of the European Communities approved a Communication to the Council of ministers presenting a global approach to testing and certification issues, a new Community policy designed to facilitate mutual recognition. It has been drawn up in the context of the European Single Act's deadline of 31 December 1992 for the completion of the Internal Market, as the missing link in the Community's policy for ensuring the free movement of goods, which already provides for an information procedure on draft technical regulations and standards, a new approach to technical harmonization and a reinforcement of European standardization activities.

The global approach to testing and certification completes the new approach by putting in place the elements to develop confidence in the product, as well as in the manufacturers, the testing laboratories and certification mechanisms for conformity assessment and setting the ground rules for the private sectors to make activities in these areas acceptable to national and Community public authorities without legislative intervention being necessary.

Based on international documentation and activities, recourse has been made to such techniques as quality assurance which encourage economic operators to control their design and manufacturing activities properly, whilst mechanisms for the evaluation of testing laboratories and certification bodies have been devised so that the quality and competence of all concerned is not only good but can be demonstrated.

### 1.1 The manufacturers

The first element of the global approach is to reinforce the confidence which can be placed in a product by reinforcing the credibility of the manufacturer himself. This can be done by inciting manufacturers to make greater use of quality assurance techniques and by making these attractive to the manufacturers. This is greatly facilitated by the CEN/CENELEC adoption of the EN 29000 series of standards relating to quality assurance (which is in fact the adoption at the European level of the ISO 9000 series of standards).

The Commission has recognized the important role such techniques can play and so will not only promote their use but will encourage the use of third party certification of quality systems when it is carried out in a transparent manner, inducing mutual recognition throughout the Community. Moreover, the Commission considers that such techniques can also be used in legislation to enhance the credibility of a manufacturer's declaration or to complement, reduce or bring more flexibility to third party intervention in respect of the products themselves.

1.2 Testing Laboratories and certification bodies

It is however, also necessary to enhance the credibility and confidence which can be placed in testing laboratories, inspection and certification bodies whether these operate in the private or in the public sectors. In order to facilitate such an enterprise the Commission will also encourage testing laboratories, inspection and certification bodies to follow the EN 45000 standards setting the criteria of competence of these bodies and providing for the means for their assessment and when possible to have them use third party intervention to demonstrate their conformity.

1.3 Accreditation techniques

Lack of confidence of Member State authorities in the laboratories and bodies of other Member States has been one major factor leading to barriers within the Community.

This has lead the Commission to support accreditation as being an efficient and tranparent manner for laboratories and certification bodies to demonstrate their competence. Accreditation entails a third party evaluation of these bodies exclusively on the basis of the relevant EN 45000 standards and therefore presents advantages of neutrality, transparence and independence, giving guarantees as to technical competence.

In the future, Member States who have to notify to the Commission the bodies responsable for implementing Community Directives, will be held to notify only bodies which can demonstrate conformity to the EN 45000 series. Should the bodies in question not be formally accredited the national authorities will have to produce documentary evidence of such conformity.

1.4 Information on testing and certification systems

Transparency also means easy access to information concerning technical regulations, standards and procedures relating to testing and certification. The Commissionn is therefore financing the creation of a European Database (CERTIFICAT) containing information on products which have to be certified, by whom, how and following which procedures. There will also be information on the third party bodies involved. This database will be placed on the public networks and will be open to consultation by all concerned also non-community operators.

## 1.5 New Techniques for certification - The modular approach

In trying to bring about greater transparency, the Commission proposes that more clarity and coherence be brought to the Community legislation. Thus the Community by putting forward what it has called a "modular approach" to conformity assessment procedures. This is based on the principle that full certification procedures can be broken down into functions or modules. These different modules (e.g. type examination, type test, quality control of the production, final product verification by a third party,...), like building blocks, may be put together to form more flexible and more appropriate procedures.

Thus, the Council of Ministers, in approving EC legislation will be able to leave a certain amount of choice to the manufacturers on the basis that the various combinations of modules should give the public authorities, not identical but equivalent technical solutions, in terms of ensuring an appropriate level of safety for the products in question. The Council will then lay down the combination of procedures which it considers to be appropriate for a given Directive, will set the conditions of application and will then leave the final choice as to which specific procedures to follow to the economic operators themselves (manufacturers or importers).

Member States will be encouraged by the Commission to follow the same approach in their national legislation. Such developments will certainly help to reinforce the possibilities for mutual recognition of test reports and certificates directly on the basis of national legislation without requiring Community legislation. The modules can also be used by the private sector when drawing up private certification systems.

## 1.6 A European organization for testing and certification

It has become evident that it is necessary to dispose of an organization at European level to pull all these elements together into a coordinated and coherent whole for those activities which are carried out in the private sector. The setting up of an infrastructure at European level in the testing and certification field complements the existence of CEN/CENELEC in the standarization field. It will bring together all the interested parties in testing and certification (manufacturers, consumers, users, public authorities, testing laboratories, certification bodies) with a view to generating agreements and common voluntary disciplines for mutual recognition. The proposed organization will not assess products, manufacturers and third parties, but will be a focal point where interested parties can find or develop the necessary instruments for drawing up mutual recognition agreements. It will in other words constitute a home for all these activities, in which basic principles rules of competence, openness and transparency will be safeguarded.

## 2 WHAT DOES ALL THIS MEAN FOR PRODUCTS ORIGINATING IN THIRD COUNTRIES ?

### 2.1 A more transparent market

First of all, it means that third country manufacturers, and economic operators in general, will be faced with a more coherent, more organized market place as well as a more open and transparent one.

### 2.2 The principle of non discrimination in access

Clearly the Court's Jurisprudence (Cassis de Dijon and Biologische Producten) operates for products orginating in third countries as do the principles of the GATT Agreement on Technical Barriers to Trade. This means that products originating in third countries must be granted access to certification systems on an equal footing to products originating in the Community whether the system be voluntary or mandatory (at national or Community level). Third country products can only be refused for the same reasons as products originating inside the Community viz : non- conformance and lack of safety.

Where Community legislation exists according to the modular approach, the choice of modules and procedures left to the economic operators applies to third country economic operators as to Community economic operators. Where Community legislation allows the manufacturer to use a manufacturer's declaration of conformity, and to self-affix the CE mark, third country operators will also be permitted to use the declaration and to affix the CE mark. If the manufacturer has the choice between third party procedures, this choice is also open to third country manufacturers, as long as they can respect the conditions laid down in the Community legislation.

The bodies notified by the Member States to carry out the third party intervention are bound to apply the rules in a non discriminatory fashion.

### 2.3 More flexibility

A considerable advantage to economic operators stemming from the implementation of the new approach to technical harmonization and confirmed and reinforced in the global approach is that manufacturers are held to respect the essential requirements set out in Community legislation, but are not obliged to follow the

European standards referred to in the legislation. Should a manufacturer follow the European standards, he benefits from simplified certificiation procedures. Should he not do so, he is obliged to process his product through a notified body which assesses the product directly to the essential requirements and gives a certificate valid throughout the Community. This brings greater flexibility for manufacturers situated inside and outside the Community in that it reduces the need to produce special goods for the Community market to unnecessarily detailed specifications. Very different types of products may conform to the essential requirements. The technical specifications spelt out in the European standards only constitute one (priviledged - it is true) technical solution. The manufacturer must only demonstrate conformity to the essential requirements, whatever technical solution he has adopted.

## 2.4 Mutual recognition agreements

Once the Council of Ministers approves the principles of the global approach, the Community will be ready to open negotiations on the conditions for mutual recognition.

The global approach clarifies the issues and allows the Community to put forward some clear basic principles to guide such negotiations.

Any negotiation must take into account that the Community must be bound by the level of protection it guarantees on its market and that therefore, the essential requirements will be taken as the level for negotiation.

The EN 45000 and 29000 series of standards will have to constitute the basic technical instruments that partners in non-Community countries will have to respect. Moreover, it will be necessary to ensure that such technical competence be maintained on a continuous basis. Mutual recognition agreements must provide each party to the agreement equivalent and equal access to the markets in question, without there being any supplementary conditions imposed.

Moreover as such agreements must be based on a level of protection as well as on the technical competence of the specific economic operators involved (manufacturers, testng laboratories and certification bodies), agreements and their benefits should be limited to those participants, on the understanding that these agreements operate in an open manner and may allow for further negotiations to widen their scope of operation. However, such agreements cannot be extended unilaterally by one of the parties to others, without the consent of the parties to the original agreement.

## 2.5 Negotiating procedures

These conditions must apply to both private and public sector activities. The procedures for the negotiations are, however, different.

When such agreements are to be developed in the private sector it is up to the private sector itself to negotiate and finalise the agreements. Once the global approach has set the ground rules and helped to organize an open market, negotiations in this area will be simpler to launch. Moreover the creation of the European infrastructure for testing and certification should contribute considerably by setting up a framework in Europe in which such negotiations can be carried out and managed efficiently.

Where mutual recognition agreements prove to be necessary in areas where there is legislation either at the national or Community level, and where either national or Community public authorities are involved in controlling the placing of products on the market, the negotiation of the agreements fall to the Community by virtue of Article 113 of the EEC Treaty.

This rule applies to those areas where there is unharmonized national legislation because of the obligation on Member States under the Community law, to accept products, including third country products, legally marketed in another Member State, implying that the conditions of access to an integrated Community market cannot be determined by agreements with third countries concluded by individual Member States.

## 2.6 Priorities

The size of the problem, however, should be put into perspective. Those products which fall in the private sector remain the vast majority, products covered by mandatory certification representing a relatively small proportion of the market. The Community negotiating process will therefore not have to take all industrial sectors into account.

The Community will in the first place have to restrict itself to those areas where there is Community legislation, especially in those sectors where it has been drawn up according to the new approach (because they cover whole industrial sectors as opposed to classical legislation which restricts itself to certain types of products).

It is of course, as of yet difficult to set precise priorities either as to the product sectors or as to the geographic areas concerned.

However as already indicated, Community policy is very much in favour of third countries (as much as for Community operators). Access to the market is assured whilst awaiting the more complex solutions of mutual recognition.

# 43 BASIC RULES AND OPERATING PRINCIPLES OF PRODUCT CERTIFICATION IN THE COMECON COUNTRIES

K. KOVÁCS
ÉMI, Institute for Quality Control of Building,
Budapest, Hungary

The organizing principles have been stated by the document "CONVENTION on the quality evaulation and certification system of products mutually supplied by the COMECON countries (SZEPROSZEV)", which was signed by the governments of the member countries in 1987. The undersigners have been nominated in the CONVENTION as the "CONTRACTING PARTIES".

Main purposes of the CONVENTION have been determined as the followings:

- the further development of the cooperation and economic integration among COMECON member countries,
- the assurance of a growing effectiveness in the cooperation among COMECON member countries,
- the steady increase of product exchange among COMECON member countries within the framework of international work distribution,
- the consequent increase of the technical level and quality in the mutually supplied products and their competitiveness in the world market; the reasonable use of products, energy and labour source, and the elimination of the repeated tests of products.

The Contracting Parties have agreed in the formation of "The quality evaluation and certification system of products mutually supplied by the COMECON countries", in the framework of which they undertake the following :

- assurance of the system's operation, according to the CONVENTION,
- determination of the list of products to be certified,
- accreditation of testing laboratories (centres) of the Contracting Parties, for the attainment of legal rights neccessary for the accomplishment of tests on products to be certified,
- checking the existence of those conditions in the manufacturing company of the product to be certified, which make possible the assurance of steady quality and effective quality control,
- examination of the products to be certified in the accredited testing laboratories (centres),
- issuing of feasibility certificates and certification marks,

- realization of the mutual recognition of "feasibility certificates" and certification marks,
- realization of the supervision on the system's operation and the quality of products to be certified and their governmental control.

"The quality evaluation and certification of the mutually supplied products, the CONVENTION, and "The basic rule of quality evaluation and certification system based on the COMECON standards for products mutually supplied by the COMECON member countries" will be accepted at a later date.

The documents of the System must be worked out with the consideration of principles accepted by other international certification bodies, if they are not in disagreement with the purposes and tasks of the COMECON economic integration.

According to the CONVENTION, the certification of the mutually supplied products must be accomplished by COMECON standards, other international and national standards, and the requirements of other technical standard documents, which are suitable for the scientific-technical world standard and which have been mutually agreed and accepted by the national bodies of the Contracting Parties.

Each contracting body nominates a responsible governmental body and provides it with an authority ensuring its participation in the System. According to the national legislation, this body accomplishes the coordination of the certification activites in its own country, and represents the given country in connection with the realization of the CONVENTION, regarding the authorized governmental bodies of other countries.
(Notice: "the concerned governmental body" is the standardization body of the member countries).

Concerning the nomination of the authorized governmental body, data are transferred to the consignatory of the CONVENTION.

Including also raw materials and complete units, mainly those products are to be qualified and certified, which are significant for the national economy of the member countries; products, which can be dangerous in their use for the life and health of people, their environment; moreover products, which must be tested obligatory before their use, according to the legislation of the Contracting Parties.
The list of the above products is agreed during the preparation and signature of the bilateral and multilateral agreements on certification by governmental bodies authorized for the assurance of participation of the contracting parties countries in the System.

"Feasibility certificates" are published by the authorized governmental bodies, or - with their permission, - testing laboratories (centres), accredited within the System.

The testing laboratories, (centres) of the Contracting Parties, which are supplied with unique testing equipments, on the basis of an agreement made among the Contracting Parties' authorized governmental bodies, can be specialized for the testing of suitable product groups or particular types of testing.

The accreditation of testing laboratories (centres) is accomplished by the authorized governmental bodies of the country locating the laboratories (centres) of the Contracting Parties, on the basis of control exercised by the expert group of the authorized governmental bodies in the other countries of the Contracting Parties, and with the consideration of the consequences; if the bi-, and multilateral agreements do not state differently.

If the testing laboratory (centre) has been accredited by the rules of other national, or international certification systems, which are in agreement with the rules of SZEPROSZEV, then this accreditation is wholly recognised within the framework of SZEPROSZEV.

The "feasibility certificates" are given on the basis of positive results of the product testing, if the manufacturing firms have those conditions, which ensure steady product quality and effective quality control.

The manufacturing firm can use the feasibility mark on the product to be certified, if it has a "feasibility certificate".

The existence of conditions ensuring the steady quality of products to be certified is stated by the authorized governmental body, with the production control of these products.

In case of mutual agreement, the authorized governmental bodies authorize the representatives of the governmental bodies of the importing country to get knowledge of the conditions of the manufacture and quality control of the new product types to be certified.

If the authorized governmental body of the importing country, during the product control, states that the supplied product is not in agreement with the certificate, then it can suspend the acceptance of the certificate in its own country, and is obliged to report it immediately to the authorized body of the exporting country.

Countries of the Contracting Parties take the necessary measures to ensure objectivity of testing results in the accredited testing laboratories (centres), and for the objectivity of the certifying results, on the basis of the uniform organizational-methodological principles and documents accepted by the bodies of COMECON.

The Contracting Parties take the necessary measures for the organization and accomplishment of information exchange in connection with the System's operation.

Costs in connection with certification are borne by the manufacturing firm (organizations, companies) of the exporting country, if it is not stated differently in the agreements between the exporting and importing members.

Debates on civil rights between the bodies of the Contracting Parties, influencing the product to be certified, and debates originating from the economic and scientific-technical cooperation, are to be settled by the court of arbitration, according to the CONVENTION.

The CONVENTION is not affecting those rights and obligations of the Contracting Parties, which originate from agreements and contracts made among them, or other bodies of their countries, or them and a third country.

The CONVENTION will be ratified and agreed (accepted) according to the legislation of the undersigning countries. The ratifying documents, or documents on agreement (acceptance) are recorded by the consignatory, who is fulfilling the consignatory functions of the Convention.

The CONVENTION becomes effective on the ninetieth day counted from the date, when the fifth ratified document or agreement (acceptance) document was given to the consignatory for filling in.

After its putting into force, with the agreement of the Contracting Parties, other countries can join the CONVENTION, having transferred the suitable documents on entering the consignatory. Entering is valid in 90 days from the date when the consignatory is given the last document on the agreement about entering.

The CONVENTION is agreed for an undetermined period. Giving a written notice to the consignatory, any Contracting Party is allowed to disrupt the participation in the CONVENTION. Disruption is operative in 12 months following the date, when the consignatory receives the notice. Disruption of the participation in the CONVENTION is not affecting those contractual and legal obligations, which have been taken by the country in connection with the formation of the System, and which are in force on the day, when the disruption becomes effective.

The consignatory immediately informs the undersigners of the CONVENTION and the entered countries on the date when the ratifying papers, agreement (acceptance) or entering documents are given for filling, when the CONVENTION is put into operation and when other reports are given, as a result of the CONVENTION.

According to the agreement of each Contracting Party, the CONVENTION can be supplemented, or modified. The supplements and modifications must be stated in minutes.

After putting into operation, the CONVENTION will have been registered by the consignatory in the Secretariat of the United Nations, according to Article 102 of the UN Charter.

The original copy of the CONVENTION will be given to the consignatory for registration, and he will send the verified copies to the ungersigners and members.

## Main principles of the Hungarian certification system

In compliance with the CONVENTION, the member countries undertake the establishment of their own national certification system, according to the principles of the CONVENTION. Thus, in 1988, the Hungarian Government updated the order of the Executive Council on the STANDARDIZATION AND QUALITY, where separate chapters state the system, the order, and the legal conditions of Quality Certification.

In the interest of the international competitiveness and the compatibility with other certification systems, the quality certification system, which is based on unified principles, has three levels.

In the certification system, the application of the principles of the first level, is obligatory. The manufacturers' participation in the certification according to the second and third levels, is voluntary, but, in connection with particular products, the obligatory application of these certification levels can be ordered by the minister concerned.

The first level certifies those properties of the products, which are significant in use for the client (buyer), according to the governmental order on the quality certification of products.

The second level - standard certification - certifies that the product is in agreement with the requirements stated in the standard (product qualification), and the manufacturer is in the possession of those conditions, which ensure the steady high level production of the product (production qualification).

The third level - differentiating qualification - certifies that the product satisfies further quality requirements (e.g. energy conservation, environment protection, ergonomics, aesthetics, fashion, etc.), in addition to the requirement of the second level.

In the case of the second and third levels, tests giving the basis for certification, are accomplished by the Hungarian Standardization Office (MSZH), on the basis of the certification given by the testing body, certifies the attainment of the second or third levels and gives permission for the use of the suitable feasibility mark.

The validity of certification according to the first level is controlled by a control body designated for this purpose, and the attainment of requirements given in the second and third certification levels is controlled by MSZH.

If MSZH discloses any lack during its control, it can suspend or withdraw the permission for use of the feasibility mark.

## Accreditation of the testing body

The accreditation of a testing body (laboratory

accreditation) is the recognition that the testing body, which is independent from the manufacturer or dealer, is ready for the accomplishment of determined tests or testing types.

The accreditation must be asked from the MSZH by the testing body.

If the conditions for accreditation are given, the MSZH is entering the testing body on the list of the accreditated testing bodies.

The accreditation made within the international and foreign certification systems, if it is comparable to the prescriptions of the national system, must be accepted by the organization, which is operating the national system.

For the certification of feasibility, marks can be applied, indicating that the product or production process is in agreement with the respective prescriptions.

In the certification systems, feasibility marks should also refer to the level of certification.

It must be noted that the initiation of the system's operation is not easy. First, the testing laboratories applied for an accreditation (on the basis of the well known ISO/EN standard series - ISO 9000 and EN 29ooo basic rules). At present,it seems that the Hungarian manufacturers are still not under such a significant market influence, as they would freely undertake more than their essential obligation on certification (first level) for obtaining the buyers' and users' confidence, e.g. with the application of a quality mark.

## Main principles of the Hungarian product approval system

For those construction products, for which there is no accepted national standard or technical specification, and the use of which is significantly dangerous for health, environment and property, the "FEASIBILITY CERTIFICATE" OF THE BUILDING INDUSTRY (ÉAB), which is equal to the TECHNICAL APPROVAL according to the new European terminology, must be acquired.

In the procedure:

PRODUCT:
- building material, structure or accessory,
- product, equipment or construction for use in building,

NEW PRODUCT:
- not applied yet,
- applied, but its important properties have been modified significantly, and
- there is no standard or normative document for it.

INNOVATIVE BUILDING and CONSTRUCTION METHOD
- not applied yet

THE WHOLE OR A PART OF AN INNOVATIVE CONSTRUCTION
- not applied yet

In Hungary, within the scope of construction products, the authorized body is the Institute for Quality Control of Building (ÉMI).

The occurence of innovative product in ÉMI:
x freely - by the manufacturers or dealers
x on the basis of building inspectors
The "obliged": x manufacturer
x importer, dealer
x user

There is also a group of innovative products, for which not a "FEASIBILITY CERTIFICATE" but only "OBLIGED REPORTING" is required. These products are:

- imported products, which are in agreement with the Hungarian or nationalized international standard concerning domestic products intended for the same purpose,
- products, the feasibility of which can be stated without any particular test (on the knowledge of field of application, standards and requirements valid in the mentioned area).

Neither a "FEASIBILITY CERTIFICATE", nor "REPORTING" is necessary for

- the experimental, single use,
- equipment, which has been given a manufacturing license by another authority, considering also the requirements of use in building.

DOCUMENTS for the "ÉAB" REQUEST, or "REPORTING"

a) description of the innovative products, their technical properties, technical planning document,
b) scope and method of application,
c) experiences on the trial application or other tests,
d) method of transport,packing and storing,
e) quality control and certification of the product,
f) draft on the prospectus or manufacturing catalogue to be published for the application of the product
g) the manufacturer's name and location
h) the applicant's data.

Summarizing the above, it can be stated:

- the CONVENTION OF COMECON countries is set up and valid, but it is not operational the known political and economic changes of the member countries are rather strengthening the intention of joining the European systems,
- there is no any universal bilateral obligation or agreement on acceptance among the member countries; there are occasional or interinstitutional agreements,
- there are relatively well developed and operating certification and approval systems in Hungary, Czechoslovakia, Poland, East-Germany, but they are not "compatible",
- the accepted view in Hungary is the adherence to activities in CEN, CEN-CER and EOTA.

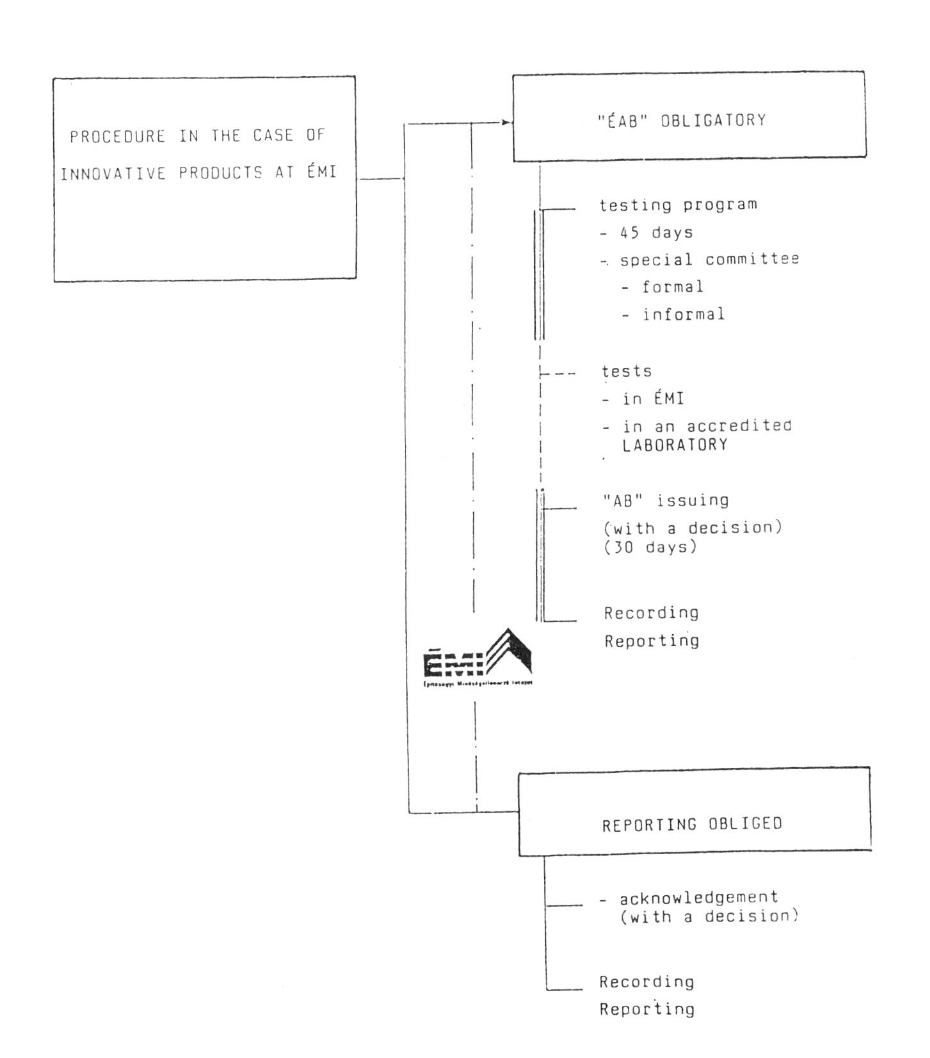

# 44 TRANSNATIONAL RECOGNITION OF TEST RESULTS IN NORDIC COUNTRIES

B. LINDHOLM
Nordtest, Esbo, Finland

Abstract
An important task in promoting free trade of goods in Europe is the creation of an European infrastructure on testing and certification.

The Nordic countries have a close collaboration within the framework of the Nordic Council and the Nordic Council of Ministers. In the field of technical testing Nordtest is the responsible organization for Nordic cooperation. This paper describes the Nordic activities guided by Nordtest on:

- Development and implementation of laboratory test methods
- Quality assurance of laboratory testing
- Development of competence

The experiences gained by the Nordic cooperation are useful also in a wider European context.

## 1 Region

The Nordic countries are Denmark, Finland, Iceland, Norway and Sweden. Four of these are members of EFTA and one, that is Denmark, is a member of EC.

## 2 Nordic cooperation

The five Nordic countries have for nearly 40 years cooperated within the Nordic Council (NR). This has lead to close contacts between these countries on cultural as well as on technical and economic level. Most important in this respect is of course their common historical background and similarities in language. The Nordic Council of Ministers (NMR) is the executive body. The NMR is today the head of more than 50 Nordic organizations. The cooperation is based on agreements on governmental level.

## 3 Trade and industry

The Nordic Committee of Senior Officials for Trade Policy Questions (the ministers of trade) is in charge of the collaboration on trade and questions concerning technical barriers to trade.

Correspondingly responsible for their own fields are the Nordic Committee of Senior Officials for Industrial Policy and the Nordic Committee of Senior Officials for Cooperation in the Building Sector. Permanent Nordic organizations under these committees are:

- Steering Committee for abolition of Technical Barriers to Trade
- Nordtest
- Nordic Fund for Technology and Industrial Development
- Nordic Committee on Building Regulations

In addition to this formal collaboration, there are also close contacts between companies, trade associations etc.

## 4 Nordtest

### 4.1 Background

Nordtest is the Nordic organization for cooperation in the field of technical testing. The agency was set up in 1973 by the decision of the Nordic Council of Ministers on the initiative of the Nordic Council. Nordtest subordinates to the Senior Officials Committee for Industrial Policy, and it also cooperates with the Senior Officials Committees for Trade and Building.

One of the principal objectives of Nordtest is to promote "free trade through abolition of technical barriers to trade". The task of the organization is to ensure that technical evaluation, testing, metrology and control are developed and carried out in a competent, reliable, efficient and uniform way. Nordtest should also contribute to the enhancement of the level of quality referring to testing and metrology and promote quality assurance for the testing activity. Further it should endeavour to create confidence in test results.

### 4.2 Organization

The activity of Nordtest is directed by the Board. The members are appointed by the Government of the respective country. The work is mainly carried out through projects. This activity is handled by working groups consisting of members from the Nordic countries. Within the field of building and construction, Nordtest has four technical groups: Building, Fire testing, Heating, Ventilation and Sanitation and Building Acoustics. The everyday work is handled by a Secretariat situated in Finland.



 

### 4.3 Fields of activity

The principal fields of activity are:

- Development and implementation of laboratory test methods
- Quality assurance of laboratory testing
- Development of competence, technical testing and testing personnel.

### 4.4 Test methods

Nordtest selects test methods - Nordtest methods - suitable for common Nordic use. These methods are in the first place selected among existing standards and other available publications. If no test methods of acceptable quality are available, Nordtest develops new test methods.

Methods issued by other organizations and institutions are not reprinted by Nordtest. New methods developed on the initiative of Nordtest are issued as original Nordtest methods, and printed in English. The total number of Nordtest methods is more than 900.

These methods have been recommended by Nordtest for common use within the Nordic countries. They are also offered as basic information and material to groups working in those areas where development of European or other international methods is in progress.

Of these methods, about 500 have been selected among existing methods after close examination and, as occasion may require, after comparison with alternative methods. Some methods have been tested by means of comparative laboratory tests. From projects partly or totally financed by Nordtest, 400 test methods have been developed. All these test methods are registered in the annually published register "Nordtest Register of Test Methods".

Through the financial support of Nordtest, experts from the Nordic countries have been given the opportunity to participate in the work of CEN/CENELEC. In this work, Nordtest methods have been offered as a Nordic contribution.

### 4.5 Quality assurance of laboratory testing

Quality assurance is a fundamental principle for mutual recognition of test results in Europe. No matter what formal structures will be chosen for the various sectors, the basis is that laboratories have their own internal systems to ascertain the quality of test results, verification reports etc. A separate European standard (EN 45001) has been developed to specify the quality requirements for testing laboratories.

Nordtest started early working on quality assurance. In 1979 the first guideline for development of quality manuals was issued which also was used as background material in the work of ILAC (International Laboratory Accreditation Conference) and ISO. A revised guideline based on the new international standards was published in 1987. Practical examples of quality manuals have been prepared for all technical groups of Nordtest.

Nordtest has also carried out interlaboratory test comparisons for certain test methods and given guidelines for conducting interlaboratory test programmes. Studies of how to meet the demands on traceable calibrations in testing laboratories and of the consequences of authorization of testing laboratories have also been carried out.

Again we can conclude that the efforts of Nordtest have not only lead to documentary results, but indeed to a much more rapid and professional development of quality assurance in laboratory testing in the Nordic countries, than would otherwise have been the case. Quality assurance is now extensively used in testing laboratories in the Nordic countries.

### 4.6 Development of competence

The activity comprises two subdivisions, viz. activities designed to improve the technical competence within specific areas and the certification of personnel working with non-destructive testing (NDT).

The first-mentioned area is developed through projects which have been initiated and partly financed by Nordtest. It comprises Fire testing, Mechanics and NDT. The work is followed and supervised by the respective Technical Group.

In the second area, certification of testing personnel, Nordtest has established a certification arrangement for non-destructive testing operators. The system is widely accepted in the Nordic countries and more than 2.500 NDT-operators are now certified.

The system was revised 2-3 years ago in order to incorporate a 3 level structure and other changes in the international development. The new system has been introduced to ISO as well as to the CEN standardization work and has been well received.

Agreements on mutual recognition have been signed with the Federal Republic of Germany and with France. At present, negotiations are under way with Great Britain, the Netherlands and Switzerland.

The practical operation of the system is delegated to examination centers in each of the Nordic countries. In Denmark, Finland and Norway the Nordtest system is the most commonly used system for NDT operators.

## 5 Mutual acceptance of test results

### 5.1 Nordic acceptance

The Nordic Council of Ministers has in 1987 accepted a programme aiming at the abolition of technical barriers to trade between the Nordic countries. There are several initiatives promoting this decision.

Due to the activities of Nordtest creating close contacts and similar test procedures in Nordic laboratories, it has already been possible to establish systems for acceptance of test results between the Nordic countries. Today there are two different ways in use.

1. Agreements between testing institutions. Within the field of building and construction there are several examples of mutual agreements concerning acceptance of test results. One of these agreements covers Denmark, Finland, Norway and Sweden.

2. Formal systems. The Nordic Committee on Building regulations has established rules, valid in all Nordic countries, covering specific building products. A product approved in accordance with the rules in one Nordic country is automatically accepted in the other countries. Sweden accepts unilaterally building products approved in any of the other Nordic countries. A creation of systems for accreditation of test laboratories and certification of products is in progress, if not already existing. There are already current national systems and principal agreements on mutual acceptance signed between them. The next step should be to put them into effect.

**5.2 International acceptance**

Decisions made by EC and EFTA are published in several documents regulating and influencing the activities in testing and certification in order to promote the free movement of goods. Mutual acceptance of test results is an important step in this process. This work is promoted by many international organizations.

The Nordic countries have always participated very actively in the work of ISO, IEC and ILAC. Denmark is a member of EC and the four other countries are members of EFTA, which guarantees close contacts with these organizations. The national standardization organizations of the Nordic countries are members of CEN/CENELEC. There is Nordic participation in EOTC (European Organization for Testing and Certification), EQS (European Committee for Quality System Assessment and Certification), Eurolab (European Organization for Testing), EURaCHEM (European Organization for Analytical Chemistry), EGOLF (European Group of Official Fire Laboratories), WECC (Western European Calibration Cooperation), WELAC (Western European Laboratory and Accreditation Conference) and other European bodies. This leads to the conclusion that the Nordic countries are well prepared to join different systems for mutual acceptance of test results when needed. They follow and influence the international development trends and they have introduced quality assurance on a professional level in their laboratories.

# 45 TRANSNATIONAL RECOGNITION OF TEST RESULTS – NETWORKS, COOPERATION AND CONDITIONS IN THE SOUTHERN HEMISPHERE

A.M. GILMOUR
Private Consultant, Sydney, Australia

**Abstract**

The Southern Hemisphere consists of Australia, Indonesia, New Zealand, Papua New Guinea, all the Island States of the South Pacific Ocean such as Fiji and Tonga, the southern African countries and most of South America.

Without extensive research it is not possible to study this topic on such a broad geographic and political basis. I am attempting some research in most of these areas and will report on my findings to the symposium. For obvious reasons I will focus my attention on the situation in Australia and New Zealand which is a reasonably unified market and quite sophisticated in the area of testing, certification and conformity assessment generally. Both countries have also established international reputations for research and development in construction and building materials, particularly with respect to resistance to cyclonic conditions and for construction in earthquake prone areas.

From a trade point of view it would also be useful to address the same questions in South East Asia and therefore the networks and cooperation should include reference to countries such as Korea, Hong Kong, China, Taiwan, Singapore, Indonesia, Malaysia, the Philippines and Thailand.

**Building Regulations in Australia and New Zealand**

Both Australia and New Zealand have similar approaches to the development and utilisation of building regulations and standards, although Australia probably has a more sophisticated (and certainly more complicated) system of regulation and conformity assessment.

**Australia**

Australia is a federation of six sovereign states two territories. The Commonwealth (Federal) Government is responsible for Australia's external relationships, for defence and trade policy. The States, on the other hand, have responsibility for health, education welfare etc. including Building Regulations. Traditionally, there have been different regulations for each state and another set for the Territories administered by the Commonwealth.

During recent years serious attempts have been made to establish a uniform set of technical requirements and standards for the design and construction of buildings and other structures throughout Australia.

A new comprehensive Building Code of Australia was published in 1988. It has been adopted by one State and is expected to be adopted by the other States during 1990.

**New Zealand**

New Zealand has a single central Government and a uniform Building Code with provisions for, for example, construction in active earthquake areas.

Traditionally, the Australasian building regulations prescribe technical requirements whereas there is now a trend to performance requirements accompanied by voluntary technical specifications compliance with which is "deemed to comply with the regulatory requirements" but with the provision that a manufacturer can choose other means of demonstrating compliance. The new Building Code of Australia is, therefore, philosophically similar to the EC's *"New Approach to Technical Harmonisation and Standards"*.

**Testing for Building and Construction**

By tradition, the infrastructure for testing and certification in Australia differs from Europe, the United States and the older industrialised countries in that it has never had major national testing laboratories or any significant third-party certification body such as UL in the United States. The major Government laboratories have concentrated on research, development and assistance to manufacturers.

Both Australia and New Zealand have national building research facilities - the National Building Technology Centre, (NBTC) in Australia and the Building Research Association of New Zealand, (BRANZ). They are both primarily for research more than testing and certification.

For testing, Australia has relied on manufacturer's own testing resources and some specialised independent laboratories. Since the 1950s a great deal of reliance has been placed on the accreditation of manufacturer's laboratories by the National Association of Testing Authorities (NATA). There is virtually no provision for third-party certification and reliance is placed on declarations of conformity from manufacturers. New Zealand is following a similar path through accreditation of laboratories by the Testing Laboratory Registration Council of New Zealand (TELARC).

NATA has always had a strong presence in the building area - from foundation engineering and soil mechanics, through the raw materials sector (steel, cement, concrete, etc.), structural members and components (trusses, frames, windows, roofing etc), safety factors (fire resistance, electrical safety) to performance factors such as acoustics and air conditioning, durability of surfaces.

Among the more than two thousand NATA accredited laboratories the largest industry grouping is the building and related sector. Accredited laboratories are to be found in manufacturing plants, service organisations, consultancy practices, academic institutions and government departments. For the most part, test reports from any accredited laboratory have equal status although, clearly, some laboratories have particular expertise which is recognised.

The New Zealand system is of more recent origin but is following the Australian pattern.

**Proofs of Conformity**

**Australia**

In Australia there are different ways of demonstrating that the building materials or a proposed form of construction meet specific requirements.

According to the current regulations in New South Wales, for example, the documentary evidence may be in one of the following forms:

a) a report issued by a competent testing authority, (= accredited laboratory);

b) an Accreditation Certificate issued by the Director, Commonwealth Experimental Building Station at the Commonwealth Department of Works; or

c) any other form of satisfactory documentary evidence that, in the opinion of the Local Building Council, correctly describes the properties and performance of the material or form of construction and adequately demonstrates its suitability for use in the building as proposed. (Where testing is required, most local Councils will require that the laboratories performing the tests are accredited by NATA).

But these provisions do not apply to evidence of fire resistance ratings.

Where a structural section of a building is required to have a fire resistance rating, the document evidence must be an official report issued by one of the following Australian or overseas testing authorities:

a) Commonwealth Experimental Building Station, Commonwealth Department of Housing and Construction;

b) Fire Research Station, Building Research Establishment, Department of the Environment, Great Britain;

c) Fire Insurers' Research and Testing Organisation, Great Britain;

d) National Institute of Standards and Technology, United States of America;

e) Underwriters' Laboratories Incorporated, United State of America;

f) National Research Council, Canada;

g) Underwriters' Laboratories of Canada;

h) Building Research Association of New Zealand;

i) A laboratory approved for that particular test by -

(i) the National Association of Testing Authorities, (NATA) Australia;
(ii) the Department of the Environment of Great Britain, (should now be read as National Measurement Accreditation Scheme, [NAMAS]); or
(iii) the Testing Laboratory Registration Council of New Zealand, (TELARC)

The new Building Code of Australia requires that every part of a building must be constructed in a proper and workmanlike manner to achieve the required level of performance, using materials that are not faulty or unsuitable for the purpose for which they are intended.

Evidence to support the use of a material, form of construction or design, may be:

a) a report issued by a Registered Testing Authority, (= accredited laboratory);

b) a current Certificate of Accreditation, (= Product or System Certification);

c) a certificate from a professional engineer or other appropriately qualified person;

d) a StandardsMark Certificate issued by Standards Australia; or

e) any other form of documentary evidence that correctly describes the properties and performance of the material or form of construction and adequately demonstrates its suitability for use in building.

The above applies to all other parts of a building, except those building elements for which a fire-resistance level is required. In this case, the results of required fire tests shall be confirmed in a report from a Registered Testing Authority.

The definition of a Registered Testing Authority according to the new Australian Building Code is:

a) the National Building Technology Centre (NBTC);

b) the CSIRO Division of Building, Construction and Engineering (CSIRO-DBCE);

c) an authority registered by the National Association of Testing Authorities (NATA) to test in the relevant field; or

d) an organisation outside Australia recognised by NATA through a mutual recognition agreement.

**Regional Networks and Cooperation**

Within the region, the Australian laboratory accreditation system is the most highly developed and is regarded as a model for the future. At present, however, testing is a serious trade barrier in the building products area. Lack of acceptance of test data is used by almost all countries but there are signs of this breaking down.

Trade barriers between Australia and New Zealand have all but disappeared. This has certainly been facilitated by the existence of NATA and TELARC and the very close relationship between those two bodies. Testing is no longer a trade barrier issue. Similarly, closer relationships between the Standards Associations of both countries and the development of joint standards will also help.

Some of the South East Asian countries have been in the past quite unashamed in their use of testing as trade barriers but in recent times, under pressure from the GATT and EC policies, there has been widespread recognition that these trade barriers will have to be reduced.

Hong Kong, Papua New Guinea (PNG), The People's Republic of China (PRC) and Singapore have functioning laboratory accreditation systems which have been designed with international trade in mind. Indeed, HOKLAS in Hong Kong has mutual recognition agreements with NATA in Australia, NAMAS in the UK and A2LA in the USA The system in PNG is managed through a very close relationship with NATA.

Other countries which have indicated intentions to establish similar bodies or have already taken the first steps include, Indonesia, Korea, Malaysia, Taiwan and Thailand.

Lack of immediate recognition of some of these systems has led to some laboratories in the region seeking external accreditation. NATA, for example, has accredited laboratories under its own program in Singapore, The Philippines and Taiwan and has an application from a laboratory in PRC.

Within the region, NATA and TELARC are seen as the models. All of the systems mentioned have relied on these more mature organisations for advice and assistance with training for both staff and assessors and, indeed, for provision of assessors.

**Mutual Recognition Between Accreditation Systems**

Mutual recognition between accreditation bodies means that each party recognises the technical equivalence of accreditation systems operated by the other parties, i.e. they recognise laboratories accredited under the other systems as if they were accredited under their own.

It is important to understand that even within one accreditation system, not all laboratories are of equal capability, the differences are recognised by means of the terms in which the accreditation is expressed in the certificate granted by the accreditation body.

For instance, one laboratory may be accredited for simple compression tests on concrete, while another may have the capability to conduct all standard tests on both wet and hardened concrete. This will also apply between accreditation systems and indeed, the two accreditation bodies may not cover an identical range of work. For instance, not all systems offer accreditation for fire tests on building products.

An acceptance body utilising an agreement between accreditation bodies to facilitate acceptance of foreign test data is therefore able to identify laboratories in the foreign country, which meet the desired level of capability by looking at the terms of accreditation granted to laboratories. This information is included in the national directories of accredited laboratories.

Agreements between accreditation bodies do, therefore, provide transparency between national systems but complete correlation is only possible where extensive intercountry laboratory proficiency testing programs are conducted.

**Conditions for Acceptance of Test results**

A general condition for acceptance of foreign test results is that the importing country can rely on the test results produced in another country. That means that the foreign test results must carry the same degree of validity as if the tests had been performed in the importing country.

In a recent ILAC document, *"The Role of Testing and Laboratory Accreditation in International Trade"*, the quality of test results have been described with four basic terms. These terms are accuracy, precision, repeatability and reproducibility.

Accuracy expresses the degree of departure or coincidence between the test result and a "true value".

The term precision is used to express the degree of departure from or coincidence with several results from repetitions of the same test.

Repeatability and reproducibility are subsets of the term precision. Repeatability is the measure of variation between test results that may be achieved within one laboratory, whereas reproducibility is a measure of variation of test results from similar tests carried out in different laboratories. Problems connected with reproducibility of test results are compounded when tests are performed in different countries.

When discussing confidence in test results a major concern is assuring a sufficient degree of reproducibility.

Ideally, all test methods should produce accurate results with a high degree of precision and be described in such a way as to ensure high levels of repeatability and reproducibility. Unfortunately, test methods possessing all these characteristics are rare and considerable effort is required in the standards writing area to improve matters.

In terms of international trade, problems with reproducibility lead to the most serious disputes.

Reproducibility of test results require that:

1. the tests must be based on test methods which do not allow different procedures or interpretations;

2. the accuracy and/or precision of the test methods must be known in advance. This requires the adoption of test methods which have been properly validated;

3. some technical competence of the laboratory must be demonstrated, which requires some kind of evaluation;

4. interlaboratory trials (proficiency tests) are required on a continuing basis to ensure that laboratories are maintaining satisfactory levels of performance and may be used to generate reliable reproducibility data.

The solutions to some of these problems can only be provided through scientific progress and better definition of standard methods of test.

However, the assessment and evaluation process used in laboratory accreditation provide assurance as to technical competence, consistency of interpretation of standards (at least within one country) and proficiency test programs. Another side effect is to identify many problems associated with the implementation of standards by a working laboratory.

**Proficiency Testing**

Laboratory accreditation provides a good framework within which laboratories can gain international recognition. Ultimately international acceptance of test data will depend on the building up of a high level of confidence at the acceptance level - regulatory authorities, certification bodies and industrial purchasers.

These acceptance bodies must be confident that the test results from foreign testing laboratories carry a sufficient degree of reproducibility in comparison with the national laboratories from which these acceptance bodies accept test results.

Accreditation is but one element in this process of establishing confidence. Another is by interlaboratory comparisons - proficiency tests - in which laboratories actually demonstrate their competence, preferably on a continuing basis.

Laboratory accreditation bodies can also be major contributors in this area and are particularly well placed for international cooperation with regard to proficiency testing programs.

NATA operates a wide range of proficiency testing programs for its own purposes but offers these to laboratories in the region. To date, laboratories in Australia, Hong Kong, New Zealand, PNG, Singapore, Taiwan and The Philippines have participated in programs for cement, steel, paint, early fire hazard of building boards and calibration of microphones organised by NATA. NATA, of course has many other proficiency testing programs in the non-building area.

**Other Networks and Cooperation**

This paper has focused on networks and cooperation between laboratory accreditation systems but, of course, full international cooperation and transnational recognition of test results involves many other participants at the technical level.

There is considerable scope for cooperation in research and between the major laboratories. Australia's major building research organisation, the National Building Technology Centre and the Building Research Association of New Zealand already have a high level of cooperation.

Australia is widely perceived as providing leadership for primary standards of measurement and much calibration in the region is traceable to Australia.

The Pacific Area Standards Congress, (PASC) provides a network for communication between standards writing bodies and for those operating product certification schemes. This later activity is not widely used in the building products area. While PASC provides a forum for discussion, the only active cooperation is between Standards Australia and

**Conclusion**

This paper has mainly focused on the advantages of networks and cooperation between accreditation systems in the efforts to achieve transnational acceptance of test results. It has given a broad view of the current activities within and between the countries in the South East Asian and South Pacific regions. These activities provide a good starting point, but it will be a number of years before the network is complete and able to be utilised effectively within broad product areas.

# INDEX

This index uses keywords provided by the authrs of the papers as its basis. The numbers are the page numbers of thefirst page of the relevant papers.